THERMAL MANAGEMENT OF ELECTRONIC SYSTEMS II

THERMAL MANAGEMENT OF ELECTRONIC SYSTEMS II

Proceedings of EUROTHERM Seminar 45,
20-22 September 1995, Leuven, Belgium

Edited by

E. BEYNE
IMEC,
Leuven, Belgium

C.J.M. LASANCE
Philips B.V.,
Eindhoven, The Netherlands

and

J. BERGHMANS
Catholic University Leuven,
Leuven, Belgium

SPRINGER-SCIENCE+BUSINESS MEDIA, B.V.

A C.I.P. Catalogue record for this book is available from the Library of Congress.

ISBN 978-0-7923-4612-8 ISBN 978-94-011-5506-9 (eBook)
DOI 10.1007/978-94-011-5506-9

Printed on acid-free paper

CONTENTS

PREFACE

For the second time, the Eurotherm Committee has chosen Thermal Managment of Electronic Systems as the subject for its 45th Seminar, held at IMEC in Leuven, Belgium, from 20 to 22 September 1995. After the successful first edition of this seminar in Delft, June 14-16, 1993, it was decided to repeat this event on a two year basis. This volume constitutes the edited proceedings of the Seminar.

Thermal management of electronic systems is gaining importance. Whereas a few years ago papers on this subject where mainly devoted to applications in high end markets, such as mainframes and telecommunication switching equipment, we see a growing importance in the "lower" end applications. This may be understood from the growing impact of electronics on every day life, from car electronics, GSM phones, personal computers to electronic games. These applications add new requirements to the thermal design. The thermal problem and the applicable cooling strategies are quite different from those in high end products.

In this seminar the latest developments in many of the different aspects of the thermal design of electronic systems were discussed. Particular attention was given to thermal modelling, experimental characterisation and the impact of thermal design on the reliability of electronic systems.

The Organising Committee succeeded in assembling an interesting conference program consisting of 38 oral and poster presentations, four invited lectures, two pre-conference tutorials, ten company exhibits and a post-conference meeting on thermal measurement standardisation. The 130 attendees from 17 different countries clearly illustrate the high interest in the subject of this seminar. Therefore it was decided that this seminar will be organised for the third time in 1997. This seminar will be hosted by ISITEM of the Univeristy of Nantes in France, from September 24 to 26, 1997.

The Organising Committee thanks the authors for submitting papers to this conference as well as the Scientific Committee in selecting and editing the papers. Special acknowledgements are due to the conference secretary Mrs. Chantal Deboes and ir. Filip Christiaens for their hard work in making this conference happen.

Eric Beyne, C.J.M. Lasance and J.Berghmans, editors

ACKNOWLEDGEMENTS

exhibition participants

Cambridge Accusence
900 Mt. Laurel Circle
Shirley MA 01464
U.S.A.

FEM CONSULT
Torenstraat 36
3384 Attenrode-Wever
Belgium

Flowmerics Limited
81 Bridge Road
Hampton Court, Surrey KT8 9HH
England

FLUID DYNAMICS INTERNATIONAL
Neuhofstraße 9
64625 Bensheim
Germany

IMEC
Kapeldreef 75
3001 Leuven
Belgium

OPTILAS
Postbus 222
2400 AE Alphen a/d/ Rijn
The Netherlands

PARKER HANNIFIN - CHOMERICS
Parkway Globe Park
Marlow, Bucks SL71YB
England

SDRC
Rivium Quadrant 81
2409 LC Capelle a/d IJssel
The Netherlands

SEMI DICE INTERNATIONAL
Middel 6
1551 SP Westzaan
The Netherlands

ZUKEN-REDAC
3 Avenue du Canada
LP 803-91974 Les Ulis Cedex
France

1. INVITED LECTURES

THE NUMERICAL MODELLING OF HEAT TRANSFER IN ELECTRONIC SYSTEMS : CHALLENGES AND IDEAS OF ANSWER

J.B. SAULNIER
Laboratoire d'Etudes Thermiques URA 1403 CNRS
ENSMA FUTUROSCOPE - 86960 CEDEX, France

1. INTRODUCTION

For the past 25 years , there has been an ever increasing interest for computerized analysis, and it would be difficult today to find a field in which searchers or engineers do not make profit of computer codes efficiency in their every day life.

Of course, heat transfer has obeyed this rule, and we can say that the development of modelling has greatly helped for the understanding of basic or coupled heat exchange phenomena like in:

- solar system, heat storage in the ground (with large time and space scales) already simulated for more than 15 years ago, nuclear waste storage today
- 3D pure convection problems (air conditionning in buildings)
- transfer in semi-transparent materials (glass processing)...

With the introduction of inverse techniques, modelling has also induced today new and efficient developments in metrology, allowing measurements which were not possible before (accessibility, intrusive captors, interface reconstruction in problems with change of phase...)

Modelling also offers design tools for heat transfer, in specific domains like spacecraft thermal control, heat exchangers, electrochemical batteries..., and it is also used with some success in electronics.

Concerning heat transfer modelling in electronic systems, we can yet observe that the complexity of the new problems to be dealt with (coupled phenomena,geometry a bit complicated, larger and larger physical systems to be simulated), combined sometimes with the difficulty to model the physics itself (multiphase flow, turbulence, ultra short heat pulse on a solid,...) still constitute some interesting challenges. It also leads very often to big size models that are difficult or even impossible to simulate with today computers. These different aspects contribute to the emergence of new limitations.

In face of this classical demand of "computer power", progresses may come both from the modelling techniques (how to represent the physics itself ?), from the simulation procedures (how to manage the equations which generally constitute the model ?), and from the technology of the computers themselves (micros, recent workstations, vectorisation and parallelisation). We could thus from now, center the discussion about a group of ideas covering the physics modelling, the simulation, techniques and the computers new features.

But we shall gain a complementary insight of the effective and potential evolutions in heat transfer modelling, thanks to a second direction crossing over the previous one.

E. Beyne et al. (eds.), Thermal Management of Electronic Systems II, 3-16.

We propose for this, to follow the different possible scales of analysis for the setup of a thermal model.

Let us first mention a very classical way of getting heat transfer models by considering matter as a continuum, which leads to write conservation equations (mass, momentum, energy, electric charge...). This approach can apply to many phenomena, involving fluids and solids as well. Radiative exchanges can also, in principle, be mastered at this level, even in presence of a semi-transparent material, and many coupled heat transfers are now currently modelled in this frame. To illustrate some typical applications that can be handled with this "continuum mechanics" approach, let us mention the example we shall discuss further in the paper, of the conjugate heat transfer in electronics boards. We call this first level the **macroscopic one.**

If we perform now a zoom backwards, we discover problems for which the models are to be stated at a system level, and for which the modelisation tools no longer rely upon the same formulation of the conservation laws. This is for instance the case when people try to implement a thermal network to represent IC_S on a board, instead of solving the previous conservation equation of the macroscopic level. This can be performed by using experimental datas and by deciding a priori upon the nature of the network. A parallel approach can use model reductions techniques, coming from the field of system analysis. We shall call this level of modelling the **megascopic** one.

Starting from the continuum mechanics frame, let us finally zoom forwards. We enter then the **microscopic** world and observe that heat transfer analysis is also attainable by an appropriate description of interactions between "particules", like molecules, atoms, electrons, or phonons. Associated with statistical treatments, this representation has provided good means for the evaluation of transport properties. But thanks to new concepts, this microscopic level also appears today to be a very promising way for modelling heat transfer, both in fluid and in solids (Monte Carlo simulations, cellular automata, molecular dynamics). This approach seems to be a good way to simulate for instance the thermal behaviour of chips at very short time scale ($< 10^{-9}$s). The paper uses this "scanning direction", through the macro, mega and micro levels, and first begins with some examples of challenges. We then examine some ideas of answer successively at the macro level (continuum mechanics is probably the most common frame today, for heat transfer modelling, and this is the reason to start from here) and extend our reflections to the mega and micro levels.

The coverage of our paper is of course limited and selective and it uses both personal results and others coming from the literature.

2. SOME EXAMPLES OF CHALLENGES ENCOUNTERED

After the development and intensive use of the first "general purpose" simulation tools, the heat transfer engineer is nowadays faced with a new kind of challenge : because of the efficiency the simulation tools offers to describe a complex model (easier description languages like in a network analyzer, or improvement of the man/machine interface to input complex geometries), because of the progresses of the resolution algorithms and of computers, the demand has grown for large models simulation.

But even if we can "enter" these large size models in the machine, we are not sure that there will be a computer enough powerful to solve the huge number of equations involved. Let us first explain some reasons for this increase of the size of the models, and we shall explore other tracks that can bring difficulties in modelling, not necessarily when the model is written, but already backwards, that is during the conceptual phase which translates the physics into a particular representation (let us call it "physical modelling").

2.1. COMPLEX GEOMETRIES

A first source of difficulty lyes in the fact that heat transfers very usually obey 3D geometries, both for conduction, convection and radiation. For instance, integrated circuits on a board are actually 3D dissipating obstacles implying flow separation, recirculation zones, eventually instabilities, which requires a lot of nodes to retrieve accurately the description of these particularities. Classical assumptions can be used to transform the 3D problem of figure 1 into a 2D one, allowing in theory to handle a stack of such boards (see figure 2).

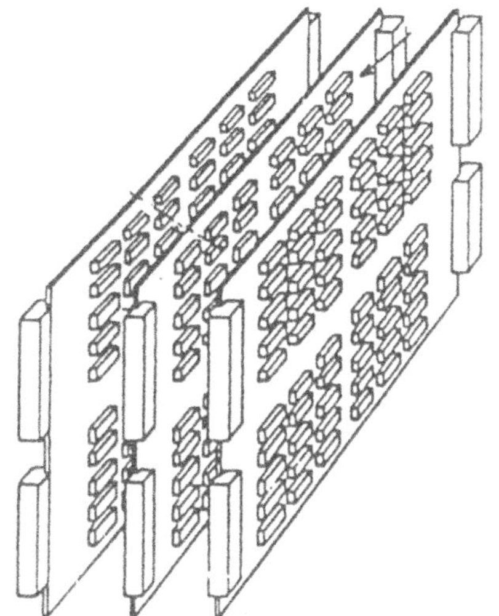

Fig. 1 - A stack of electronic Printed Circuit Boards with IC_S

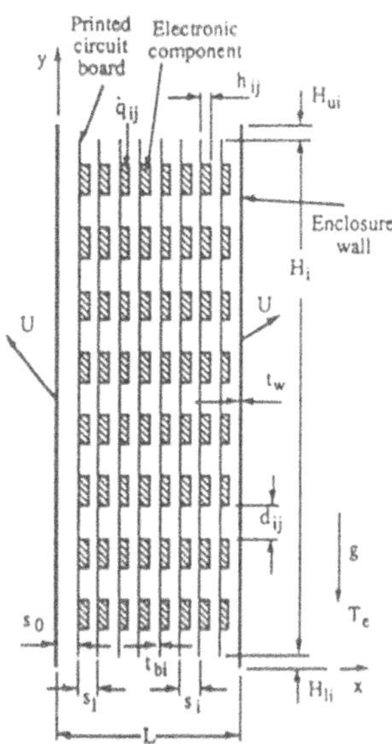

Fig. 2 - 2D representation of the 3D problem of figure 1

We have performed the simulation of a single channel (2 boards) with IC_S as obstacles both in a 2D and in a 3D representation /1/. We discovered of course strong differences in the structure of the flow, in the Nusselt numbers, but we could also show that there was a serious difficulty to choose the heat flux dissipation density in the 2D model, starting from the 3D datas. Finally, big discrepancies could also appear between the 2 models at the chip temperature level. Let us mention that this simple exercise requires about 100.000 nodes for the size of the model (conjugate).

Industrial applications, often begin with the simulation of subsystems, that must be finally integrated in wider systems. Because of the necessity of observing the interactions at the final integrated level, a simple addition of all the subsystems models may lead to a dramatic number of nodes in the final model. This could be the case for instance, when building a complete model for a PC (fig. 3) or a television receiver (fig. 4).

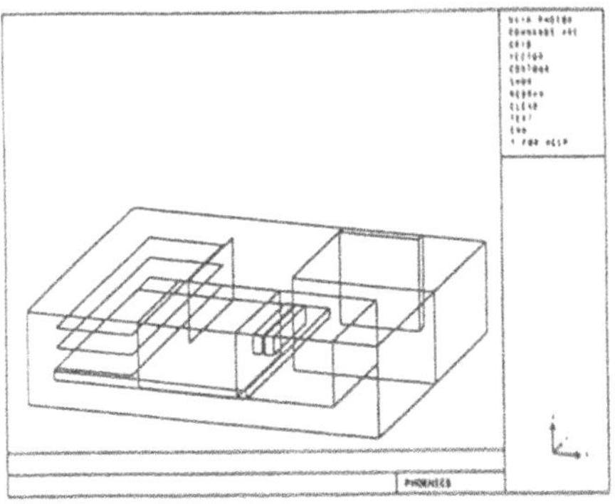

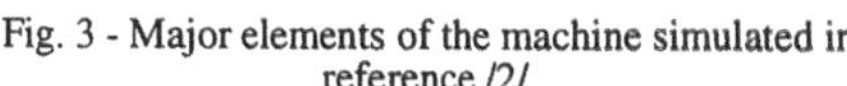

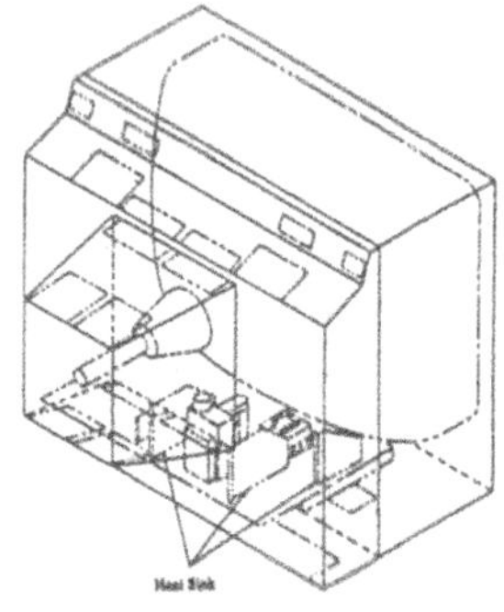

Fig. 3 - Major elements of the machine simulated in reference /2/

Fig. 4 - Schematic view of a television receiver /3/

2.2. COMPLEX INTERACTION BETWEEN COUPLED PHENOMENA

Problems have already appeared for a while, in which it is no longer possible to simulate a given mode of heat transfer without taking account of others. More complex situations are encountered when heat transfer is combined with other phenomena.

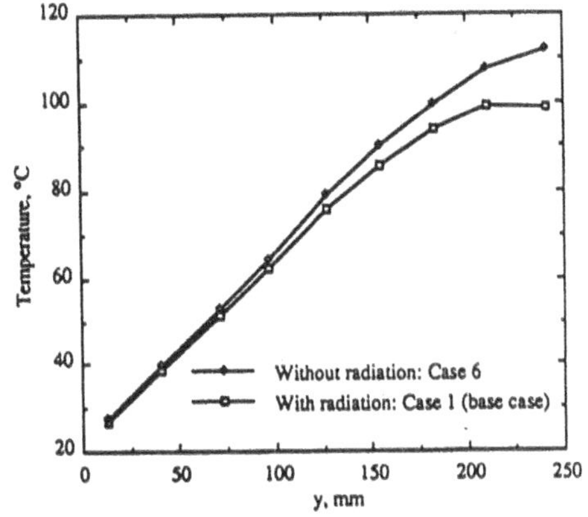

Fig. 5 - Local board temperature of illustrating effect of radiation for the configuration shown on Fig. 2 /4/

The modelling of electronic boards (Fig. 1) illustrates the case of coupling only heat transfer modes. The density of subsystems, here the integrated circuits, is still at the origin of very large models, and particularly in configurations where the dissipated power is not uniform in all the IC s, which is very often the case. But another reason, for the big size of the models comes here from the conjugated nature of the transfers : conduction inside heterogeneous solids (the ICs, the boards) with localized heat sources (the chips inside the IC's), convection in the space between the boards where the ICs appear like many obstacles, and with the presence of radiation too (see fig. 5 for the influence of radiation).

A last example is presented by Figure 5, concerns an electroluminescent diode, where the coupling between heat transfer and electrical potential, the small dimension of the object and the still smallest dimension of the heat source also dictates a large size for a model using discrete techniques (finite volumes, differences...).

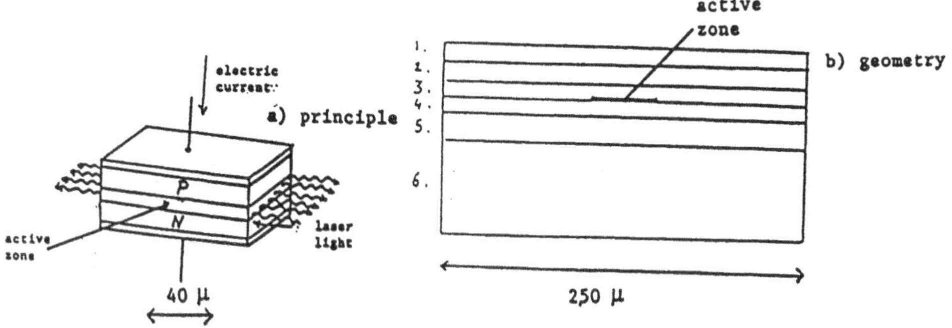

Fig. 6 - An electroluminescent diode /5/

2.3. UNSATISFACTORY MODELLING OF THE PHYSICS

There is for instance still a big challenge for the representation of turbulence. No physical model seems today to be universal (mixing length, k - ε, higher order closure equations, low Reynolds number...), and one must necessarily be aware of the intrinsic limitations of this physical modelling.

Another example concerns the heat transfer in the solid, at very short time scales (short pulses of high heat flux). This is connected to the fact already recognized in the fifties, that the Fourier law generates an infinite velocity for the heat propagation. At that time (1958), Vernotte /6/ and Cattaneo /7/ proposed a complementary term, to introduce the concept of time delay, in the basic Fourier law, establishing a link at a macroscopic level between the heat flux density $\vec{q}$ and the local gradient, as :

$$\vec{q} + \tau \frac{\partial \vec{q}}{\partial t} = -\lambda \, \overrightarrow{\text{grad}}\, T \tag{1}$$

The FOURIER heat diffusion equation, from its classical parabolic version,

$$\frac{\lambda}{\rho c} \Delta T = \frac{\partial T}{\partial t} \tag{2}$$

became then :

$$\frac{\lambda}{\rho c} \Delta T = \frac{\partial T}{\partial t} + \tau \frac{\partial^2 T}{\partial t^2} \tag{3}$$

This new version, analog to the propagation equation on an electrical line, can be solved easily, but the problem remained until today to identify the coefficient.τ Some attempts were done with statistical physics calculations, whereas the experimental estimation remains still difficult today.

Here the fundamental understanding of phenomena probably lies in a representation at the level of the energy vectors, atoms, electrons and phonons, and finally the best representation, that is what we call the physical modelling, is probably to be the provided by this microscopic approach (see development in § 5).

Let us also mention that a similar problem concerns the definition of the thermal conductivity for thin films of material used for instance in optoelectronics.

3. THE POINT WITH THE MACROSCOPIC LEVEL

The general organization of the modelling starts here from the phenomenological conservation equations (usually partial differential equations, becoming integro-differential ones in presence of radiation in a semi-transparent material) which are generally first discretized for every dimension of the problem. Classical approaches split into finite differences, elements or volumes for which a still up to date evaluation may be found in Patankar /8/. A significant difference between finite element method and finite difference lies, according to Patankar, in the ability of the first one to handle irregular shaped domains. An alternative is to formulate finite differences method in a curvilinear grid (body fitted coordinates), which also applies for finite volumes. Some attempts have also been developed to combine finite volumes with finite elements as for instance in Prakash and Patankar /9/, Baliga /10/.

In any case, a non negligible contribution to the efficiency of a code will lie in the automatic installation of the mesh, and in what is called the friendly features of the input of the datas. The same comments may also apply to the output visualizations. Aside from these input/output problems one can identify at least two other crucial steps in the development of a code:

- the appropriate choice of a "global" algorithm, depending upon the nature of the transfer (SIMPLE generation algorithms for the fluid, Sn, Pn, their variants and the spectral representations for gases, in case of radiation in semi-transparent materials).
- inner algorithms dedicated to local tasks like the resolution of linear systems. Gauss-Seidel point by point or line by line has been commonly used for this purpose, but new methods have recently been proposed and evaluated, particularly in the context of vector or parallel computers: Benbouta et al /11/, Sidi et al /12/. Some classical techniques have been found to be inefficient in the parallel context (see LU decomposition).

Let us come now to less general observations and mention 3 particular examples of efficient numerical techniques, the objective of which is to decrease the dimensionality of the problem. We just present the general lines of these techniques, and more information can be found in /13/.

3.1. BOUNDARY ELEMENTS

The phenomenological equations are supposed to be solved on a domain Ω, limited by a boundary Γ.

Boundary element method consists in the transformation of the partial differential equations describing the phenomena inside and on the boundary of Ω, into an integral equation relating only boundary values; then it finds out the numerical solution of this last equation by discretization of Γ only.

If values are required at internal points, they can be calculated afterwards, from the boundary datas already computed. Since the numerical calculations take place only on the boundary, the dimensionality of the problem is reduced by one, and a smaller system of equations is obtained, in comparison with full domain resolution techniques. This technique has been intensively impulsed by the group of Brebbia in Southampton /14/.

3.2 THE INTEGRAL-TRANSFORM TECHNIQUE

The integral-transform technique can be considered as another way of decreasing the dimensionality of a problem, and we find here a good opportunity to mention, how useful it can still be today to combine the resources of analytical mathematics and those of the "classical" discretization approach of computerized resolution, to efficiently improve the CPU consumption.

This technique is a systematic approach of the solution both of steady state and transient boundary value of heat conduction. The fundamental theory is given for instance in Sneddon /15/, but the most popular reference for its application to heat transfer is Ozisik /16/. An extension to convection problems has recently been presented by Cotta/17/.

The integral transform technique derives from the classical method of separation of variables, in the sense that it needs integral transform pairs that are obtained by considering the representation of an arbitrary functions in terms of the eigenfunctions of the corresponding eigenvalues problem. The main step in the solution are the following ones :

- development of appropriate an integral transform pairs (direct transform and inversion) ;
- the partial derivatives are then removed from the heat equation, which is reduced to an ordinary differential equation (here is the decrease of the dimensionality of the problem) ;
- this ordinary differential equation is solved, and the physical temperature is retrieved thanks to the inversion transform.

A presentation using this integral transform technique to perform the characterization of multistripe heterostructure laser is given bu Scudeller in /5/.

3.3. THE MULTIGRID ALGORITHM

Another way of improving the consumption of CPU time with a large model, is to keep its initial extension (that is to remain and solve the problem in the domain Ω of § 3.1), but to perform a decrease of the dimensionality in a dynamic sense, throughout the iterative process of resolution. Starting from a fine grid (1, 2 or 3D) on which we whish to get the solutions, a series of coarser and coarser embedded grids is built.

The general idea lies in the study of the spectral analysis of the error attached to a given level of grid.

On a given grid, the contribution of spacial high frequencies disappears quickly with iterations, whereas after many iterations there still remains an influence of the low frequencies attached to that grid. The trick is that a set of low frequencies for a grid becomes higher frequencies for a coarser grid: the process has then to perform few iterations on a fine grid, to smooth out the local high frequencies contribution, and then to switch to the coarser grid, where it can still be quickly efficient.

The extension to several level of grids is very simple in principle, and the global procedure involves then several "cycles" forth and back from the finest to the coarsest grids. This have been proved to converge efficiently ant to generate a gain of time (relatively to single grid procedures) by a factor 10 to 20 in usual heat transfer problems. Another advantage is that it gives access to the detailed knowledge attached to the finest grid.

Due to Brandt /18/, popularized by Hackbusch /19/, it has been successfully applied to the conjugate heat transfer problem in PCB'$_S$ (see /1/ and figures 7 and 8).

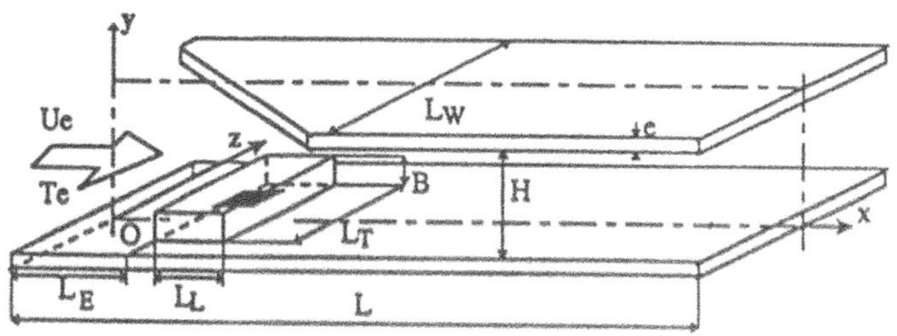

Fig. 7 - A rectangular channel with an integrated circuit (3D conjugate problem)

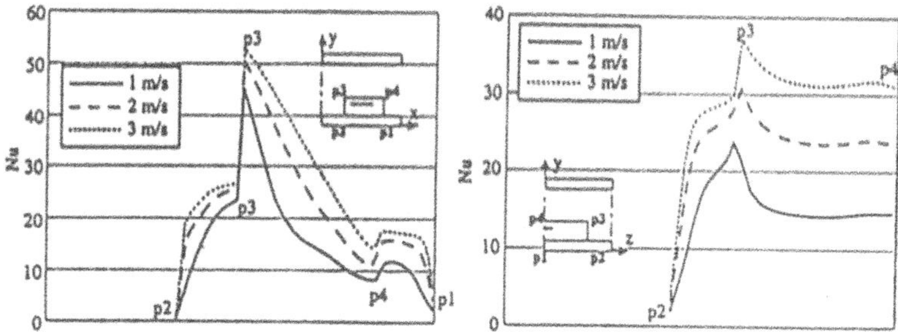

Fig.8 - Example of Nusselt number variation around the IC

A last important feature of the multigrid algorithm (similar to the integral transform) is the linear variation of the CPU consumption with the number of nodes: this characteristic is of course very interesting when compared to the corresponding polynomial law of classical simple grid techniques

4. THE DISCUSSION AT SYSTEM LEVEL

If the progresses in the previous macroscopic description result from exchanges with the field of numerical analysis, modelling at the megascopic level imports concepts rather from automatic control, and systemics.

An efficient way of representation of a system is the concept of network which we shall discuss first and then we shall deal with model reduction techniques.

4.1. NETWORK REPRESENTATION

The thermal characterization of an electronic device by a reduced network is an old concept. Initially a very coarse representation just involved a resistance between the junction and ambiant. The value of this resistance was shown to depend on the power dissipated, on the "ambiant", and on the board temperature... Extensions were introduced, with more refined networks, that are assumed by the a priori knowledge of the critical parts of the device. A good example is presented in that conference by Vinke/20/ in which the technique consists in:

- assuming the network topology of a compact model
- identifying the values of its components by an optimization procedure which minimizes a cost function involving the discrepancies between "full model" and "compact model" temperatures or fluxes.

One of the default of this method, the dependance of the components values versus the boundary condition, seems to have been mastered. Another one remains, in the fact that the compact model is valid only for steady states.

4.2. MODEL REDUCTION TECHNIQUES

The idea is to start here from the initial full model, either at system or subsystem level, and to eliminate the unknowns or nodes that are not of final interest. The common reduction techniques are rather well suited for state variable representation (state variables are generally the temperatures in heat transfer).

Concerning the steady states, a classical elimination (Gauss type) can be applied for linear systems. The works are still in progress for non linear ones, and a contribution has been presented in /21/.

For the transient case, only the linear problems seem to have been considered, and a good contribution came from the field automatic control, where the formulation of the reduction was stated as follows : starting from a set of ordinary differential equations representing the initial complete model as :

$$X'(t) = A\,X(t) + Bu(t) \qquad Y(t) = C\,X(t) \quad ,$$

how can a reduced system of equations (reduced model) be installed, like :

$$Z'(t) = F\,Z(t) + Gu(t) \qquad \hat{Y}(t) = H\,Z(t) + Ku(t)$$

where the dimension of the new state vector Z is much smaller than that of X and the output vector $\hat{Y}(t) \simeq Y(t)$? Very good references present a review of the techniques performing this operation (Michailesco /22/, Petit /23/). Let us briefly summarize the main ideas that have given development to particular reduction procedure in heat transfer modelling :

* analysis of the eigenvalues of A, and identification of what are called the dominant modes (aggregation technique : Aoki /24/, Michailesco /22/, Neveu /25/).
* minimization of a quadratic criterion between the complete and reduced model outputs, of state equations (Eitelberg /26/, Saulnier /27/, Merour /28/).
* serial development with a perturbation technique (Sadat /29/).

An interesting example of such reduction technique applied to the thermal characterization of a power transistor can be found in Petit et al /30/ where the technique is very similar to the principle presented above (see fig. 9, 10, 11).

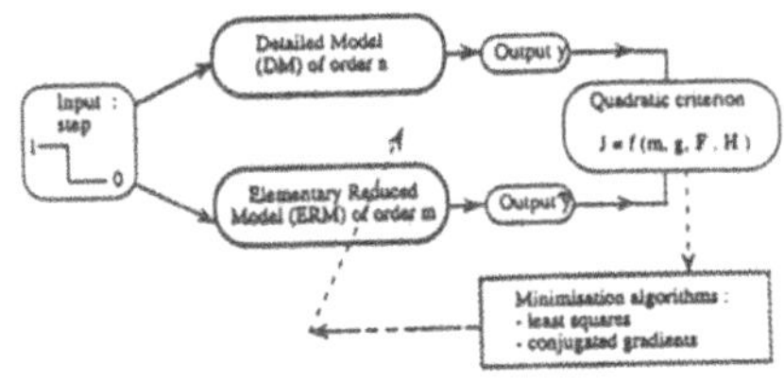

Fig. 9 - Principles of reduction

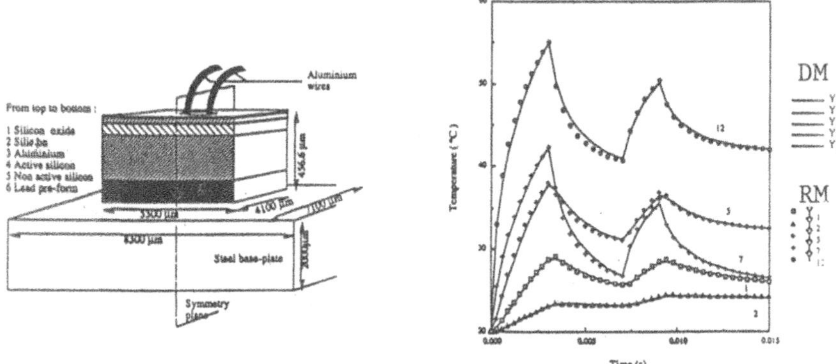

Fig. 10 - Schema of the power transistor

Fig. 11 - Comparison between full (DM) and reduced (RM) models

5. HEAT TRANSFER MODELLING AT THE MICROSCOPIC LEVEL

A simple introduction to the microscopic level simulation could refer to the Monte Carlo modelisation of radiative exchange. Instead of reasoning in a macroscopic sense with the view factor concept, one just describes the exact nature of radiation by following a sample of rays which simulate the photons emitted or reflected by a surface.

Simulation by the molecular dynamics method we refer to, seems in fact to be a good way to explore a new field of heat transfer, precisely proposed by electronics: micro scales in time connected to short heat pulse, or in space, in thin films.

A basic introduction to the philosophy of molecular dynamics can be found in Verlet /31/, who considered a volume of fluid as composed of a relatively small amount of 864 "hard spheres" representing the molecules. The equations of motion of this system of particules, interacting through a Lennard Jones potential was integrated. As the model is a set of second order ordinary differential equation, a very simple algorithm was used. The equilibrium caracteristics (temperature, pressure, critical values,...) obtained, were found in good agreement with the properties of the gaz simulated (Argon).

Recent progresses in this field are presented by Mareschal in /32/. We present here interesting results we have obtained /33/ concerning the answer of a finite wall submitted to a step of temperature on its boundaries (see fig. 12, 13). We clearly identify in this answer corresponding to some ps, a preliminary signal corresponding to an acoustic wave, followed later on by a classical heat diffusion phenomena.

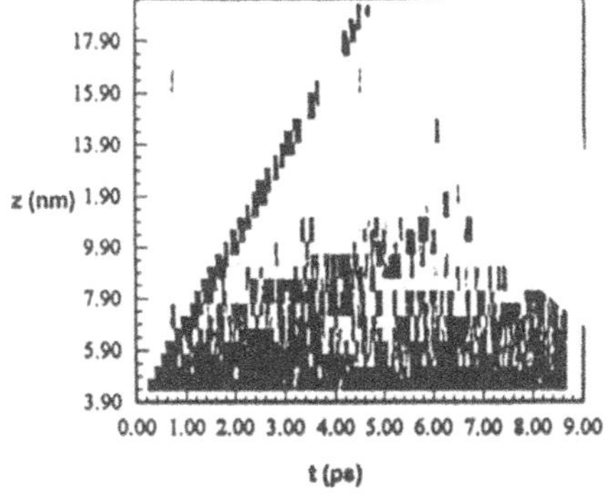

Fig.12 - Answer of the wall in the time/space domain

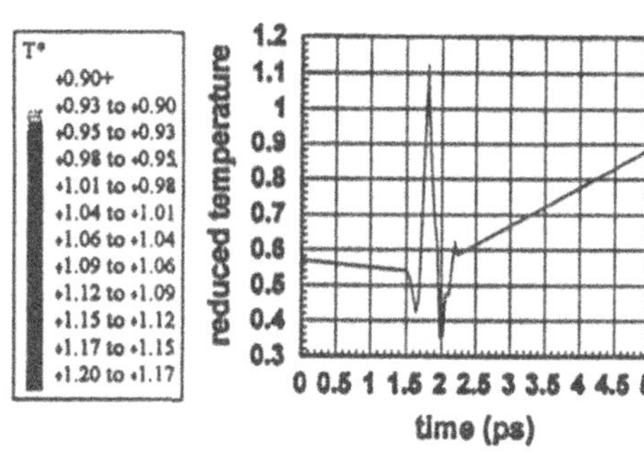

Fig. 13 - Example of local observed signal (z=9,9 nm)

6. CONCLUSIONS

Electronic system is a good example to demonstrate the new challenges appearing in heat transfer modelling. Complexity may be due to the geometry (probably simple, but compact, with heterogeneities), to the presence of combined modes of heat transfer, and with eventually the emergence of very small scales both in time and in space. When facing these challenges, what could be expected for the coming years ?

Modelling should be able to tackle to the 3 levels of analysis we have proposed :

- microscopic both to help to understand, and to provide strong basis for other levels of modelling,
- macroscopic, for which development of numerical analysis can still bring some improvements. On may think of multigrids combined with finite volume, finite element particularly suited to complex geometries. Of course, this level will probably be the first for which massively parallel machines will have an efficient impact concerning both the size, and the combined aspects of the models,
- megascopic for which, there will be yet an evident interest to deal with reduction techniques of any kind (network, automatic methods...), which will be to mix with the increase of power of the machines. Why could not we hear about a new kind of multigrids, at the mesh level, at a reduced network level, and at processors levels ?

There will be probably also the necessity to improve the switching forth and back through the different levels, and to develop adequate interfaces for this.

Nevertheless, much remains to be done to provide efficient tools not only available for analysis, but also for sensitivity, for optimization, which are in fact necessary for efficient design purpose.

REFERENCES

1. Saulnier, J.B., Wang, H.Y. and Fourka, B. : *Thermal Management of Electronic* Systems, Eurotherm Seminar n° 36, pp. 6.1, Delft, 1993
2. Linton, R.L., Agonafer, D. : *Thermal model of a PC*, Journal of electronic packaging Transaction of the ASME, 134/Vol. 116, June 1994
3. Ahmed, I., Krame, R.J., and Parsons, J.R. : *A preliminary investigation of the cooling of electronic components with flat plate heat sinks,* Journal of electronic pachaging Transaction of the ASME, 60/Vol. 116, March 1994

4. Beckermann, C., Smith, T.F., and Pospichal, B. : *Use of a two-dimensional simulation model in the thermal analysis of a multi-board electronic module*, Journal of electronic packaging, Transactions of the ASME, 126/Vol. 116, June 1994
5. Scudeller, Y., Val, C. : *Thermal management of electronic systems'95*, Eurotherm seminar n°36, Sep. 20-22, Leuven Belgium, 1995
6. Vernotte, P. : *Les paradoxes de la théorie continue de l'équation de la chaleur*, C.R. Acad. Sci., 246, pp. 3154, 1958
7. Cattaneo, C. : *Sur une forme de l'équation de la chaleur éliminant le paradoxe d'une popagation instantanée*, C.R. Acad. Sci., 247, pp. 431, 1958
8. Patankar, S.V. : *Recent Developments in Computational Heat Transfer*, Journal of Heat Transfer, Vol. 110, pp. 1037, 1988
9. Prakash, C., and Patankar, S.V. : *A Control-Volume Based Finite-Element Method for Solving the Navier-Stokes Equations Using Equal-Order Velocity-Pressure Interpolation*, Numerical Heat Transfer, Vol. 8, pp. 259-280, 1985
10. Baliga, R.B. : *An overview of control-volume finite element methods for fluid flow and heat transfer,* Advanced concepts and techniques in thermal modelling, Sept. 21-23, Eurotherm seminar n°36, Poitiers, 1994
11. Benbouta, N., Ferrand, and Leboeuf, F. : *Convergence acceleration for linear systems iterative resolution and application to computational fluid mechanics*, High Performance Computing, Elsevier, pp. 55, 1989
12. Sidi, A., Ford, W.F. and Smith, D.A., *Acceleration of Convergence of Vector Sequences*, Siam J. Numer. Anal., Vol. 23, N°1, pp. 64, 1986
13. Saulnier, J.B. : *Some gateways to optmization of heat transfer modelling*, 10th International Heat Transfer Conference, Brighton England, August 1994
14 Brebbia, C.A., Telles, J.C.F., Wrobel, L.C. : *Boundary Element Techniques : Theory and Applications in Engineering*, Springer-Verlag, Berlin, Heidelberg, New York, Tokyo, 1984.
15. Sneddon, I.N. : *Fourier Transforms*, Mc Graw-Hill Book Co., New York, 1951
16. Ozisik, M.N. : *Heat Conduction*, Chapt 13, ed. John Wiley, 1980
17. Cotta, R.M. : *The integral transform method in computational heat and fluid flow*, 10th International Heat Transfer Conference, Brighton England, August 1994
18. Brandt, A. : *Multi-Level Adaptive Solutions to Boundary-Value Problems*, Mathematics of computation, Vol. 31, Num. 138, pp. 333-390, 1977
19. Hackbusch, W. : *Multigrid Methods and Applications*, Springer-Verlag, 1985
20. Vinke, H., Lasance, C. : *Thermal management of electronic systems'95*, Eurotherm Seminar n°45, Sept. 20-22, Leuven Belgium, 1995
21. Lemonnier, D., Sadat, H. and Saulnier, J.B. : *A New Reduction Technique for Non Linear Thermal Models with Conductive and Radiative Couplings*, 10th IHTC, Brighton, 1994
22. Michailesco, G. and Duc, G. : , *L'approche de la Réduction des Modèles en Automatique : Classification et Simulation en Thermique*, Journées d'Etude ENSMA, Poitiers, 1984
23. Petit, D. : *Réduction de Modèles de connaissance et Identification de Modèles d'Ordre Réduit, Application aux Processus de Diffusion Thermique*, Thèse d'Etat, Marseille, 1991
24. Aoki, M. : *Control of Large Scale Dynamic Systems by Aggregation*, IEEE, Vol. AC13, N°3.

25. Neveu, A. and Flament, B. : *Traitement de Grands Systèmes Linéaires par Synthèse Modale, Journées d'études sur la modélisation des champs thermiques*, SFT, 1991
26. Eitelberg, E. : *Interactive Model Reduction by Mimizing the Weighted Equation Error*, IFAC, 79, Zurich, Computer aided design of control systems, 1980
27. Saulnier, J.B. *: La Réduction des Modèles en Thermiques*, AI 83 IASTED Symposium, Lille, 1983
28. Mérour, P. : *La Réduction des Modèles en Thermique : Application à l'Etude d'un Circuit Electronique*, Thèse de l'Université de Poitiers, 1986
29. Sadat, H. : *Une Nouvelle Méthode de Modélisation des Transferts Thermiques en Régime Graduellment Varié*, Thèse de l'Université de Poitiers, 1988
30. Petit, D., Hachette, R., and Veyret, D. : *A modal identification method to reduce a high order model : application to heat conduction modelling, Submitted International Journal of Modelling and Simulation, 1995*
31. Verlet, L. : *Computer Experiments on Classical Fluids. I. Thermodynamical Properties of Lennard-Jones Molecules*, Physical Review, Vol. 159, N° 1, pp. 159, 1967
32. Mareschal, M. : *Microscopic Simulations of Complex Flows,* NATO ASI Series, Series B : Physics, Vol. 236, 1990
33. Volz, S., Lallemand, M. and Saulnier, J.B. : *Etude du comportement thermique des solides aux temps courts par la méthode de la dynamique moléculaire*, Congrès de la Société Française des Thermiciens 17-19 Mai, Poitiers France, 1995

DELPHI - A STATUS REPORT ON THE ESPRIT FUNDED PROJECT FOR THE CREATION AND VALIDATION OF THERMAL MODELS OF ELECTRONIC PARTS

H I ROSTEN
Technical Director, Flomerics Limited
81 Bridge Road, Hampton Court, Surrey, KT8 9HH, UK
Tel +44 81 941 8810 Fax +44 81 941 8730 E-mail: harvey@flopc.demon.co.uk

Abstract

The accurate prediction of the operating temperatures of critical electronic parts at the component-, board- and system-level is seriously hampered by the lack of reliable, standardised input data. This paper reports on the status of work-in-progress in the 3-year European collaborative project, named DELPHI, whose goal is to solve the aforementioned problem.

1. The DELPHI Project

DELPHI stands for DEvelopment of Libraries of PHysical models for an Integrated design environment. The DELPHI project is a three-year project awarded under the micro-electronics domain of the Esprit III workprogramme. The project commenced on November 1, 1993 and is due finish in November 1996. It involves 278 man-months of work. The cost is 4 MECU, 50% funded by the European Union. The project Consortium Partners consists of a mix of industrial companies manufacturing a range of electronic equipment, a software supplier and a university-linked research institute, namely: Alcatel BELL (Belgium) which specializes in telecommunications equipment; Alcatel Espace (France) which makes electronic systems for space applications; Flomerics (UK) which makes the electronic thermal CAD software FLOTHERM; the National Microelectronic Research Centre (University of Cork, Ireland); the consumer electronics company Philips (Netherlands); and Thomson CSF (France) which manufactures electronics for professional applications like radar for civil airports. The project coordinator is Flomerics.

The project is concerned with the development and experimental validation of thermal models of a variety of electronic parts. The models created are generic to certain classes of electronic parts (termed 'species'). The parts studied are tabulated on the next page.

The creation of thermal models of these parts, correct for the intended range of thermal environments they are likely to encounter in applications, is a non-trivial exercise. The possession of thermal CAD software is only a small part of what is needed to create such

E. Beyne et al. (eds.), Thermal Management of Electronic Systems II, 17-26.

models. DELPHI is therefore creating the knowledge of the best methodologies for model building.

Family	Species
Plastic mono-chip	PQFP, PLCC, PPQFP, PBGA, PDIP, TSOP, SSOP, TSSOP,T0220, T03
Ceramic mono-chip	CPGA, CDIP, CERQUAD, LCCC, CQFP, CBGA
Specific parts	MCM, chip-on-board
Passive parts	electrolytic capacitor,transformer
Air flow parts	perforated plates, axial and radial fans
Heat transfer enhancers	flat heat sink, extruded heat sink, pin fin heat sink, enhanced PCB
Interfacing materials	thermal paste, adhesives, die attach

For all parts considered, two kinds of thermal model are constructed: a detailed (or full) model and a compact model. For example, for a mono-chip package the detailed model consist of a detailed conduction model containing several thousand computational nodes whereas its compact model equivalent would be a thermal resistor network containing no more than 10 (say) nodes. This terminology also extends to recognize that even detailed models contain compact models within them: an example of this is the representation of the lead frame of a PQFP by a layer of uniform property material and the validation of this assumption by a detailed model of the lead frame by finite-element software (Rosten, et al [1]).

The project requires that the generic models developed are made available to all the thermal CAD tools in use by the Consortium partners, namely: the board-level thermal tools produced by the major EDA vendors, viz. THERMAX (from Cadence), VTAT (from Zukon-Redac) and AUTOTHERM (from Mentor Graphics) and the computational fluid dynamics (CFD) tool FLOTHERM (from Flomerics). The schematic on the next page shows how the system will work.

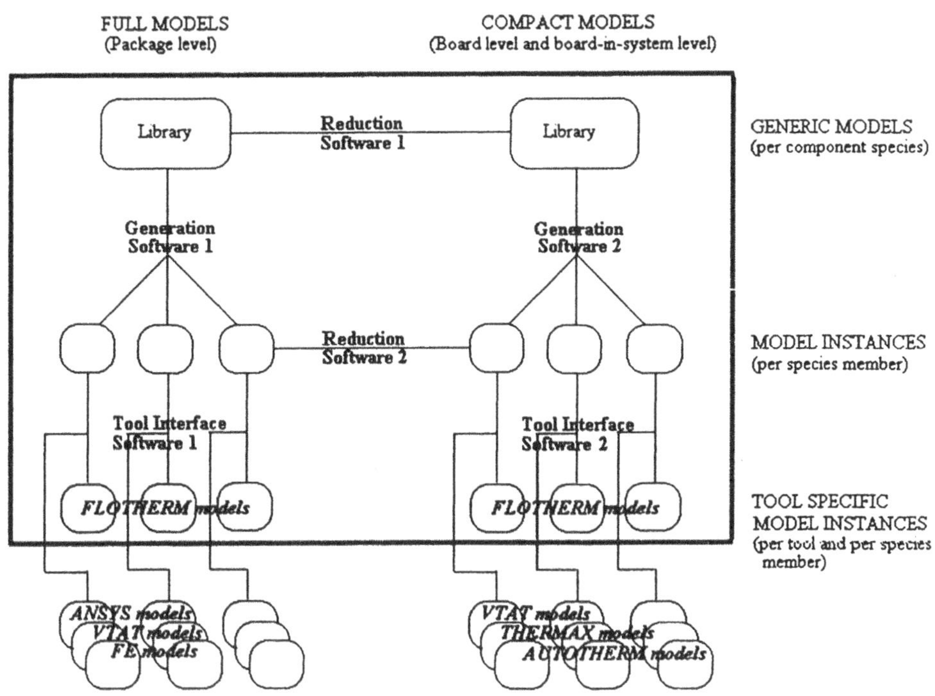

The adoption of this scheme will lead to more reliable thermal designs for electronic systems because of the use of validated thermal models of the constituent parts of the systems, the validation being for the range of the thermal environments encountered in the system. This result, significant though it is, can only achieve its full potential when the suppliers of the parts, ie. the component manufacturers, supply with their hardware thermal models of their parts in some generic format determined by the DELPHI project. The project specifically allocated a portion of funds for developing links with component manufacturers and international standardization bodies.

A crucial outcome of DELPHI is therefore a methodology, for adoption by component manufacturers, which will enable them to produce validated thermal models for passing on to their customers. The precedent for this already exists in that component manufacturers have for many years provided electrical models of their parts for use in circuit simulations algorithms. Indeed, the following quotation from Bar-Cohen et al [2] made several years ago pointed the way ahead:

> *"The thermal precision required in the development of a competitive packaging design could best be served by vendor delivery of a validated numerical model...for*

each chip package in its inventory"

On the global scale, what is required is the setting of an international standard. Since the start of the project, the DELPHI consortium has been actively represented on the JEDEC committee JC15.1 which has been concerned with defining new standards for thermal measurement for mono-chip packages. Much of the work of this committee over the past two years been in preparing a new standard for still-air thermal-characterization measurements of SMT packages. The following quote from 'Environmental Conditions - Natural Convection (Still Air)' is, in the context of DELPHI, revealing:

> *The intent of the θ_{ja} measurements is solely for a thermal performance comparison of one package to another in a standardized environment. This methodology is not meant to and will not predict a package's performance in an application-specific environment.*

This is recognition of the fact that the thermal resistance of a package is a function of its thermal environment, as demonstrated by Hardisty et al [3]. A detailed model of a chip package will always account for its thermal environment through the boundary conditions which are applied at its surface and leads. However, the development of a compact model that produces satisfactory predictions of junction temperature for any thermal environment it might encounter in practice, is a very difficult problem. At least for steady-state applications, this problem has been solved in DELPHI.

Component manufacturers should take heart from the fact that DELPHI defines a very clear boundary to separate their thermal responsibilities from those of their customers. Thus DELPHI enjoins a 'divide and conquer' philosophy whereby the component manufacturer is responsible for the thermal model of the part and nothing else, and the end user is responsible for the specification of the thermal environment to which the part is exposed (ie. the end user sets the heat transfer coefficient and the thermal links that connect the part to the board on which it is mounted). For the component manufacturer, DELPHI provides: generic detailed models for certain species of part; a methodology for generation of compact models from the detailed models to produce compact models valid for a very wide set of thermal environments; and a new experimental procedure for the validation of the detailed conduction models. The diagram on the next page illustrates the foregoing statements.

It should be stated that the methodologies for compact model generation are intended to be independent of any particular CAD tool. The tool most often found at component manufacturers is the finite-element package ANSYS. Many component manufacturers are already in the habit of building analysis models with ANSYS principally for the purpose of evaluating thermo-mechanical behaviour. The thermal conduction model implicit in this is well suited for the task of generating data for the construction of a compact model, more about which will be described in the next section of this paper. The tool-independence of the methodologies for treating mono-chip packages is of course a pre-requisite for any international standardization. However, it should be stated that results for some of the other parts studied in DELPHI will be exclusively for the use in FLOTHERM.

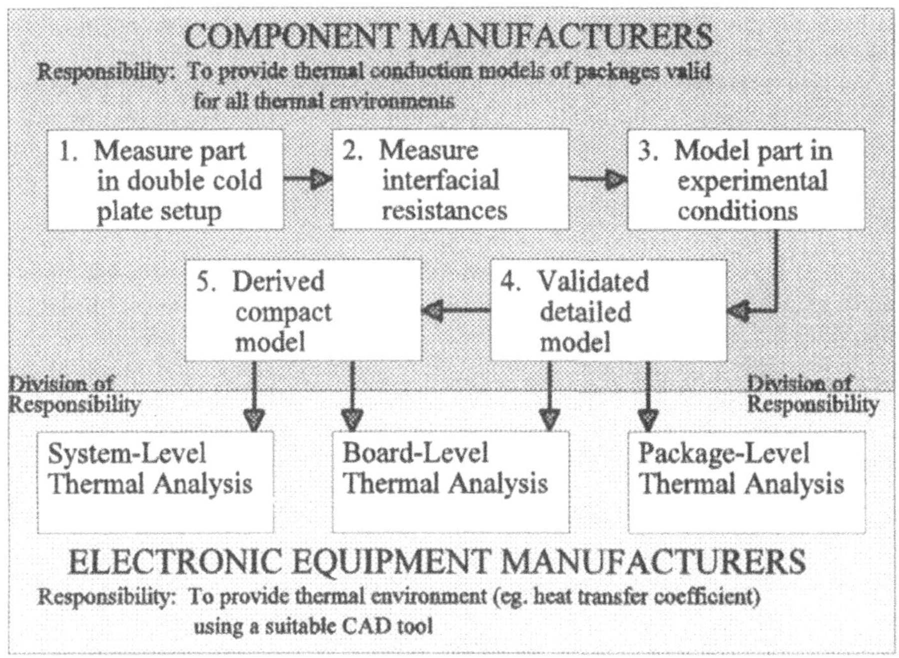

2. DELPHI Bibliography

In the first two years, the project Partners have published a number of papers [1], [4] - [16]. A commentary on the most significant contributions is provided below, with special emphasis on the DELPHI papers published at the EUROTHERM conference.

A comprehensive review of the deficiencies of the pre-existing methods of thermal resistance characterization of mono-chip packages is provided in Rosten and Lasance [7]. This provides many references to the literature of this subject. A careful analysis is provided of the circumstances under which the term 'thermal resistance' has a precise meaning at all: many of the apparent paradoxes present in the literature are attributed to the application and measurement of this quantity in situations where its physical meaning is not well defined. Special attention is devoted to the method of Bar-Cohen et. al. [2], which extends the single thermal resistor model to include a star-shaped network resistor topology, viz. a node for the junction from which radiates resistors to the top-, bottom-, sides- and leads of the package.

This theme is taken up in the paper by Lasance et. al. [6] in which a direct method of computing the Bar-Cohen resistors is demonstrated and in which it is demonstrated that this model fails when non-uniform boundary conditions are applied at the package surfaces. A novel method of solution to this problem is proposed and demonstrated for a PQFP (the same package as documented by Rosten et. al. in [1]). The novelty involves extension of the Bar-Cohen topology to allow for direct thermal connections between nodes on the package surface (eg. top node to leads node, etc), with the values of the thermal resistors determined by an optimization procedure. In the EUROTHERM paper by Vinke and Lasance [8], the method is refined and extended to deal with a CPGA (using the Pentium model described by Rosten and Viswanath [4]). Meanwhile, unpublished work has been done by Flomerics on compact model generation for an SSOP and a CBGA (the 603/4 Motorola processors).

The project is as much about experimental methods and measurements for validation of thermal models of electronic parts as about building thermal models of them. Much of the project has focused on model building for chip packages and methodologies by which conduction models of them can be validated. A comprehensive review of the existing standards including SEMI, MIL and JEDEC standards (for still-air, natural and forced-convection, cold plate and fluid bath) revealed that none of them were suited to the task of conduction model validation. The outcome has been the development of two novel experimental methodologies, viz. the double-cold plate method (DCP) and the submerged double-jet impingement method (SDJI).

In the investigation of various experimental options, attention was given to both 'repeatability' and 'reproducibility' of measurements for any particular technique. The repeatability test questions whether the same results are obtained when the experiment is simply repeated once the part is mounted in the test equipment. The reproducibility test questions whether the same results are obtained if the test equipment is dis-assembled and re-assembled. Clearly, the passing of these tests must be a requirement for any methodology that can be considered as a contender for standardization.

Temmerman et. al. [9] give a full discussion of the issues relating to the DCP method and SDJI method. They show that the DCP method provides repeatability, reproducibility, in situ calibration of the thermal test die, and well-defined ('hard') boundary conditions for validation of a conduction simulation. The DCP method provides measurements for several thermal boundary condition configurations, namely:

- package up and down for an isothermal case with leads insulated;

- a non-isothermal temperature distribution at the case top (which approximates the bell-shaped profile that occurs in natural convection), achieved by introduction of a low conductivity spacer between the package and the top cold plate; and

- heat extraction exclusively through the leads to the bottom cold plate by specially designed test boards.

What the DCP method does is to measure the package under conditions which exercise its principle heat flow paths. In each case, the DCP experiment provides well-defined boundary conditions for input to the thermal conduction model for model validation. All Consortium Partners have now adopted the DCP method which Alcatel BELL are manufacturing.

An added bonus of the DCP method is that it may be possible to derive the thermal conductivity of the packaging encapsulant by the method of 'parameter extraction', a method already used for electrical models. The accurate determination of encapsulant thermal conductivity is extremely important for any thermal model. The values provided by the mold-compound manufacturers is to be treated with due caution. When the manufacturer's value (0.62 W/mK) of the thermal conductivity of the mold compound was used in the PQFP model (Rosten et. al. [1]) very wrong results were obtained. Only the measured value provided good results. This is hardly surprising in view of the fact that the major contribution to the thermal resistance within a plastic package is from the encapsulant, but it does draw attention to the need for a simple and reliable method for the determination of the encapsulant conductivity. Because of the importance of this issue and also because of the importance of interfacial resistances generally, two of the Partners (Philips and Alcatel Espace) are investigating a method which uses a transient cooling method from which an accurate thermal conductivity can be extracted. This method has been investigated by modelling and now the experimental rig is being built. It is planned to publish this work later at another venue.

The EUROTHERM paper by Christiaens et. al. [10] examines the existing standards for fluid bath measurements for thermal characterization of packages. The basic idea behind the fluid-bath method is that the liquid-side thermal resistance is so low that the case of the package is isothermal, under which circumstance the lowest theoretical value of junction-to-case thermal resistance is attained (the advantages of having such a number are discussed in Rosten and Lasance [7]). In practice, the existing standards nowhere near achieve this condition. The problem is that the convective heat transfer from the package depends heavily on the stirring speed of the fluid, the circulating pump, the bath dimensions, the position of pump inlet and outlet, the size of the test board, etc. The authors showed the measurements failed the reproducibility requirement even within the same laboratory: a standard should allow different investigators in different laboratories to obtain the same values of thermal resistance for the same batch of parts. The authors investigated a double submerged jet impingement system (SDJI), in which the device under test is suspended in a fluid bath and subjected to impinging jets on both sides. Their results show that this method gives both accuracy and reproducibility, with such high fluid-side heat transfer as to make the package case very close to the isothermal condition. They tested both Flourinert and de-ionized water for the bath fluid on a PQFP and Flourinert on a CPGA (but not water due to serious current leakage problems).

The advantage of the SDJI method compared with the DCP method is that there are no interfacial contact resistances present. This is especially important for ceramic packages where the internal thermal resistance is very low. Christiaens et. al. find that θ_{ja} for the PQFP and the CPGA are around 7.4°C/W and 2.5°C/W respectively. However, the SDJI method can only ever measure the isothermal condition whereas the DCP method can measure the part under several different boundary conditions as explained above. The two methods should be regarded as complementary to one another, and maybe both of them should form part of the final standard.

Another theme of DELPHI is taken up by Christiaens et. al., in that they carefully treat and report the experimental errors. The method described by them has been adopted by all

Consortium Partners. It considers both systematic error and random errors in the analysis, with the random error calculated for a 95% confidence interval. For further details, their paper should be consulted.

Temmerman et. al. [14], will be reporting to an American audience on both the DCP method and the SDJI method at the IEPS conference to be held in San Diego in September 1995.

The poster session at EUROTHERM, contains three papers that report DELPHI work. The first by Van Es and Lasance [11] investigates the extent of thermal similarity between geometrically dissimilar objects under conditions of natural convection. In consumer electronics (Philips), proper representation of cylindrical components (like electrolytic capacitors) in the simulation of natural convection cooling is important. In FLOTHERM, the simplest approximation to a cylinder is a cuboidal block. Preliminary results for the axis vertical shows that heat transfer coefficients for both shapes differ by no more than 6%. Questions regarding accuracy of the temperature with an IR-spot meter are being investigated by replacement of it by a very small thermocouple.

The second poster paper by Belache et. al. [12] investigates one of the other DELPHI parts, viz. an MCM. One of the parts considered consists of four die mounted in a conventional PQFP consisting of 160 ferro-nickel leads. The second part considered is a 224-pin CPGA. As a preliminary step, the parts were measured under conditions of natural and forced convection. Results are compared between: measurement, detailed-model FLOTHERM simulation and a compact resistor network. These parts are now due for measurement in the DCP system with the thermal resistors of the network generated by the DELPHI procedures.

The third poster paper by Henissen et. al. [13], reports on the modelling of axial fans used in electronic equipment. This is part of a larger piece of work which includes measurements of axial and tangential velocity profiles of the air downstream of the fan using LDA techniques.

O'Flaherty et. al. [16] report work on a batch of CPGA parts mounted with test die. The experimental method used pre-dates the DCP system, in that it consists of a single temperature-controlled cold plate to which the part is mounted via a copper pedestal. Another metal plate is present on top, separated from the package top by a layer of nylon insulation. The temperature of this top plate is monitored by a thermocouple and is used as an input boundary condition for a FLOTHERM simulation. This set-up satisfied the repeatability test, but their were some questions about the reproducibility test. In the simulations, the authors used the die-attach thickness (50μm) quoted by the manufacturer which produced quite good agreement between measurement and simulation of the junction temperature. However, to avoid destructive physical analysis to check the thickness of the die attach, they tested the sensitivity of the simulation to three different values of die-attach thickness, and found best results were achieved with a value of 60μm. So, in this sense the model was calibrated against the measurement. Note also that they assumed a 50μm thickness for the thickness of the thermal paste that was applied at both the top and bottom surfaces of the package.

The use by O'Flaherty et. al. of a system in which the top plate temperature find its level by the balance of natural convection at it top surface with the heat conducted in to its lower surface, and the use of this value in the simulation as an applied constant temperature boundary condition is example of what we have termed a 'soft' boundary condition. A 'soft' boundary condition is in contradistinction to the 'hard' boundary condition given when the top plate is also temperature controlled. The authors are now adopting the double cold plate system and will repeat their measurements.

3. Concluding Remarks

This paper has discussed the mid-term status of the DELPHI project. Clearly, the project has made considerable progress in all its main areas of focus. At the next EUROTHERM conference, the project will have been completed, and the author hopes then to give a final presentation on the achievements of the project. However, even when the DELPHI project is finished its work will not be over, and in a sense its work has only just started. The true success of the project can only be gauged by the extent to which the methods developed are adopted by the thermal community at a whole. Certainly, the achievement of an international standard for preparation of thermal models of chip packages will take more time, for the component manufacturers must be appraised of the advantages of the method and their end users must express (vehemently) to the component manufacturers that they will cease to accept the current situation in which they receive from the manufacturers thermal measurement resistance data that is valid only in the idealized conditions under which they are measured (eg. real systems do not sit in wind tunnels, real packages do not sit isolated on test coupons, etc).

There is another sense in which the work started by DELPHI will not be complete. Accurate determination of junction temperatures is a means to an end, the end being more reliable electronic products. In this respect it is noteworthy that amongst other things, transient thermal characterization has not been considered in the project. No doubt the reader will be able to add his own list of things which urgently need attention to achieve the goal of greater reliability.

4. Acknowledgements

The author wishes to acknowledge the contributions of the following Consortium Partners: Willem Temmerman and Wim Nelemans (of Alcatel BELL), Celine Lacaze and Patrick Zemlianoy (of Alcatel Espace), John Parry and Bilgin Ali (of Flomerics), Ciaran Cahill (of NMRC), Clemens Lasance (of Philips CFT), Thierry Gautier and Yannick Assouad (of Thomson CSF). FLOTHERM is a registered trademark of Flomerics Limited. THERMAX, AUTOTHERM and VTAT are trademarks of Cadence Design Systems, Mentor Graphics and Zukon-Redac, respectively.

5. References

1. H. I. Rosten, J. D. Parry, M. Davies, E. Fitzgerald and R Viswanath, "Development, Validation and Application of a Thermal Model of a PQFP", pp. 1140-1151 of Proc. 45th Elect. and Comp. and Tech. Conf., May 1995, Los Vegas, Nevada.

2. A Bar-Cohen, T Elperin and R Eliasi, "R_{jc} characterization of chip packages - Justifications, limitations and future", IEEE Trans. Comp, Hybrids, Manuf. Technol., vol. 12, pp 724-731, Dec. 1991.
3. H. Hardisty and J. Abboud, "Thermal Analysis of a Dual-in-Line Package Using the Finite Element Method", IEE Proceedings, Vol. 134, Part 1, No. 1, pp. 23-31, February 1987.
4. H. I. Rosten and R. Viswanath, "Thermal Modeling of the Pentium Processor Package", pp. 421-428 of Proc. 44th Elect. and Comp. and Tech. Conf., May 1994, Washington, DC.
5. H. I. Rosten and C. J. M. Lasance, "The Development of Libraries of Thermal Models of Electronic Components for an Integrated Design Environment", pp. 138-147 of Proc. of Inter. Elect. Pack. Conf. (IEPS), Sept 1994, Atlanta, GO.
6. C. J. M. Lasance, H. Vinke, H. I. Rosten and K-L Weiner, "A Novel Approach for the Thermal Characterization of Electronic Parts", pp1-9 of Proc. of 11th SEMITHERM Conf., Feb. 1995, San Jose, CA.
7. H. I. Rosten and C. J. M. Lasance, "DELPHI: the Development of Libraries of Physical Models of Electronic Components for an Integrated Design Environment", pp63-90 of Model Generation in Electronic Design, eds. J-M. Berge, O. Levia and J. Rouillard, 1995, Kluwer Academic Press.
8. H. Vinke and C. J. M. Lasance, "Thermal Characterization of Electronic Devices by Means of Improved Boundary Condition Independent Compact Models", Proc. of EUROTHERM Seminar no. 45, Sept., 1995, Leuven, Belgium.
9. W. Temmerman, W. Nelemans, T. Goosens and E. Lauwers, "Experimental Validation Methods for Thermal Models", Proc. of EUROTHERM Seminar no. 45, Sept., 1995, Leuven, Belgium.
10. F. Christiaens, E. Beyne, W. Temmerman, K. Allaert and W. Nelemans, "Experimental Thermal Characterization of Electronic Packages in Fluid Bath Environment", Proc. of EUROTHERM Seminar no. 45, Sept., 1995, Leuven, Belgium.
11. R. Van Es and C. J. M. Lasance, "Natural Convection Experiments with Cuboids and Cylinders of Equal Area", Poster at EUROTHERM Seminar no. 45, Sept., 1995, Leuven, Belgium.
12. A. Belache, T. Gautier, N. Leveau and G. Paulet, " Thermal Characterization and Modelization of Multichip Modules using Standard Electronic Packages", Poster at EUROTHERM Seminar no. 45, Sept., 1995, Leuven, Belgium.
13. J. Henisson, J. Berghmans, W. Temmerman and K. Allaert, "Modelling of Axial Fans for Electronic Equipment", Poster at EUROTHERM Seminar no. 45, Sept., 1995, Leuven, Belgium.
14. W. Temmerman, W. Nelemans, T. Goossens, E. Lauwers and C. Lacaze, "Validation of Thermal Models for Electronic Components", to be published in Proc. of Inter. Elect. Pack. Conf. (IEPS), Sept 1995, San Diego, CA.
15. C. J. M. Lasance, "About the Validation of a Numerical Model of an Electronics System", Proc. of Third International FLOTHERM User Conference, Sept., 22-23, 1995, Guildford, UK.
16. M. O'Flaherty, C. Cahill, K. Rodgers and O. Slattery, "Validation of Numerical Models of a Ceramic Pin Grid Array Packages", presented at the THERMINIC International Workshop on Thermal Investigations of ICS and Microstructures, Sept. 25-26, 1995, Grenoble.

TEMPERATURE AS A RELIABILITY FACTOR
"We have a headache with Arrhenius"[1]

Michael Pecht
CALCE Electronic Packaging Research Center
University of Maryland
College Park, MD 20742

Pradeep Lall
Motorola Inc.
Radio Products Group
Plantation, FL 33322

Edward Hakim
U.S. Army Research Laboratory
Component Reliability Branch
Fort Monmouth, NJ 07703-5601

Many reliability engineers and system designers consider temperature to be a major factor affecting the reliability of electronic equipment. Unfortunately, in an effort to improve reliability, design teams have often lowered temperature without fully understanding the impact on cooling system reliability, in dollars, weight, and size, and the extent of any actual reliability improvement.

In this paper, various modeling methodologies for temperature acceleration of microelectronic device failures are discussed, as are situations in which some current methodologies give misleading results. The aim is to raise the level of understanding of the impact of temperature on reliability and to define the objectives of a new temperature modeling and design methodology.

1. Background

Reliability, defined as the ability of a device to fulfill its intended function, is often expressed in terms of number of years of useful life. Reliability-related failures render the device non-operational due to damage caused by a failure mechanism, actuated generally by external and internal stresses. Failure mechanisms determine device

---

1 Takehisa Okada, Senior General Manager of Sony Corporation, when asked about Sony's perspective on reliability prediction methods during a U.S. Japanese Technology Evaluation Center visit [Kelly et al. 1993]

E. Beyne et al. (eds.), Thermal Management of Electronic Systems II, 27-41.

reliability; most often, some failure mechanisms will dominate and cause device failure before others.

A device may also fail to fulfill its intended function when its application to operating and environmental conditions lies outside the specification limits. Performance malfunctions may commonly arise due to a threshold voltage drift, a large leakage current, or unacceptable propagation delay or noise margins, although normal operation is often resumed once the operating and environmental conditions return within specifications. Performance problems generally indicate either the need for a system design change or the unsuitability of the device technology for a beyond-specification, high-temperature application. For example, Figure 1 shows how the minimum output voltage for an output high (1) changes with respect to temperature. Clearly, different devices have different values; in this case, the lower the temperature, the smaller the margin for safe operation. The results are similar for the maximum output voltage for an output low (0) (see Figure 2). Of course, other electrical device technologies may work the other way, with noise margins becoming worse at high temperature.

It is important that product engineers, be knowledgeable of the effect of temperature on system performance requirements (as specified in the device catalog). This discussion will focus only on the influence of temperature on reliability.

2. Activation Energy-Based Models

Steady-state temperature, temperature cycles, temperature gradients, and time-dependent temperature changes all have the potential to affect the reliability of modern electronic devices and equipment. However, because of the required use of reliability prediction methods such as Mil-Hdbk-217 [1991] and progeny [HRD5, 1995; CNET, 1983; Siemens, 1986], steady-state temperature has often be considered the only stress parameter affecting reliability.

Current methodologies are based on the work of Savante Arrhenius a Nobel prize winner in chemistry in 1889. He published the results of an experimental study of inversion of sucrose, in which the steady-state temperature dependence of a chemical rate reaction was fit to the form

$$r_r = r_{ref} e^{-E_{chrt}/K_B T} \tag{1}$$

where r_r is the reaction rate (moles/meter^{2}second), r_{ref} is the reaction rate at a reference temperature (moles/meter^{2}second), E_{chrt} is the activation energy of the chemical reaction (eV), K_B is the Boltzman's constant (8.617×10^{-5} eV/K), and T is steady-state temperature (Kelvin). Equation 1, now called the Arrhenius equation, has been used to assess the temperature dependence of a wide variety of reaction rate constants and

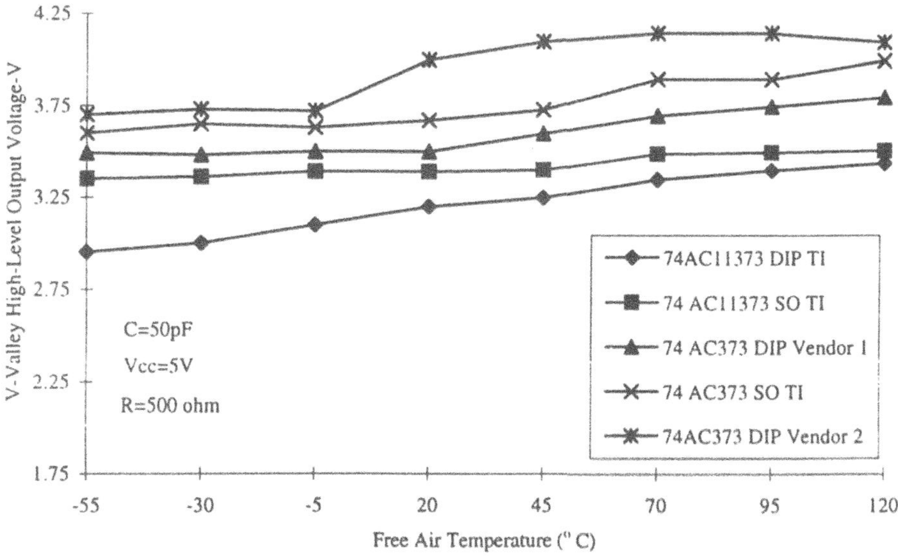

Figure 1. Valley high-level output voltage vs. free-air temperature (74AC11373 compared to end-pin product) [Texas Instruments, 1990].

diffusion coefficients - often crudely, but sometimes quite accurately [Wong, 1990; Blanks, 1990; Witzmann, 1991, Klinger, 1991; Berry, 1980; Clark, 1979][2].

The Arrhenius-based models have been reformulated to predict the influence of steady-state temperature on electronic device reliability. In this case, the mean time to failure, *MTF* (hours), for a given steady-state temperature is represented as

$$MTF = MTF_{ref} e^{E_{dev}/K_B T} \tag{2}$$

where MTF_{ref} is the mean time to failure at a specified reference temperature and E_{dev} is the device activation energy (eV). Figure 3 shows an Arrhenius plot in which the activation energy is obtained from curve-fitting and extrapolating experimental data to an Arrhenius equation.

In the modeling formulation, activation energies of the various failure mechanisms that arise in the device are lumped together to generate a weighted average activation energy for the device [LSI Logic, 1990; Setliff, 1991]. That is,

---

2 Theoretical work in kinetic theory, thermodynamics and statistical mechanical treatments has developed forms that contain exponentials similar to the Arrhenius form [Eyring, 1980; Wigner, 1938; Evans, 1938]. At their core is the assumption that a state of equilibrium exists between the reactants and the products of a reaction, which are separated by a finite energy difference [Reif, 1965].

$$E_{dev} = \sum_{i=1}^{n} m_i E_i \quad 0 < m_i < 1 \tag{3}$$

where m_i is the weight assigned to the activation energy of each failure mechanism (dimensionless), and E_i is the activation energy of the i-th failure mechanism.

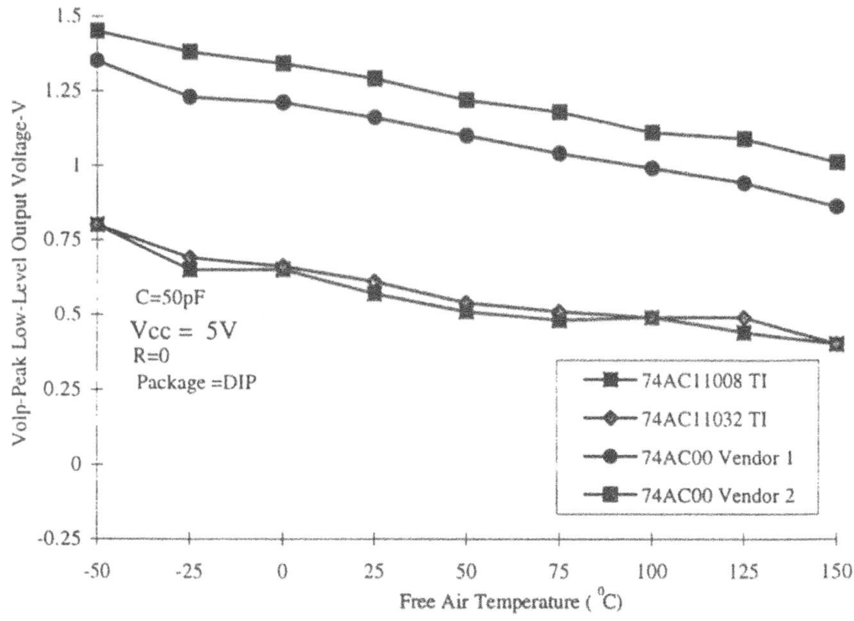

Figure 2. Evaluation of temperature effects (peak low-level output voltage vs. free-air temperature) [Texas Instruments, 1990].

The use of an activation energy to describe a device failure rate is extremely complex and often misleading [O'Connor, 1990; Hakim, 1990]. First, the weighted activation energy approach given in Equation 3 is so sensitive to the variability in the relative dominance of the failure mechanisms (i.e., the assigned weight) that useful conclusions cannot generally be drawn[3]. For example, the effect of even a 0.1 eV variation in the value of activation energy on the *MTF* predicted by the Arrhenius model at a temperature of 70 °C is:

---

[3] Table 1 demonstrates the extreme variability of the dominant device failure mechanisms for different manufacturers of VLSI devices. Considering that activation energies for different failure mechanisms can have values ranging from - 0.06 eV (for hot electrons) to 2 eV (for intermetallic growth), reliability is highly sensitive to the weighing factors.

$$F = \frac{e^{\frac{E_{dev}+0.1}{K_B T}}}{e^{\frac{E_{dev}}{K_B T}}} \approx 29.5 \quad (4)$$

This means that a variation of 0.1 eV at 70 °C results in an error of 29.5 times. This error can be orders of magnitude larger at lower temperatures. Figure 4 shows the sensitivity of a change in activation energy on the mean time between failures (MTBF) as a function of temperature [LSI Logic, 1990].

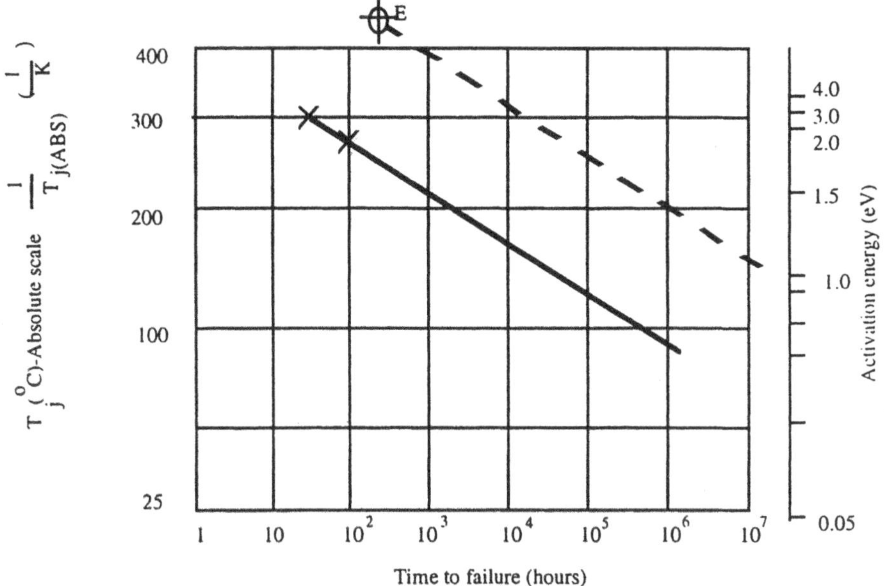

Figure 3. The Arrhenius plot depicts the inverse of junction temperature vs. time to failure. To determine the activation energy, E_{dev}, for a failure mechanism, draw a straight line through the experimental data points, then draw a parallel line from the reference point E. The intercept of this line (dashed line in the figure) on the right-hand scale gives the activation energy [Pollino, 1987].

The mean time to failure predicted by the Arrhenius model is also very sensitive to the value of the activation energy even when the specific device failure mechanism is known. In other words, activation energies for any one failure mechanisms also vary over a wide range (Table 2) and depend on the materials, geometries, manufacturing processes, and quality control methods. For example, the activation energy for electromigration varies from 0.35 to 0.85 eV, even for the same metallization system. Predicted reliability using this approach will have little useful meaning.

Another problem with the use of an activation energy is indicated by studies such as those observing failure rate and steady-state junction temperature for various semiconductor devices (See Figure 5). Finally, the effects of temperature on electronic

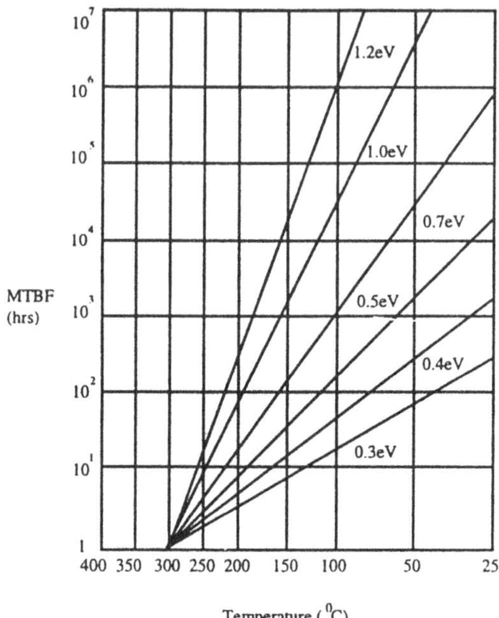

Figure 4. The sensitivity of a change in activation energy on the mean time between failures as a function of temperature. [Note: Device lifetime shown does not refer to any specific device. Times are given only as an aid in determining acceleration factors.]

devices are often assessed by accelerated tests carried out at extremely high temperatures. For example, electromigration tests are generally conducted at temperatures above 250 °C and at current densities ten times those applied in actual operation; the test results are then extrapolated to operating conditions to obtain a value for the thermal acceleration of device failures.

Often implicit in the test strategy is the assumption that the failure mechanisms active at higher temperatures are also active in the equipment operating range, and that the Arrhenius relationship holds. Problems arise when the failure mechanisms precipitated at accelerated stress levels are not activated in the equipment operating range[4].

In particular, many failure mechanisms have temperature thresholds below which failure will not occur. In some other cases, high temperature can actually inhibit or de-accelerate a failure mechanism that will occur at a lower temperature (e. g. hot electron failure mechanism) [Hakim, 1989]. Often, threshold information provides a more effective way to design and test a device and to provide stress management.

4 A NIST study noted that, "there is ample evidence that a straight forward application of the Arrhenius equation, with activation energies determined from high temperature accelerated stress testing, is not strictly valid for predicting real device lifetime" [Kopanschi, et al. 1991].

Table 1 Dominant VLSI failure mechanisms based on survey response [IITRI, 15 January 1988, RADC]. Clearly, no single activation energy can be assigned to the device because the failure depends on manufacturing processes and the actual failure mechanism experienced by the device can vary. Furthermore, few failure mechanisms remain dominant, or even significant, for very long [Witzmann, 1991]

Failure model/mechanism	Survey response					
	1	2	3	4	5	6
Electromigration						13%
Dielectric breakdown	X	50%	< 0.1%	98%		2%
Soft errors						
Parametric drift	X		1%			38%
Hot electrons	X					
Latch-up	X	10%	0.1%		X	
Electrical overstress		20%		2%	X	
Package related		20%	< 0.1%		X	28%
Other					X	19%

3. Reliability Prediction Methods

As noted previously, Arrhenius-based models have been incorporated into some reliability prediction methods. This section reviews these methods and the impact of temperature-dependent models on system effectiveness.

The reliability of electronic devices has often been represented by an idealized plot called a bathtub curve (Figure 6), which consists of three regions. Region 1, in which the failure rate decreases with time, is called the infant mortality or early-life failure region. Region 2, in which the failure rate reaches a constant level, is called the constant failure rate or useful life region. Region 3, in which the failure rate increases, is called the wear-out region.

Modern semiconductor designs, manufacturing processes, and process controls have improved to the point where the infant mortality and useful life regions of the semiconductor devices have failure rates so near zero that the bathtub curve "no longer holds water" [Wong, 1990; Beasley, 1990]. Furthermore, for a device operated within specification limits, the wear-out portion of the curve has been delayed well beyond the useful life of most products [Hakim, 1990, McLinn, 1990, O'Connor, 1990]. In fact, Pecht and Ramappan [1992] found that the majority of electronic hardware failures over the past decade were not component failures, but were attributable to interconnects and connectors, system design, excessive environments, and improper user handling [Table 3]. Nevertheless, various non-scientific attempts to predict the failure rate of devices [HRD5, 1995; CNET, 1983; Siemens, 1986; Mil-Hdbk-217, 1991] are being used, even though they have been proven inaccurate, misleading, and damaging to cost-effective and reliable design, manufacture, testing, and support [Cushing et al. 1993, 1994]. An overview of these reliability prediction models can be found in Bowles [1992]. The models are:

$$\lambda_{dev} = \lambda_{base} \prod \pi_i \tag{5}$$

Table 2 Activation energies for common failure mechanisms

Failure mechanisms	Activation energy	Reference
Die metallization failure mechanisms		
Metal corrosion	0.3 to 0.6 eV	[Hakim, 1989; Jensen, 1982; Amerasekera, 1987]
	0.77 to 0.81 eV	[Peck, 1986]
Electromigration	0.5 eV (small-grain Al)	[Black, 1982]
	0.43 eV (Al)	[Ghate, 1981; Towner, 1983]
	0.35 to 0.85 eV (Al)	[Lloyd, 1987]
	1.0 eV (large-grain glassivated Al)	[Nanda, 1978; Jensen, 1982]
	0.24 to 0.57 eV (Al)	[Reimer, 1984]
	0.7 eV (Al)	[Saito, 1974]
	1.67 to 2.56 eV (Al-1%Si)	[Suehle, 1989]
	0.58 eV (Al-1%Si)	[Schafft, 1985]
	0.96 eV (Al-1%Si)	[Fantini, 1989]
Metallization migration	1 eV	[Abbott, 1976]
	2.3 eV	[Jensen, 1982]
Stress-driven diffusive voiding	0.4 eV	[McPherson, 1987]
	1.0 to 1.4 eV	[Tezaki, 1990]
Device and device oxide failure mechanisms		
Ionic contamination (surface, bulk)	0.6 to 1.4 eV	[Amerasekera, 1987]
	1.4 eV	[Jensen, 1982]
Hot carrier	-0.06 eV	[Hakim, 1989]
Slow trapping	1.3 to 1.4 eV	[Jensen, 1982]
Gate-oxide breakdown		
ESD	0.3 to 0.4 eV	[Baglee, 1984]
	0.3 eV	[Crook, 1979]
TDDB	1 eV	[Hokari, 1982]
	0.3 eV	[Crook, 1979]
	2.1 eV	[Anolick, 1979]
	0.3 to 1.0 eV	[McPherson, 1985]
EOS	2 eV	[Anolick, 1979]
Surface-charge spreading	1.0 eV	[Hakim, 1989]
	0.5 to 1.0 eV	[Jensen, 1982; Amerasekera, 1987]
First-level interconnection failure mechanisms		
Au-Al intermetallic growth	0.5 eV	[Irvin, 1978]
	1.0 eV	[Hakim, 1989; Jensen, 1982]
	1.1 eV	[Mizugashira, 1985]
	2.0 eV	[White, 1978]

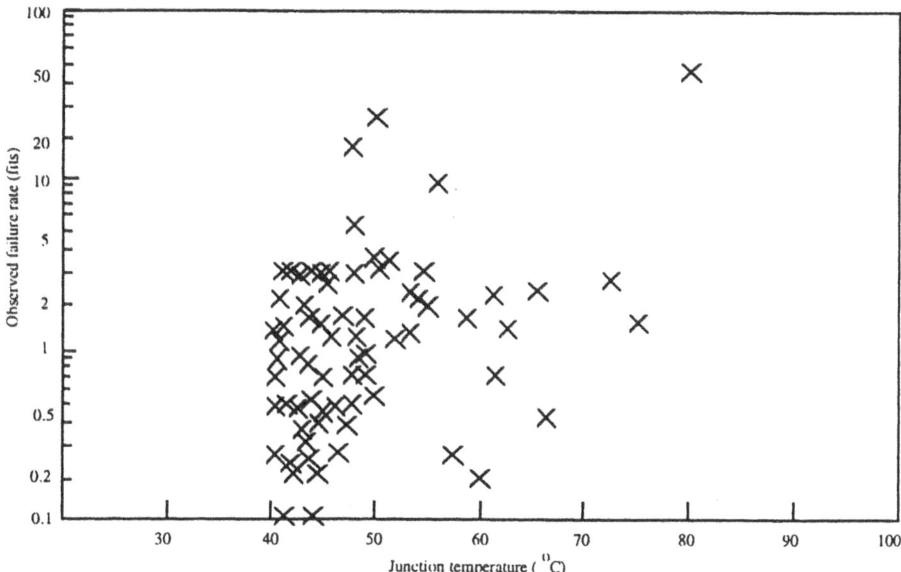

Figure 5. Scatter diagram showing the lack of correlation between observes failure rate and junction temperature of different types of bipolar logic ICs [Hallberg 1994].

where λ_{dev} is the device failure rate (failures/10^6 hours), λ_{base} is the base failure rate (failures/10^6 hours), and π_i are various functional factors for device technology, complexity, package type quality, temperature, and voltage (dimensionless). The temperature acceleration factor, π_T, generally has the form of an Arrhenius equation (see Table 4). Because steady-state temperature is the only temperature factor, or for that matter, generally the only stress parameter (i.e., temperature cycling, vibration, moisture, voltage, and current are usually not applicable), system designers often use temperature reduction as the primary means to improve reliability, often without understanding the actual reliability or the hidden costs associated with the temperature reduction.

As an example of the damage on product design and tradeoffs caused by current reliability prediction methods and the steady-state temperature relation, Figure 7 shows a plot of mean time between failures (MTBF) as a function of package case temperature for a Boeing E-3A multiplexer hybrid. The Mil-Hdbk-217 prediction is an order of magnitude incorrect which could easily lead to the non-use of this electronic device.

As another example, Figure 8 comes from the U.S. Joint Inter Agency Working Group (JIAWG), which developed reliability requirements for such new military systems as the F-22 and the Comanche (light helicopter). Figure 8 was developed, using the values and temperature relations from Mil-Hdbk-217, to provide guidance for reliability allocations of the new systems. To meet system reliability requirements, the maximum component junction temperature was determined to be 65 °C. For the Comanche, this dictated the development of a super-cooling system pumping air at -60 °C in order to lower the temperature outside the sealed electronic boxes enough to get component temperatures to 65 °C . Initially there was no consideration of the reliability impact. In particular, on a hot day with 43 °C outside ambient, cooling is started first; the electronic box will cool to around -40 °C, then rise to around 60 °C when the electronics are turned on. This extreme temperature cycling would occur every time the helicopter is started and stopped. In addition to fatigue damage, Boeing engineers

Table 3 Historical perspective of dominant failures in microelectronic devices [Pecht, 1992]

Source of data	Year	The dominant causes of failure
Failure analysis for failure rate prediction methodology [Manno, 1983]	1983	Metallization (52.8%); oxide/dielectric (16.7%)
Westinghouse failure analysis memos [Westinghouse, 1989]	1984-1987	Electrical overstress (40.3%)
Failure analysis based on failures experienced by end-user [Bloomer, 1989]	1984-1988	Electrical overstress and electrostatic discharge (59%); wirebonds (15%)
Failure analysis based on Delco data [Delco, 1988]	1988	Wirebonds (40.7%)
Failure analysis by power products division [Taylor, 1990]	1988-1989	Electrical overstress damage (30.2%)
Failure analysis on CMOS [Private correspondence]	1990	Package defects (22%)
Failure in vendor parts screened per MIL-STD-883	1990	Wire bonds (28%); test errors (19%)
Pareto ranking of failure causes per Texas Instruments study [Leonard, 1989]	1991	Electrical overstress and electrostatic discharge (20%)

estimated significant standby water in the bottom of the electronic assemblies due to condensation. When was further reviewed by the army, junction temperatures were raised and the use of Mil-Hdbk-217 in general, and the temperature functions specifically, was dropped. The final statement from Boeing was that "the validity of the *steady state temperature* relationship to reliability is constantly in question and under attack as it lacks solid foundational data."

The questionable validity of Mil-Hdbk-217 has not discouraged its use by the U.S. military. In fact, the military standards body has declined to evaluate the technological merit of the document and is waiting for a replacement. Unfortunately, the military electronics industry is somewhat relieved to shed the burden of technical validity.

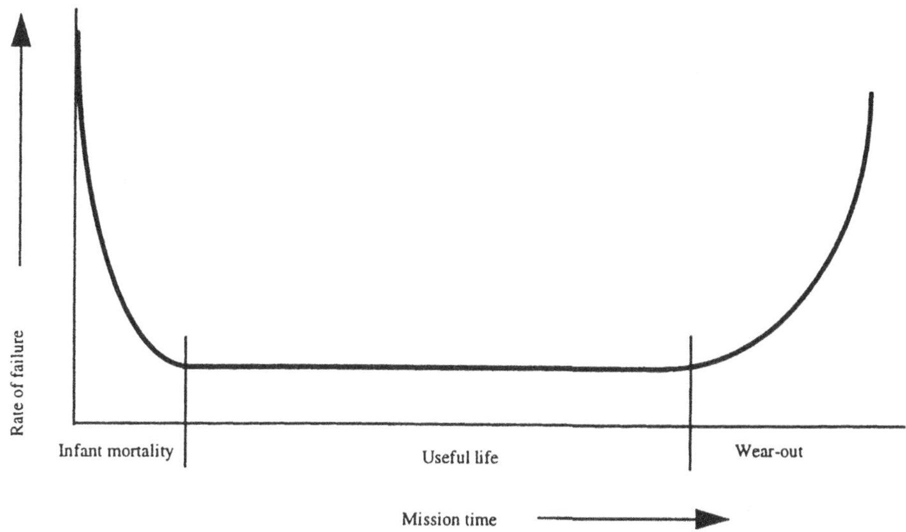

Figure 6. The reliability of electronic devices has often been represented by an idealized graphic called the bathtub curve.

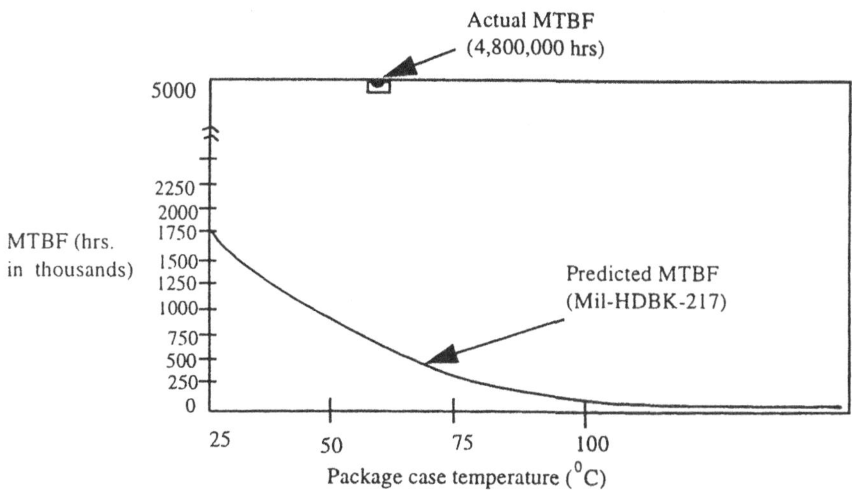

Figure 7. Plot of meant time between failures (MTBF) as a function of package case temperature for a Boeing E-3A multiplexer hybrid.

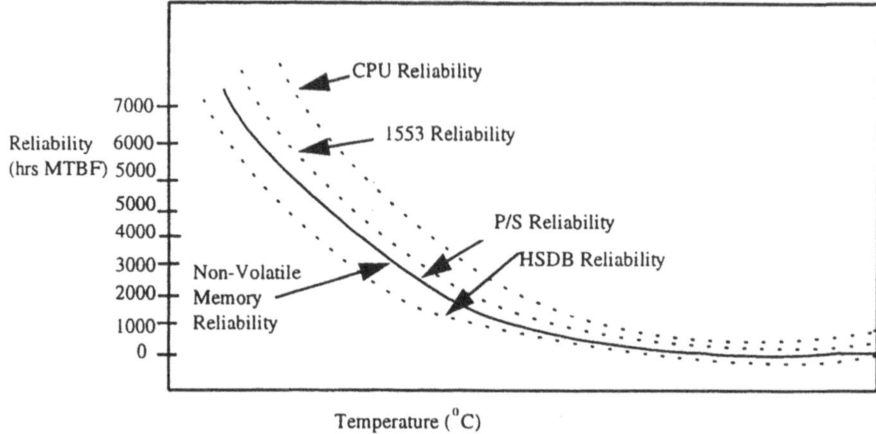

Figure 8. Guidance for reliability allocations of a system.

4. How Should Design, Thermal Management, And Reliability Engineers Work Together?

To address the actual impact of temperature, design, thermal management, and reliability engineers should work together, utilizing a physics-of-failure methodology. There are six steps to this method:

- Develop a thorough knowledge and understanding of the environment in which the equipment will operate. Usually, the customer will specify the operating environment in terms of absolute physical parameters, such as temperature ranges, or will quote the relevant chapter in some handbook or specification. While this may be a useful starting point for the designer, it rarely identifies the actual range of environments experienced by the equipment. It may be better, and from the customer's point of view, more contractually sound, to state where and how the equipment will be used. As a point of interest, consumer goods manufacturers, such as the automobile industry, have never had the benefit of a detailed environmental specification supplied by their customers (the public), but have been able to effectively ascertain the environment for themselves.
- Develop an understanding of the material properties and architectures used in the design. This involves tailoring the product design to requirements by modifying materials geometry, allowable manufacturing defects, and operating stresses.
- Learn how products fail under various degrading influences. This involves assessing the potential failure mechanisms and determining the role of stresses, including steady-state temperature, temperature cycling, temperature gradients, and time-dependent temperature changes, on the failure mechanisms.
- Examine field failure data providing information on how failures occur.

Control manufacturing to reduce those variabilities that cause failure.
Design the product to account for temperature-related performance degradation. Steady-state temperature has an influence on many electrical functional parameters, including propagation delays and noise margins.

Table 4 Temperature - acceleration factors [Bowles, 1992]

Source of procedure	Temperature acceleration factor	
HRD5 1995	$\pi_T = 1$	$for \quad T_{junc} \leq 70\ ^oC$
	$\pi_T = 2.6\times10^4 e^{-3500/T_{junc}} + 1.8\times10^{13} e^{-11600/T_{junc}}$	$for \quad T_{junc} > 70\ ^oC$
CNET 1983	$\pi_T = A_1 e^{-3500/T_{junc}} + A_2 e^{-11600/T_{junc}}$	
Mil-Hdbk-217	$\pi_T = 0.1e^{-A\left(1/T_{junc} - 1/298\right)}$	
Siemens 1986	$\pi_T = Ae^{11605 E_{siem,1}\left(1/T_{junc,1} - 1/T_{junc,2}\right)} + (1-A)e^{11605 E_{siem,2}\left(1/T_{junc,1} - 1/T_{junc,2}\right)}$	

5. SUMMARY

There are alternatives to the Arrhenius relation and the Mil-Hdbk-217 approach to reliability. In Japan, Taiwan, Singapore and Malaysia, a physics-of-failure approach is used by most companies [Kelly et al. 1995] and in the U.S., the CADMP Alliance has developed methods and software to conduct reliability assessments [Evans et al. 1995].

REFERENCES

Abbott, D. A., and Turner, J. A., Some Aspects of GaAs MESFET Reliability. *IEEE Transactions on Microwave Theory and Technology*, MTT, 24, 1976, 317.

Amerasekera, E. A., and Campbell, D. S., Electrostatic Pulse Breakdown in HMOs Devices. *Quality and Reliability Engineering International*, 2, 1986, 107-116.

Amerasekera, E. A., and Campbell, D. S., *Failure Mechanisms in Semiconductor Devices*, New York: John Wiley and Sons, 1987, 12-96.

Anolick, E. S., and Nelson, G. R., Low Field Time Dependent Dielectric Integrity. *Proceedings of the 17th Annual IEEE Reliability Physics Symposium*, 1979, 1.

Baglee, D. A., Characteristics and Reliability of 100 Oxides. *International Reliability Physics Symposium*, 1984, 152-155.

Beasley, K., New Standards for Old. *Quality and Reliability International*, 6, 1990, 289-294.

Berry, R. S., Rice, S. A., and Ross, J., *Physical Chemistry*, New York: John Wiley and Sons, 1980.

Black, J. R., Current Limitations of Thin Film Conductors, *International Reliability Physics Symposium*, 1982, 300-306.

Blanks, Henry J., Arrhenius and the Temperature Dependence of Non-constant Failure Rate, *Quality and Reliability Engineering International*, 6, 1990, 259-265.

Bloomer, C. et al., Failure Mechanisms in Through Hole Package, *Electronic Material Handbook*, 1, 1989, 976.

Bowles, J. A, Survey of Reliability Prediction Procedures for Microelectronic Devices. *IEEE Transactions on Reliability*, 41, 1, March 1992, 2-12.

Clark, I. D., Harrison, L. G., Kondratiev, V. N., Szabo, Z. G., and Wayne, R. P., *The Theory of Kinetics*, edited by C. H. Bamford and C. F. H. Tipper, Elsevier Scientific Amsterdam, 1979.

CNET Recueil de Donnees de Fiabilitee du CNET (Collection of Reliability DATA from CNET), Centre National d'Etudes des Telecommunications (National Center for Telecommunication Studies), 1983.

Crook, D. A., Method of Determining Reliability Screens for Time Dependent Dielectric Breakdown. *17th Annual Reliability Physics Symposium*, 1979, 1-7.

Delco Data. (as of 11-2-88) ETDL LABCOM-from 636 nos ICs analyzed 513 nos Bipolar & 123 nos MOS; quoted from Dicken, H. K. *Physics of Semiconductor Failures*. DM Data Inc. 6900, E. Camelback Rd. Suite 1000, Scottsdale, AZ 85251, 1988, 119.

Evans, M. G., and Polanyi, M., Inertia and Driving Force of Chemical Reaction, *Faraday Society (London) Transactions*, 34, 1938, 11.

Evans, J., Lall, P. and Bauernschub, R. A., Framework for Reliability Modeling of Electronics. *Reliability and Maintainability* Symposium, 144-151, 1995.

Eyring, H., Lin, S. H., and Lin, S. M., *Basic Chemical Kinetics*, New York: John Wiley and Sons, 1980.

Fantini, F., Specchiulli, G. and Caprile, C. The Validity of Resistometric Technique in Electromigration Studies of Narrow Stripes. *Thin Solid Films*, 172, L85-L89, 1989.

Ghate, P. B., Aluminum Alloy Metallization for Integrated Circuits, Thin Solid Films, 8 3, 1981, 195-205.

Hakim, E. B., *Microelectronic Reliability, Reliability Test and Diagnostics*, I, Norwood, MA: Artech House, 1989.

Hakim, Edward B., Reliability Prediction: Is Arrhenius Erroneous?, *Solid State Technology*, 57, August 1990.

Hallberg, Ö., Hardware Reliability Assurance and Field Experience in a Telecom Environment, *Quality and Reliability Engineering International*, 10, 1994, 195-200.

HRD5 *Reliability Prediction*, British Book. British Telecom, March 1995.

Hokari, Y., Baba, T., and Kawamura, N., Reliability of Thin SiO_2 Films Showing Intrinsic Dielectric Integrity. *IEDM Technical Digest*, 1982, 46.

Irvin, J. C. and Loya, A., Failure Mechanisms and Reliability of Low Noise GaAs FETs, *Bell System Technology Journal*, 57 1978, 2823; quoted from Ricco, B., Fantini, F., Magistralli, F., Brambilla, P., Reliability of GaAs MESFET's, in Christou, A., and Unger B. A., *Semiconductor Device Reliability*, Netherlands: Kluwer Academic Publishers, 1978, 455-461.

IITRI, 15 January 1988, RADC.

Jensen, F. and Peterson, N. E., *Burn-in: An Engineering Approach to the Design and Analysis of Burn-in Procedures*, New York: John Wiley and Sons, 1982.

Kelly, M. J., Boulton, W. R., Kukowski, J. A., Meieran, E. S., Pecht, M. G., Peeples, J. W. and Tummala, R. R., *JTEC Panel Report on Electronic Manufacturing and Packaging in Japan*, Loyola College, MD (February 1995).

Klinger, D. J., On the Notion of Activation Energy in Reliability: Arrhenius, Eyring, and Thermodynamics, *1991 Proceedings of Annual Reliability and Maintainability Symposium*, 1991, 295-300.

Kopanschi, J. K., Blackburn, D. L., Harman, G. G., and Berning, D. W., Assessment of Reliability Concerns for Wide-temperature Operation of Semiconductor Devices and Circuits. *First High-Temperature Electronics Conference* June 16-20, 1991.

Lloyd, J. R. and Koch, R. H., Study of Electromigration-induced Resistance and Resistance Decay in Al Thin Film Conductors. *Proceedings of the 25th IEEE International Reliability Physics Symposium*, 1987, 161-168.

LSI Logic, *Reliability Manual and Data Summary*, Milpitas, CA, 1990, 12-18.

McLinn, J. A., Constant Failure Rate: A Paradigm Transition. *Quality and Reliability International*, 6, 1990, 237-241.

Manno, P. T., Failure Rate Prediction Methodology-Today and Tomorrow, *NATO ASI Series*, F3, 1983.

McPherson, J. W., and Dunn, C. F. A., Model for Stress-induced Metal Notching and Voiding in Very Large-scale integrated Al-Si (1%) Metallization, *Journal of Vacuum Science and Technology*, B5(5), 1987, 1321-1325.

McPherson, J. W., and Baglee, D. A., Acceleration Factors for Thin Gate Oxide Stressing. *23rd Annual Reliability Physics Symposium, IEEE,* 1985, 1-5.

Mil-Hdbk 217F, *Reliability Prediction of Electronic Equipment*, U.S. Department of Defense, Washington D. C., 1991.

Mizugashira, S., and Sakaguchi, E., *15th Symposium on Reliability and Maintainability*, 1985, 53; quoted from Ricco, B., Fantini, F., Magistralli, F., and Brambilla, P., Reliability of GaAs MESFET's, in Christou, A. and Unger B. A., *Semiconductor Device Reliability*, Netherlands: Kluwer Academic Publishers, 1985, 455-461.

Nanda, V., and Black, J. R., Electromigration of Al-Si Alloy Films. *International Reliability Physics Symposium*, 1978.

O'Connor, P. D. T., Rel.ability Prediction: Help or Hoax?, *Solid State Technology*, August 1990, 59-61.

Pecht, M., and Ramappan, V. Are Components Still a Major Problem? A Review of Electronic Systems and Device Field Failure Returns. *IEEE Transactions on Components, Hybrids, and Manufacturing Technology*, 15, 6, December, 1992, 1-5.

Pecht, M., Lall, P., and Hakim, E., Temperature Dependence on Integrated Circuit Failure Mechanisms, *Advances In Thermal Modeling III*, edited by Avaram Bar-Cohen and Allan D. Kraus: New York: ASME Press, New York/IEEE Press, New York, 1993, 61-152.

Peck, Stewart D., Comprehensive Model for Humidity Testing Correlation, *Proceedings of the 1986 International Reliability Physics Symposium*, 1986, 44-50.

Pollino, E., Microelectronic Reliability, *Integrity Assessment and Assurance* 2, Boston: Artech House, 1987, 364.

Reif, F., *Fundamentals of Statistical and Thermal Physics*, New York: McGraw Hill, 1965.

Reimer, J. D., The Effect of Contaminants on Aluminum Film Properties. *Journal of Vacuum Science Technology*, 2, 1984, 242-243.

Saito, M., and Hirota, S., Effect of Grain Size on the Lifetime of Aluminum Interconnections. *Review of the Electrical Communications Laboratory*, 1974, 22.

Schafft, H. A., Grant, T. C., Saxena, A. N., and Kao, C. Y., Electromigration and the Current Density Dependence. *Proceedings of the 23rd IEEE International Reliability Physics Symposium*, 1985, 93-99.

Setliff, J. E. A, Review of Commercial Microcircuit Qualification and Reliability Methodology, *Proceedings of the 1991 Advanced Microelectronics Technology, Qualification, Reliability and Logistics Workshop*, August 13-15, Seattle, WA, 1991, 325-335.

Siemens Standard, SN29500 *Reliability and Quality Specification Failure Rates of Components*, 1986.

Suehle, J. S., and Schafft, H. A., The Electromigration Damage Response Time and Implications for DC and Pulsed Characterizations. *Proceedings of the 27th IEEE International Reliability Physics Symposium*, 1989, 229-233.

Taylor, D., Temperature Dependence of Microelectronic Devices Failures. *Quality and Reliability Eng. Intl.*, 6, no. 4, 1990, 275.

Tezaki, Atsumu, Mineta, T., Egawa, H., and Noguchi, T., Measurement of Three Dimensional Stress and Modeling of Stress Induced Migration Failure in Aluminum Interconnects, *Proceedings of the International Reliability Physics Symposium*, 1990, 221-229.

Towner, J. M., and Van de Ven, E. P., Aluminum Electromigration under Pulsed D. C. Conditions. *Proceedings* of the *21st IEEE International Reliability Physics Symposium*, 1983, 36-39.

Westinghouse Electric Corp., Summary Chart of 1984/1987 Failure Analysis Memos, 1989.

White, P.M., *Proceedings of European Microwave Conference*, 1978, 405.

Wigner, E. P., The Transition State Method-Effects of the Electron Interaction on the Energy Levels of Electrons in Metals, *Faraday Society (London) Transactions*, 34, 1938, 29.

Witzmann, S. and Giroux, Y., Mechanical Integrity of the IC Device Package: A Key Factor in Achieving Failure Free Product Performance, *Transactions of the First International High Temperature Electronics Conference*, June 1991, 137-142.

Wong, Kam L., What Is Wrong with the Existing Reliability Prediction Methods?, *Quality and Reliability Engineering International*, 6, 1990, 251-257. *Microelectronic Device Failures* Boca Raton: CRC Press, publication pending in 1995.

2. ANALYTICAL AND COMPUTATIONAL THERMAL MODELING

NUMERICAL SIMULATION OF COMBINED CONDUCTION-RADIATION HEAT TRANSFER IN PCB ASSEMBLIES FOR SPACE APPLICATIONS, EXPERIMENTAL VALIDATION

P.LYBAERT, C.NAVEAU, D.PETITJEAN
Faculté Polytechnique de Mons, Service de Thermique,
rue de l'Epargne, 56, B-7000 Mons (Belgium)
E.FILIPPI, A.STURBOIS
Alcatel Bell SDT, Structural/Thermal Analysis Department,
rue Chapelle Beaussart, 101, B-6032 Mont-sur-Marchienne (Belgium)

Introduction

In space applications, the cooling of electronic systems relies on two heat transfer mechanisms:

- heat conduction at the component, PCB and system (casing) levels as well as contact heat transfer between the different elements of the system;
- radiation heat transfer inside the component, between the component and its environment, from each PCB to the other ones and between the PCB's and the enclosure.

The contribution of radiative heat transfer in practical systems has received little attention (e.g. Ledac *et al.*, 1993) and is often neglected in finite element analyses. It is the purpose of this paper to evaluate the contribution of PCB/PCB and PCB/casing radiation for a typical PCB assembly placed in vacuum conditions.

1. Description of the PCB Assembly

The reference system is sketched in figure 1. It consists of two stacks of four single side epoxy PCB's placed side by side in an aluminium (Al 5754, $k = 125$ W.m^{-1}.K^{-1}) casing. The PCB's of each stack are fixed to each other and to the bottom of the casing by six screws with standoffs. The standoffs placed between the PCB's are made out of stainless steel (AISI 303, $k = 17.4$ W.m^{-1}.K^{-1}), the other ones, placed between the lower PCB and the base plate, are made out of aluminium (Al 2011, $k = 125$ W.m^{-1}.K^{-1}).

In order to accurately control power dissipation at the surface of the PCB's, surface resistances are used instead of dissipating components. These resistances are obtained by zig-zag circuits of equally spaced copper traces. Two identical circuits are traced on every PCB and provide uniform dissipation on each half of the PCB's. Conformal coating is applied on both faces of the PCB's. The « equivalent » thermal

E. Beyne et al. (eds.), Thermal Management of Electronic Systems II, 45-52.

conductivities, k_{xx} in the direction of the traces and k_{yy} in the perpendicular direction, have been calculated, they are equal to 1.2 and 0.2 $W.m^{-1}.K^{-1}$ respectively.

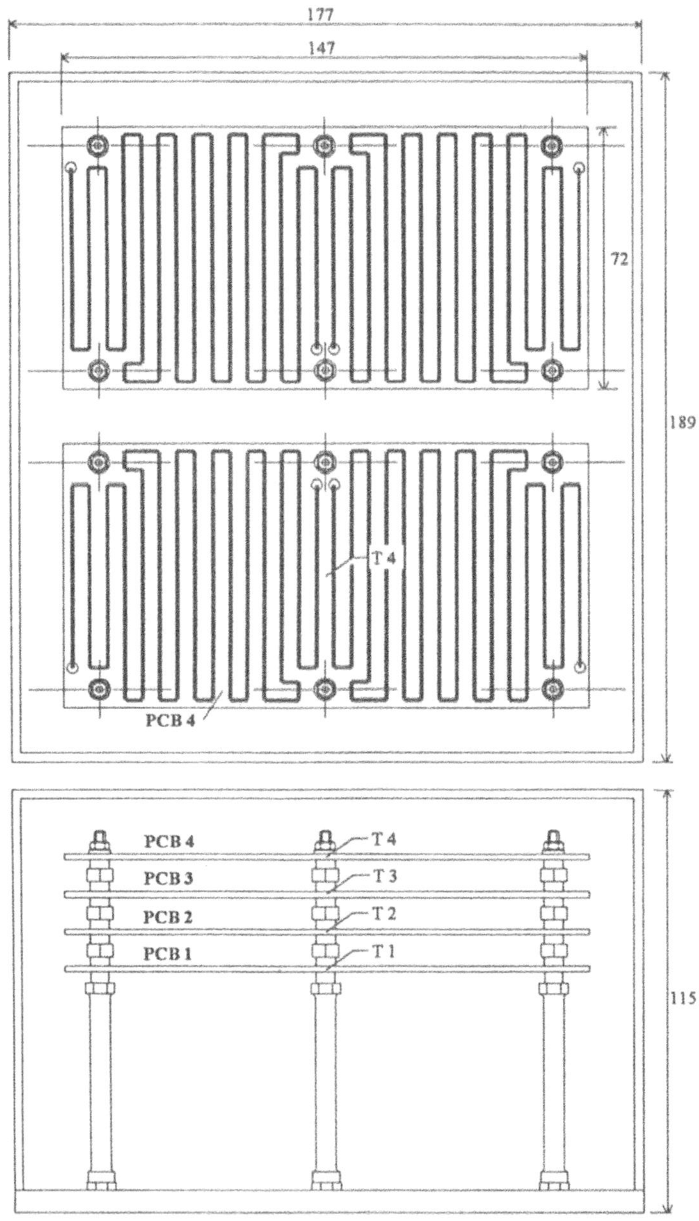

Figure 1. Sketch of the reference system

2. Experimental Study

The casing is bolted to the cooling plate of a small vacuum test facility. The system is described in a companion paper (Petitjean *et al.*, 1995). Twenty-seven thermocouples are used to measure the temperature at different points of the system : on the PCB's, in the standoffs between the different PCB's, and in the casing. The PCB temperatures are measured at the center of each PCB.

The tests have been performed under high vacuum conditions, about 5.10^{-5} Torr. During the tests, both the power dissipation of the PCB's and the cooling plate temperature have been made to vary. Symmetric heating conditions have been applied, each resistance having the same power dissipation.

The effect of the power dissipation on the measured PCB temperatures (T_1 to T_4 in figure 1) are given in table 1, for a constant cooling plate temperature of 50°C. The highest temperatures are always obtained on the intermediate PCB's (nr 2 and nr 3) and the PCB at the bottom of the stack (nr 1) is colder than the top PCB (nr 4). The difference between the top and bottom PCB temperatures is due to the additional heat flow by conduction through the standoffs between PCB 1 and the base plate.

TABLE 1. Measured PCB temperatures - Effect of power dissipation (T_{ref} = 50°C)

	P=0.1 W	*P=0.3 W*	*P=0.5 W*	*P=0.7 W*
PCB 1 (°C)	56.1	70.4	83.9	100.1
PCB 2 (°C)	57.6	75.1	91.3	110.7
PCB 3 (°C)	57.9	76.1	92.7	112.7
PCB 4 (°C)	56.9	73.5	88.9	107.1

The effect of the cooling plate temperature, for a constant power dissipation of 0.5 Watt per resistance, is given in table 2. As radiation heat transfer is a non-linear function of the temperatures, the temperature differences with respect to the cooling plate temperature decrease when the cold plate temperature increases.

TABLE 2. Measured PCB temperatures - Effect of cooling plate temperature (P = 0.5 W)

	T_{ref}=10 °C	T_{ref}=30 °C	T_{ref}=50 °C	T_{ref}=60 °C
PCB 1 (°C)	51.5	66.8	83.9	93.9
PCB 2 (°C)	59.6	74.4	91.3	101.5
PCB 3 (°C)	61.1	75.9	92.7	103.0
PCB 4 (°C)	57.3	72.2	88.9	98.9

3. Finite Element Analysis

3.1. DESCRIPTION OF THE MODEL

Using the ANSYS finite element package, detailed radiation and conduction analyses of the system have been carried out. The program uses a modified Nusselt method

(Smith and Ketelaar, 1989) to compute the radiation view factors between the elements. A hemisphere of unit radius is constructed over the radiating surface *i* and the receiving surface *j* is projected onto the surface of the hemisphere. If the emitting surface is infinitely small (i.e. differential area), the view factor is then given by (Siegel and Howell, 1992)

$$F_{ij} = \frac{1}{\pi}\int_{A_s} \cos\vartheta_i \, dA_s \qquad (1)$$

with A_s being the projected area of the receiving surface onto the hemisphere surface and θ_i being measured from the direction normal to A_i. In ANSYS, the R.H.S. of equation (1) is evaluated by numerical integration: discrete rays are projected from the center of the hemisphere to the hemisphere surface, the rays being distributed such that their density at any point on the hemisphere is proportional to cos θ_i , and the view factor is then approximated by the number of rays intercepting the projected surface divided by the total number of rays emitted. The accuracy of the procedure depends on the number of rays which are emitted. Furthermore, as relation (1) is only valid for a differential emitting surface but is applied to a finite emitting area, large inaccuracies can result if the distance between the elements is small with respect to their dimensions.

For conduction heat transfer, the experimental system has been modelled the following way:

- the casing and the PCB's are modelled by isoparametric quadrilateral shell elements,
- the standoffs are approximated by 3-D heat conducting bars,
- the contact resistances between the PCB's and the standoffs, as well as between the standoffs and the casing are modelled by convective links. A unique value of 3000 $W.m^{-2}.K^{-1}$ has been used for all the contact conductances.

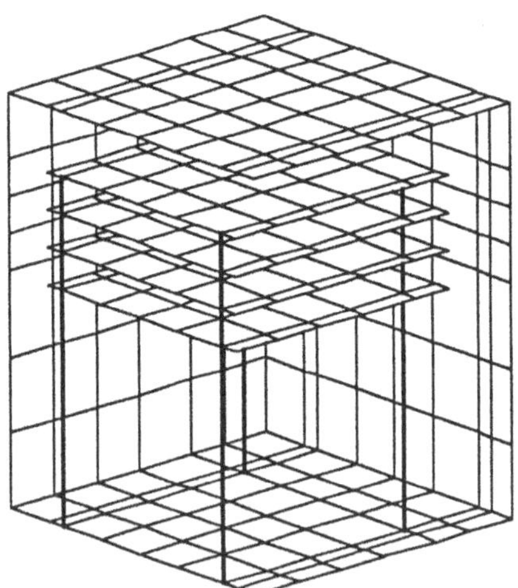

Figure 2. FE discretisation of the system

For radiation, the view factors have been evaluated using about 1300 rays. The emission factors of the PCB's and the casing have been measured., they are equal to about 0.82 and 0.13 respectively. The outer surface of the casing is assumed to radiate to a black environment which is at the same temperature as the cooling plate.

The FE model is represented in figure 2. Owing to wavefront limitations, the system has been rather roughly discretized: 370 elements and 424 nodes are used to approximate one quarter of the system.

3.2. VALIDATION OF THE MODEL

Typical PCB temperatures computed by the model for a cooling plate temperature of 50°C and 0.5 W dissipation of each resistance are compared with the measured data in table 3. A fairly good agreement is obtained, the temperature differences with respect to the cooling plate temperature differ by less than 7 percent. The temperatures in the standoffs are predicted with less accuracy, with maximum errors of about 15 percent.

TABLE 3. Measured and computed PCB temperatures
(T_{ref} = 50 °C, P = 0.5 W)

	PCB 1	*PCB 2*	*PCB 3*	*PCB 4*
T measured (°C)	83.9	91.3	92.7	88.9
T computed (°C)	86.3	91.8	93.3	90.9

3.3. APPLICATION OF THE MODEL

3.3.1. *Effect of radiation on the temperature levels of the PCB's*

Simulations performed with and without radiation allowed to assess quantitavely the effect of radiation on the temperature levels of the different elements of the system.

TABLE 4. Comparison of PCB temperature levels computed with and without radiation
(T_{ref} = 50 °C, P = 0.5 W)

	PCB 1	*PCB 2*	*PCB 3*	*PCB 4*
T_{max}-T_{min} with radiation (°C)	86.8-60.5	92.1-65.9	93.6-69.2	91.3-71.7
T_{max}-T_{min} without radiation (°C)	219.5-67.1	231.1-77.0	238.9-83.6	245.7-89.4

The PCB temperature levels computed by the model, with and without radiation, are compared in table 4. The comparison shows that:

- radiation reduces the temperature levels of the PCB's by about 140°C and of the standoffs by about 15°C;
- on each PCB, the temperature differences are much lower with radiation than without, radiation tends to equalize the temperatures of the system;
- without radiation, the upper PCB's (i.e. the most distant from the cold plate) are the hottest ones while with radiation, the highest temperatures are obtained on the intermediate PCB's (i.e. which do not see the casing surface).

These results show that radiation has a significant effect on the temperature distribution inside the system.

3.3.2. *Effect of the casing emission factors on the temperature levels of the PCB's*
The emission factors of the usual casing materials are very low, about 0.1. In order to increase radiation heat transfer between the casing and the dissipating components as well as between the casing and the surroundings, high emissivity (about 0.9) coatings can be applied on the casing surfaces. This operation is rather complicated and expensive, as the surface treatments have to be space qualified.

The effect of the casing emission factors on the PCB temperatures has also been simulated. Three cases have been considered, depending on whether no coating is applied (base case), coating is only applied on the external side of the casing (case 2) or coating is applied on the internal and external sides of the casing (case 3). The resulting PCB temperatures are given in table 5.

TABLE 5. Effect of the casing emission factors on the PCB temperatures (T_{ref} = 50 °C, P = 0.5 W)

	PCB 1	*PCB 2*	*PCB 3*	*PCB 4*
Base case - ε_i=0.12/ε_e=0.12 (°C)	86.3	91.8	93.3	90.9
Case 2 - ε_i=0.12/ε_e=0.9 (°C)	85.7	91.2	92.6	90.0
Case 3 - ε_i=0.9/ε_e=0.9 (°C)	69.3	76.3	76.9	71.2

The table shows that the most significant improvement is obtained when both the internal and external sides of the casing have a high emission factor: the PCB temperatures are about 15°C to 20°C lower when coating is applied on both sides of the casing (case 3). The effect of increasing only the external emission factor is negligible.

4. Development of a Simplified Radiation Model

Detailed modelling of conduction and radiation heat transfer requires large computational resources, CPU time as well as memory space: for 424 elements, the CPU time on a PC 486DX - 50 Mhz is about 6 hours, most of this time being spent for the computation of the radiation view factors. For rapid temperature estimates, a simpler model is required.

This simplified model has also been developed. It is based on the following discretization of the equipment:

- a single node is associated to each PCB and to the casing;
- conduction heat transfer in the PCB's and the standoffs as well as contact heat transfer between the PCB's and the standoffs are accounted for by thermal conductances;
- radiation heat transfer is computed by Hottel's zone method (Hottel and Sarofim, 1967), with a single zone for each PCB and for the inner surface of the casing. The radiation shape factors are approximated by classical analytical relationships (Siegel

and Howell, 1992) obtained for simple geometrical configurations, i.e. parallel and perpendicular plane elements.

A total of 23 resistances are needed to approximate the system. The resulting resistance network is represented in figure 3.

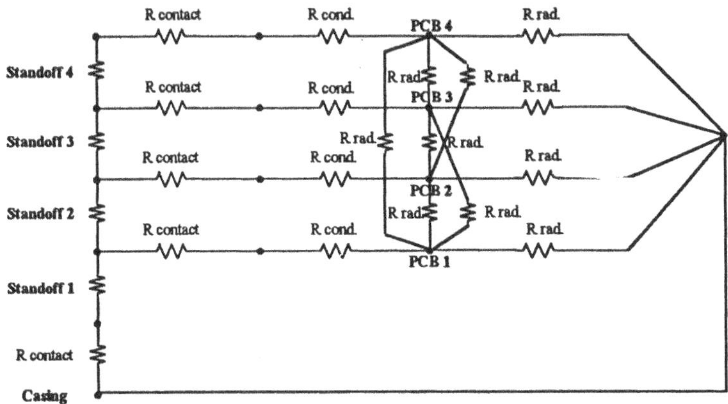

Figure 3. Simplified model - Resistance network

The PCB temperatures obtained by this simple model are compared with the measured values and the temperatures computed by the detailed finite element analysis in table 6. The table shows that very good temperature estimates are obtained by the simplified model. The temperature differences with respect to the cooling plate temperature differ by about 15 percent from the values predicted by detailed FEA. Furthermore, the temperatures computed by the simple model are on the safe side, i.e. higher than the measured and FEA values.

TABLE 6. Comparison of the measured PCB temperatures with the values computed by the simplified and detailed models (T_{ref} = 50 °C, P = 0.5 W)

	PCB 1	*PCB 2*	*PCB 3*	*PCB 4*
Measured (°C)	83.9	91.3	92.7	88.9
Simplfied model (°C)	93.5	97.6	97.8	93.7
Detailed model (°C)	86.3	91.8	93.3	90.9

The model also allows to compute the heat balances of the different elements of the system. The computed heat flows are represented in figure 4. The figure shows that, for the selected assembly, about 75 percent of the heat dissipated at the surface of the PCB's are transferred to the casing by radiation.

5. Conclusions

Combined radiation and conduction heat transfer at system level has been investigated for a typical PCB assembly for space applications. Two complementary approaches have been used : experiments in vacuum conditions and numerical simulation. Both approaches have shown that, for the selected assembly, radiation has a significant effect on the temperature levels of the PCB's.

A simplified model of the system has also been developed. The model uses the zone method with a single zone for each PCB side to compute radiative heat transfer. This model allows to evaluate the PCB temperatures with reasonable accuracy, showing that this approach could be used early in the design process of new systems in order to get realistic estimates of the key temperatures.

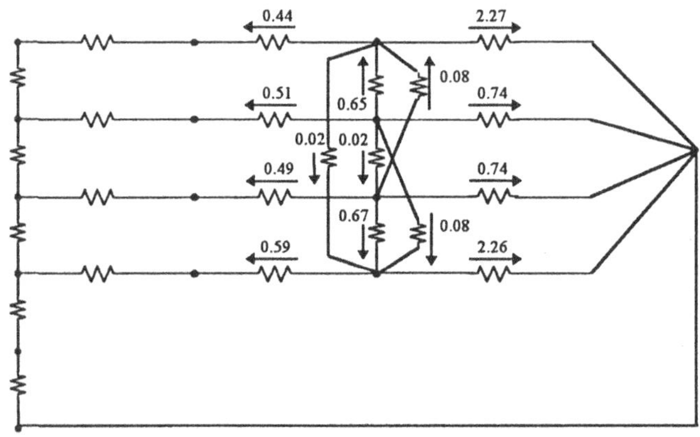

Figure 4. Simplified model - Computed heat flows
(P = 0.5 W per resistance, i.e. 2 W for each level)

6. Acknowledgments

The authors would like to acknowledge the support of the Walloon Regional Government.

7. References

Ledac, P., Pimont, V., Ratel, G. and Viault, A. (1993) Simulation of radiation in electronic equipment, in *Thermal Management of Electronic Systems*, Proceedings of the Eurotherm Seminar No. 29, Delft, 14-16 June 1993, 25.1-25.7.

Petitjean, D., Lybaert, P., Filippi, E. and Sturbois, A. (1995) Measurement of practical thermal resistance values for multicavities power hybrids (MCPH) in a vacuum environment, this seminar.

Smith, R.W. and Ketelaar, D.D. (1987) Radiation matrix generation utility, *ANSYS Revision 4.3 Tutorial*, Doc. DN-T004, Swanson Analysis Systems Inc., Houston.

Siegel, R. and Howell, J.R. (1992) *Thermal Radiation Heat Transfer*, Hemisphere Publishing Corporation, Washington D.C.

Hottel, H.C. and Sarofim, A.F. (1967) *Radiative Heat Transfer*, Mc Graw Hill, New York.

CONJUGATE MODEL OF A PIN-FIN HEAT SINK USING A HYBRID CONDUCTANCE AND CFD MODEL WITHIN AN INTEGRATED MCAE TOOL

DEREJE AGONAFER, Ph.D.
IBM, System 390 Division
522 South Road, MS P520
Poughkeepsie, NY, USA 12601-5400
Email: dereje@vnet.ibm.com

ARNOLD FREE, Ph.D.
MAYA Heat Transfer Technologies Ltd.
4999 St. Catherine Street West, Suite 400
Montreal, Quebec, Canada, H3Z 1T3
Email: arnold@mayahtt.ca

Abstract

The application of an *integrated design tool* in thermal modeling of electronic packages is presented. The methodology is first applied to a heat sink as reported by Kamath [1]. The heat sink considered is pinned-finned with a rectangular cross section. Two flow cases are considered; impinging and parallel flow. The present study agrees very well with the reported data. In this study, however, only one-tenth the number of control volumes reported by Kamath [1] are utilized. The technique is then applied to model a pinned finned heat sink with circular cross section. There is very little difference in the amount of time required to set up this model (pre-processing) as compared with the rectangular cases. This is markedly different than finite control volume codes which need to employ a body-fitted-coordinate approach. The body-fitted-coordinate usually results in a much longer time for pre-processing. In addition, several codes are limited to rectangular geometry and cannot effectively model circular pins. Results for both parallel and impinging flow cases are presented for the circular pin-fin heat sink. The paper concludes with a suggestion for integrating detailed heat sink models at the component level with package and system level simulation.

1. Introduction

As the power density in electronic systems continues to increase, there is a growing interest in the cooling design of electronic packages. In particular, since the use of *geometrically complex* heat sinks is prevalent in the cooling of all computers (from laptops to mainframes), the need for effective modeling tools which are preferably well integrated with a mechanical design tool is very important. These models, if used appropriately, will significantly reduce the lead time for new products. Most of the numerical models in the past have focused on low density rectangular fin arrays as the modeling for denser fins resulted in a grid size that surpassed the capacity of the computer system available to a practicing engineer. Also, there has been very little effort in computational modeling of cylindrical and other non-Cartesian fins as most of the available commercial CFD codes did not have body-fitted-coordinate (BFC) or unstructured meshing capability. Even for those codes that had BFC, modeling of a high density fin array was

E. Beyne et al. (eds.), Thermal Management of Electronic Systems II, 53-62.

impossible due to the high number of required control volumes. It is to be emphasized that if the problem can not be solved in a reasonable time, the model is not of too much practical value as it would be difficult to perform a parameterized study when each simulation takes a significant amount of time. Accordingly, most models have consisted of decoupling the conduction heat transfer from the convective heat transfer. This is accomplished by first coming up with a heat transfer coefficient (using a CFD code, or a correlation), and then using a finite element code (FEM) to solve the conduction problem.

In this paper a novel approach is utilized for solving the simultaneous conduction, convection (conjugate) problem. The conduction/convection problem is solved using a thermal conductance network model and the 3D flow problem is solved using an element based finite volume technique. The pre-processing is performed using I-DEAS Master Series^{TM} [2]; an integrated design/simulation MCAE tool. The technique is applied to a module cooled with a 7x7 pin-fin heat sink.

2. Numerical Method

Thermal analysis of electronic systems requires the simulation of turbulent air flow, conduction, convection, advection and often radiation. Issues include, flow non-uniformity, turbulent low Reynolds Number flow as well as complex geometry. [1][3] The I-DEAS^{TM} Electronic System Cooling [2] (I-DEAS ESC) software package was utilized to simulate the thermal and flow behavior of the heat sink models. I-DEAS ESC uses a hybrid approach. An element based finite difference method is used for simulating conduction, radiation and convection. This is coupled with an unstructured element based finite volume computational fluid dynamics (CFD) code which solves the complete Navier-Stokes equations. A unique thermal coupling method allows I-DEAS ESC to simulate convective and conductive heat paths between disjoint element meshes on arbitrary geometry. This dramatically reduces the required model mesh size since nodes and elements do not need to align between the fluid/solid mesh as well as between solid parts.

2.1 CONDUCTION NUMERICAL FORMULATION

The thermal conduction, convection and radiation simulation uses a finite control volume (finite difference) method to obtain the solution to the thermal model. It utilizes the unstructured finite element mesh to discretize the model.

2.1.1 The Finite Difference Method for Thermal Analysis

The software uses a control volume approach in formulating the discrete finite difference equations which approximate the governing partial differential equations. This method, also known as the lumped parameter method, is based on conservation of energy applied both locally and globally within the domain. [4][5] This technique involves the geometrical discretization or meshing of the model into control volume regions and establishing within each region, a calculation point called the *Finite Difference* (FD) calculation point. Heat balance equations are established for each of these control volumes. The heat flow across the boundary between adjacent control volumes is characterized by a conductance $Q_{ij} = G_{ij} \cdot f(T_i, T_j)$. The temperatures T_i and T_j are evaluated at the control volume FD calculation points. For a linear conductive conductance, G_{ij} is a constant value and the heat flow is proportional to the FD calculation point temperature difference. Radiative, advective, and other non-linear conductances involve more complicated functions of temperature.

The control volume approach utilized by I-DEAS ESC has the advantage of being compatible with finite element modeling. Simply by interpreting each geometric element as a control volume, a finite element mesh is used directly to create a finite difference model. The method is numerically concise and highly efficient. The finite difference method is well known and further information on its implementation can be obtained from references [4], [5] and [14].

2.2 FLUID FLOW NUMERICAL FORMULATION

The flow solver computes a solution to the non-linear, coupled, partial differential equations for the conservation of mass, energy and momentum for arbitrary 3D geometry. It uses an element-based finite volume method and a coupled algebraic multigrid [9][11][12] method to discretize the domain and solve the governing equations. Physical models include laminar or turbulent incompressible flow, natural convection and boundary conditions for fluid flow and heat transfer in enclosures. A good introduction to finite volume methods can be found in reference [6].

2.2.1 Governing Equations

The time-averaged Navier-Stokes equations for conservation of mass, momentum and energy, for an incompressible, Newtonian fluid are descretized and solved for 3D flow conditions. The eddy viscosity assumption is used to model the Reynolds stresses. For the fixed viscosity model, the turbulence viscosity is modeled as $\mu_t = 0.01\rho \ V_m \ L_t$. Where V_m is a mean flow velocity scale and L_t a turbulent eddy length scale

2.2.2 Discretization

In the finite volume method, the governing partial differential equations are first integrated over a control volume. Using the energy equation as an example, the result is,

$$\frac{\partial}{\partial t}\int_V \rho H \, dV + \int_A \rho \, U_j \, \tilde{n}_j H \, dA = \int_A \Gamma_{eff} \frac{\partial H}{\partial x_j} \tilde{n}_j \, dA + \int_V S_H \, dV$$

where $\tilde{n}$ is the unit outward surface normal, and A and V are the outer surface area and volume of the control volume respectively. These volume and surface integrations are next approximated numerically over a discrete finite volume defined on a computational grid or mesh. [6][8]

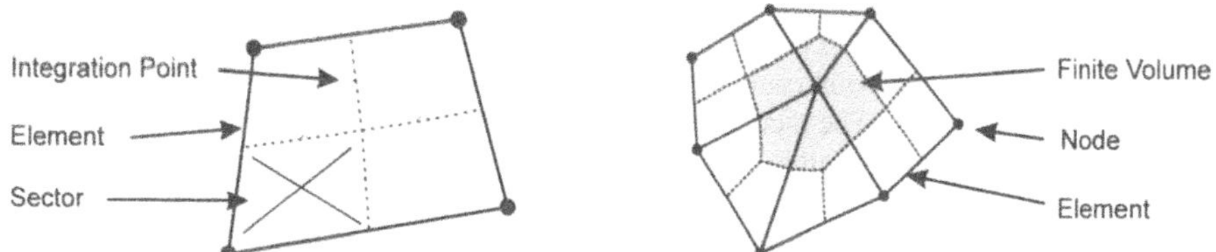

Figure 1. Element sectors and finite volume discretization

All the dependent variables, including pressure and the velocity components, are stored at the element nodes (i.e. it is an equal-order, co-located method). Problems related to pressure-velocity coupling are resolved by using a pressure re-distribution term in the mass equation.[10] Each finite volume sub-surface is an element bisector plane, as shown in Figure 1, where a complete finite volume results from 2 surrounding quad elements and 3 triangular elements.

These sub-surfaces are called *integration point surfaces*, and the integrand to be evaluated is computed at their mid-points. This method of finite volume definition extends directly to 3D. Taking one element in isolation as shown in Figure 1, the integration point surfaces and the finite volume sectors, are indicated. The discretization scheme thus entails representing discrete approximations to the volume integrals over the element sectors and discrete approximations to the surface integrals on the element integration point surfaces.

2.3 COUPLING CONDUCTION AND FLOW

Coupling of the thermal and flow solvers is performed by passing energy between the two solvers at the fluid/solid interface. I-DEAS ESC allows for the creation of *thermal couplings* between elements which are not geometrically connected; that is a path for heat to flow between elements that do not share common nodes. The couplings are established based on proximity, and are distributed to account for overlap. This methodology is also used to couple the solid and fluid. Hence the mesh used for the thermal conduction (conjugate) model does not need to align with the flow model mesh.

Local convection heat transfer coefficients are calculated using the local physically based solution and geometry. The near-wall heat transfer coefficient is modeled using the thermal log-law wall function proposed by Kader.[13] Typically, a wall function's accuracy is mesh dependent. This is a common problem with CFD simulation. I-DEAS ESC uses a semi-analytical method to compute correction factors and applies them in the wall function to adjust for varying mesh size next to the wall. This dramatically improves the reliability of the predictive approach and nearly eliminates dependency on mesh spacing and element type. This is demonstrated in this study. Additional detail about the implementation of the Kader wall functions and thermal coupling methodology can be obtained from the I-DEAS ESC User's Guide [14].

3. Model Description

Two pin-fin heat sink and module assemblies are modeled with parallel and impinging flow conditions. The two heat sinks and modules are identical except the first heat sink uses square pins and the second uses circular pins. Four cases are considered:

1. Parallel flow over a module with square pins.
2. Impinging flow over a module (fanned heat sink) with square pins.
3. Parallel flow over a module with circular pins.
4. Impinging flow over a module (fanned heat sink) with circular pins.

In all cases a 7x7 array of pins are modeled. The square pins are 4mm x 4mm, spaced at 8mm centers. The circular pins have a 5.1mm diameter in order to match the surface area of the square pins and are also spaced at 8mm centers. Figure 2 shows the module and heat sink geometry for the square pin-fin heat sink. The module contains two 1mm thick silicon chips, k=120.0 W/m·°C. The module is made of ceramic, k=25.0 W/m·°C and the heat sink is aluminum 6061, k=172.0 W/m·°C.

Two flow conditions are considered for each heat sink. Cases 1 and 3 model forced convection parallel flow. The extents of the air volume around the heat sink and module assembly are shown in Figure 3. Cases 2 and 4 model forced convection impinging flow from a fan mounted above the heat sink and module assembly. The extents of the air volume and location and size of the fan are shown in Figure 3.

A heat load of 10W is defined on each chip for a total heat load of 20W on the module. The exterior of the module and the heat sink convects to the surrounding air. The surface of the heat sink and component is assumed to be a smooth wall (no slip). Boundary conditions for fans are velocity or volume flow rate specified as shown. Outlet vents (constant static pressure) are defined as shown in Figure 3. All other exterior walls are assumed to be adiabatic and smooth (no-slip). The ambient air temperature is 20°C and pressure is 101351Pa.

The heat sink, module, and air volume parts are modeled using I-DEAS Master Series. The solid model part geometry is meshed directly with I-DEAS free (unstructured) meshing using a variety of element types; hexahedral, wedge and tetrahedral. The flow and thermal boundary conditions are defined directly on the same *master model*. This methodology permits the direct use of design geometry to model the thermal and flow phenomena in an integrated mechanical design and simulation tool. Note that the actual circular outer diameter and hub geometry of the fan was modeled since the unstructured free meshing technology in I-DEAS ESC supports the meshing of arbitrary 3D surfaces and volumes.

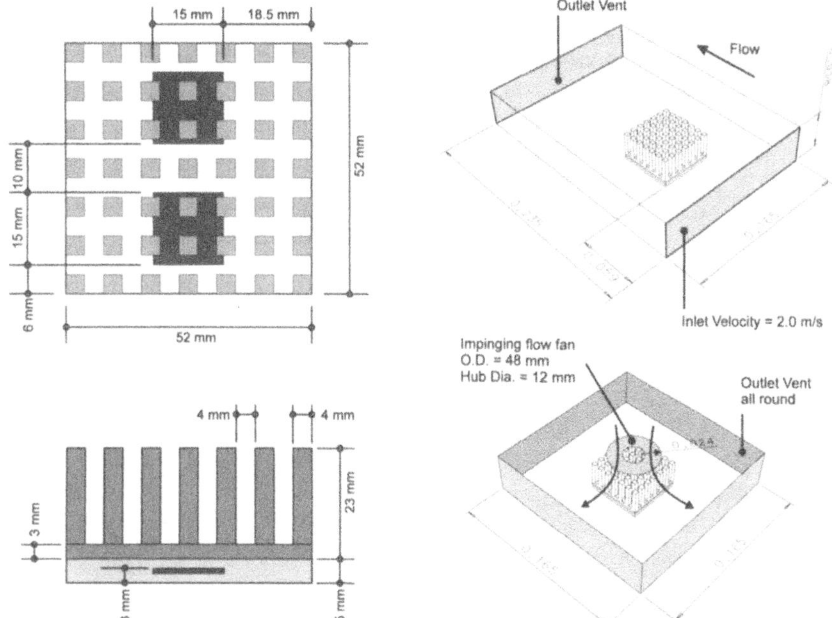

Figure 2. Module and pin-fin heat sink assembly

Figure 3. Parallel flow and Impinging flow

4. Element Mesh

The heat sink and air domain are meshed independently since the fluid and solid mesh do not need to align. The fluid mesh varies between each model and is shown for several of the cases. The mesh for the square pin-fin heat sink is the same for both case 1 and case 2 and is shown in Figure 4. The heat sink mesh for the circular pin-fin heat sink is very similar except 8 wedge shape elements are used to approximate the circular pin.

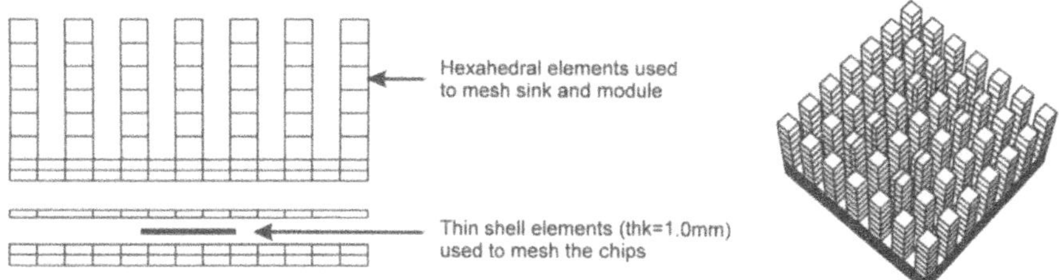

Figure 4. Exploded and perspective view of element mesh on square pin heat sink

Note that even though a coarse mesh is used to model the heat sink, a finer fluid mesh can be used to model the surrounding fluid. This instantaneous transitioning of the solid/fluid mesh is possible due to the thermal coupling approach discussed previously.

5. Results

5.1 CASE 1: SQUARE FIN WITH PARALLEL FLOW

The geometry for the parallel flow case is shown in Figure 3. A uniform flow velocity of 2.0 m/s is defined at the inlet for the parallel flow condition. Based on the projected heat sink cross-section area in the inlet plane, the volumetric flow rate over the assembly is 0.1435 m^3/min. Figure 5 shows the fluid element mesh used to model these flow conditions.

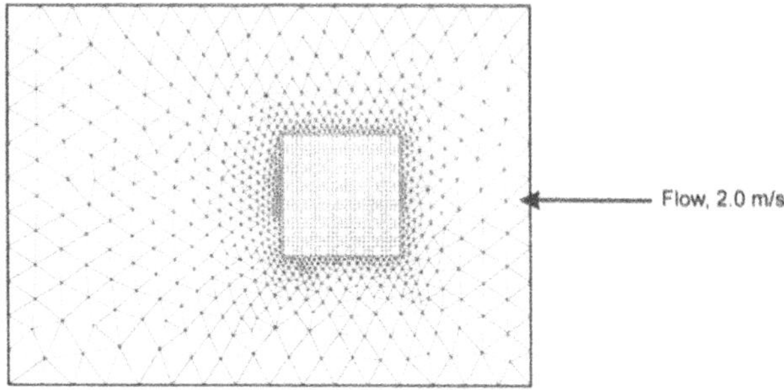

Figure 5. Fluid element mesh for parallel flow with square pin-fin heat sink showing mesh transitioning.

The mesh consists of hexahedral elements and wedge shaped elements. Two hexahedral elements (3 finite volumes) are used to model the air between each pin fin. The wedge shape elements transition the mesh from the detail at the heat sink to the boundaries of the flow domain. Transitioning dramatically reduces the number of cells (or nodes) required to model this geometry.

A solution to the steady-state flow and thermal conditions was computed using I-DEAS Electronic System Cooling. Table 1 compares the results calculated by Kamath [1] and those calculated by this study. The results for chip junction temperature and pressure drop across the device compare well with those values calculated by Kamath. Note that the predicted value of 74.4°C is approximately 5°C hotter than Kamath's prediction but this value agrees well with the measured value of 74°C [15].

Quantity	Kamath	Agonafer & Free
Chip Temp, °C	69.5	74.4
Total Pressure Drop (N/m²)	9.3	10.7
Number of Cells / Nodes	176,800	16,404

Table 1. Comparison of results calculated by Kamath [1] and Agonafer &Free for square pin heat sink in parallel flow

The convective heat flux from the surface of the pins is also computed and the heat flow (in Watts) for each pin is shown in Figure 6. The values computed by Kamath are plotted with the results from this study.

The calculations show that the heat flow from each pin decreases along the length of the heat sink in the direction of flow. The most effective pins are on the leading edge of the heat sink. Kamath predicted the total convected heat from all pins is 17.05 W and this study predicts 17.15 W. The remaining heat is convected from the heat sink base and module.

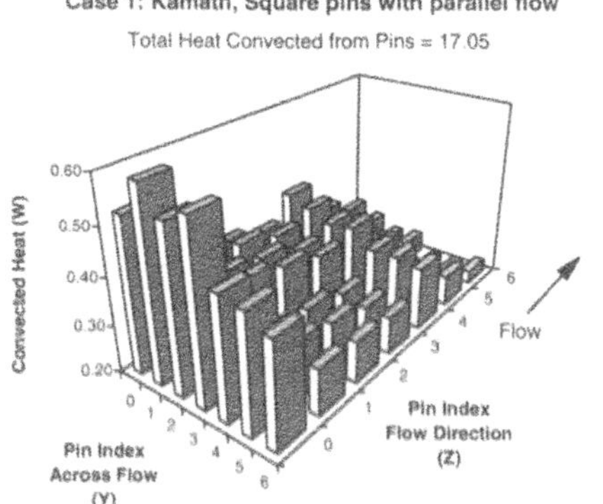

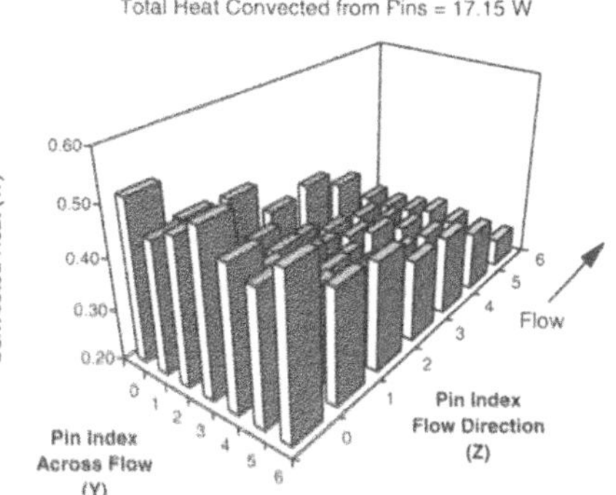

Figure 6. Heat loss (W) for each pin for square pin-fin heat sink in parallel flow conditions. Calculated results from Kamath [1] and Agonafer and Free.

Kamath also investigated the effect of grid refinement on the computed solution. He found that grid refinement had a dramatic effect on solution accuracy. This behavior is attributed to the formulation of temperature and velocity log-law wall functions. A similar study was conducted using I-DEAS ESC. The solution dependence on mesh size was dramatically reduced by the semi-analytical mesh correction factors applied by I-DEAS ESC.

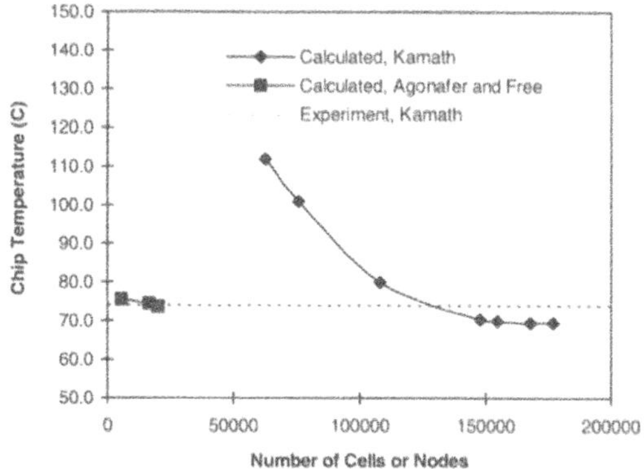

Figure 7. Comparison between Kamath [1] and Agonafer & Free on the effect of grid/node refinement on chip temperature prediction

The identical geometry was free meshed with a coarse tetrahedral mesh (2 finite volumes between pins) and a finer tetrahedral mesh (4 finite volumes between pins). The solution variation due to mesh size is clearly shown in Figure 7. Even though the mesh size used in this study is substantially less, the number of control volumes (or cells) used between pins compares closely with Kamath.

5.2 CASE 2: SQUARE FIN WITH IMPINGING FAN FLOW

The geometry for the impinging fan flow case is shown in Figure 3. The circular inlet fan and hub are modeled as shown with a specified volume flow rate of 0.1435 m^3/min. A circular *fan swirl or twist* component at 17° from the fan flow direction is also modeled. This results in a inlet velocity of 1.7 m/s at the inlet. This velocity is slightly higher that the inlet velocity of 1.64 m/s utilized by Kamath. This is due to the smaller inlet area modeled when using the actual circular fan geometry. None-the-less, the identical volume flow rate is modeled and matched to the effective flow rate imposed on the heat sink in the parallel flow condition.

Quantity	Kamath	Agonafer & Free
Chip Temp, °C	65.7	64.0
Total Pressure Drop (N/m²)	7.1	7.2
Number of Cells / Nodes	160,550	18,007

Table 2. Comparison of results calculated by Kamath [1] and Agonafer & Free for square pin heat sink with impinging fan flow.

The lower chip junction temperature and the pressure drop across the heat sink agrees with the value computed by Kamath. The junction temperature for the impinging flow is lower since the flow is utilized more effectively; there is better high velocity flow coverage throughout the heat sink. The temperature difference between the parallel flow and impinging flow cases were calculated to be higher than predicted by Kamath. This may be due to the effect of modeling the actual fan geometry whereas Kamath idealized the fan and hub as rectangular openings.

5.3 CIRCULAR PIN FIN HEAT SINK

The geometry and flow conditions for the circular pin-fin heat sink is identical to the model for the square pin heat sink shown in Figure 3. For case 3 and 4 the pin-fin heat sink uses circular pins with a diameter of 5.1 mm in order to match the surface area of the square pins. Figure 8 shows the fluid element mesh used to model the impinging flow conditions. The mesh used for the parallel flow case is similar except for the flow domain size and mesh local to the fan flow boundary conditions.

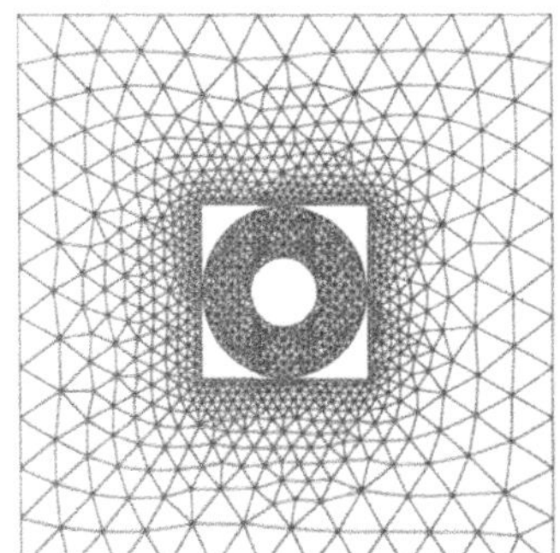

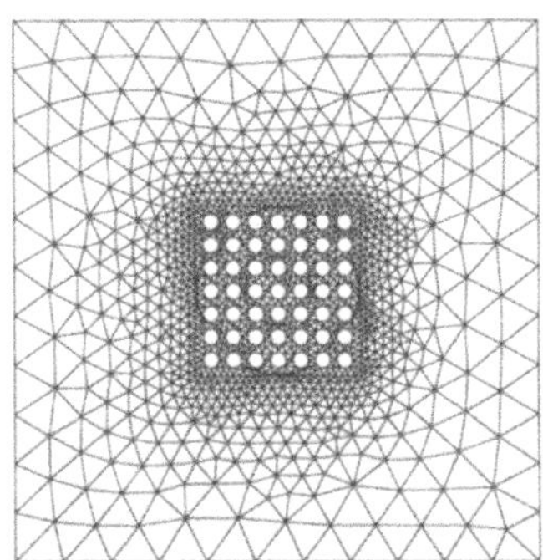

Figure 8. Fluid element mesh for impinging fan flow with circular pin-fin heat sink. Cross-sections at the fan and through the heat sink.

The fluid mesh consists entirely of wedge shaped elements. Two to three elements (3 to 4 finite volumes) are used to model the air flow between each pin fin. This mesh density was found to yield good results without an excessive number of nodes. The results for the steady-state analysis are presented in Table 3.

Quantity	Parallel flow	Impinging flow
Chip Temp, °C	69.6	62.0
Total Pressure Drop (N/m²)	11.4	11.0
Number of Nodes	28,220	34,760

Table 3. Comparison of calculated results for circular pin-fin heat sink. Parallel flow and impinging fan flow.

The chip junction temperature for the circular pin heat sink for the parallel flow conditions is 4.8°C cooler than the square pin under the same flow conditions. The circular pins, with equivalent surface area, are more effective. This is due to the improved distribution of air flow around the circular pin. The square pins promote dead zones behind each pin. The pressure loss across the circular pin heat sink is 2.1Pa higher than the square pin heat sink. This is likely due to the larger pin diameter for the circular pins which results in less space available for flow between the pins. The larger pin diameter was used in order to match the surface area of the circular and square pins.

For the circular pins, like the square pins, for impinging flow conditions the chip junction temperature is less than for parallel flow conditions. In addition, the chip junction temperature for the circular pin heat sink for the impinging flow condition is 2.0°C cooler than the square pin under the same flow conditions. This is less than the temperature difference in parallel flow conditions. Two degrees is within the bounds of solution accuracy so for all intents and purposes these two heat sinks are equally effective in impinging flow conditions.

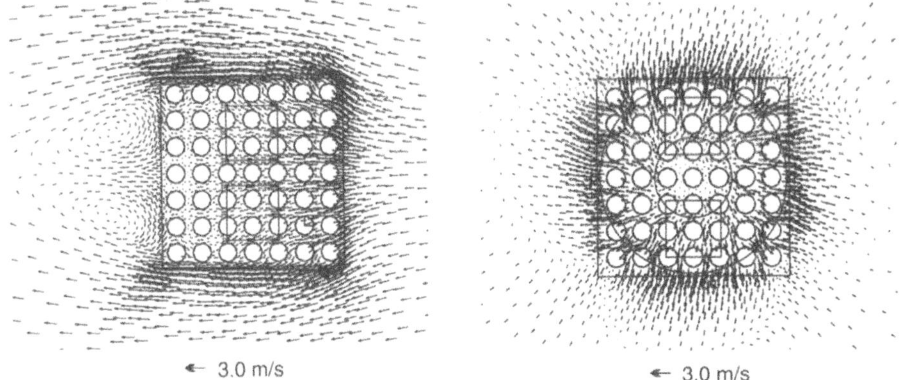

Figure 9. Velocity profiles for circular pin heat sink for parallel flow and impinging flow conditions.

Figure 9 shows the flow velocities for the circular pin-fin heat sink for both parallel flow and impinging flow conditions. Note the slight asymmetry of the flow for the impinging flow case due to the swirl component caused by the fan blades. The peak velocity between pins is approximately 2.8 m/s in both cases. However, the distribution of flow is quite different for each case. For parallel flow conditions the first two rows of pins and the side rows are most effective. This is also evident for the square pin heat sink as shown in Figure 6. For impinging flow conditions the pins directly below the fan blades are most effective whereas the pins at the corners and below the fan hub are least effective.

6. Conclusions

The ease of modeling heat sinks for electronic cooling applications using an integrated design tool is demonstrated. As opposed to finite control volume techniques which require body-fitted-coordinate for complex geometry (and a considerably increased pre-processing time), the set up time for both rectangular and non-rectangular cross-section heat sinks is about the same. In addition, due to the unique approach for modeling conjugate flow, a much faster processing time is realized since the mesh can be transitioned (considerably smaller mesh size) and the fluid/solid meshes do not need to align. In a follow up paper, a technique for integrating a detailed model of a heat sink at the module level to a coarse model at the card and system level will be presented.

The approach is to characterize the flow and thermal behavior of the heat sink and use these global characterizations to model the heat sinks within a system. The characterization captures the overall behavior of the heat sink without modeling the details of flow and heat transfer in the heat sink at the system level. Equivalent fluid pressure loss, obstruction to flow and convection heat transfer from the module is accounted for in the heat sink model. The application for this technology can be extended to cataloging common heat sink designs for packaging applications.

7. References

[1] Kamath, V., Air injection and forced convection cooling of single and multi-chip modules: A computational study, *Proceedings of the 4th ITHERM Conf*, IEEE Catalogue Number 94CH3340-7, pp.83-90.

[2] I-DEAS , I-DEAS Master Series and I-DEAS Electronic System Cooling are trademarks of Structural Dynamics Research Corporation. SDRC is a Registered Trademark of Structural Dynamics Research Corporation.

[3] Mansuria, M.S., and Kamath, V., *Design Optimization of a High-Performance Heat-Sink/Fab Assembly*, ASME 94-WA/HTD-Vol. 292, pp.95-103.

[4] Patankar, S.V., *Numerical Heat Transfer and Fluid Flow*, Hemisphere Publishing, 1980.

[5] Emery, A.F., and Mortazavi, H.R., *A Comparison of the Finite Difference and Finite Element Methods for Heat Transfer Calculations*, NASA Conf. on Computational Aspects of Heat Transfer, 1981.

[6] Patankar, S.V., *Numerical Heat Transfer and Fluid Flow*, Hemisphere, Washington, D.C., 1980.

[7] Schneider, G.E. and Raw, M.J., Control-Volume Finite Element Method for Heat Transfer and Fluid Flow Using Co-located Variables-1. Computational Procedure, *Numerical Heat Transfer*, Vol.11, pp.363-390, 1987.

[8] M. E. Thomas, N. R. Shimp, M. J. Raw, P. F. Galpin and G. D. Raithby, The Development of an Efficient Turbomachinery CFD Analysis Procedure, *AIAA/ASME/SAE/ASEE 25th Joint Propulsion Conference*, Monterey, CA, July 10-12, 1989.

[9] B. R. Hutchinson and G. D. Raithby, A Multigrid Method Based on the Additive Correction Strategy, *Numerical Heat Transfer*, Vol.9, pp.511-537, 1986.

[10] Rhie, C.M. and Chow, W.L., *A Numerical Study of the Turbulent Flow Past an Isolated Airfoil with Trailing Edge Separation*, AIAA paper 82-0998, 1982.

[11] Raw, M. J., *A Coupled Algebraic Multigrid Method for the 3D Navier-Stokes Equations*, 10th GAMM Seminar, Kiel, 1994.

[12] W. L. Briggs, *A Multigrid Tutorial*, SIAM, Philadelphia 1987.

[13] Kader, B.A., Temperature and concentration profiles in fully turbulent boundary layers, *Int. J. Heat and Mass Trans.* 21, No.9, pp 1541-1544., 1981

[14] *The I-DEAS Electronic System Cooling User's Guide*, MAYA Heat Transfer Technologies, 1995

[15] Kamath, V., Conversation with A. Free discussing experimental measurements for the square pin-fin heat sink in parallel flow conditions, June 1995.

ANALYSIS OF 3D CONJUGATE HEAT TRANSFER IN ELECTRONIC BOARDS: INTERACTION BETWEEN THREE INTEGRATED CIRCUITS

B. FOURKA and J.B. SAULNIER
Laboratoire d'Etudes Thermiques (URA CNRS 1403), ENSMA, B.P. 109
86960 Futuroscope Cedex, FRANCE

Abstract

A numerical study is performed by a multigrid technique, to simulate the combined heat transfer conduction/laminar convection in a three-dimensional channel, with three protruding heated integrated circuits (ICs). The effects of the variation of the streamwise spacing between the ICs on the conjugate heat transfer have been studied through the evolution of the velocity, the temperature, the Nusselt number and heat flux fields.

Nomenclature

B	height of the IC	L_E	distance between the entrance and the IC
e	plate thickness	L_S	transversal distance betwen the ICs
H	channel height	S_L	streamwise spacing betwen the ICs
L	channel length	U_e	inlet velocity (m/s)
L_W	channel width	Q	heat dissipation of the IC (W)
L_T	IC width	x, y, z	coordinates
lx	chip length	u, v, w	velocity components
lz	chip width	λ_f	thermal conductivity of the fluid (W/m.K)
T_p	temperature of surfaces	λ_{PCB}	thermal conductivity of the PCBs (W/m.K)
T	temperature (°C)	λ_{IC}	thermal conductivity of the ICs (W/m.K)
T_e	inlet temperature (C°)	ϕ	general variable
U_e	inlet velocity (m/s)		

1. Introduction

The miniaturization of Integrated Circuits (ICs) and the resulting increase in power density have emphasized the importance of heat transfer in electronic equipments. The detailed analysis of the heat transfer between air, integrated circuits and Printed Circuit Boards (PCBs) is in fact a challenge, because even with simple geometries, convection from complex flows, with recirculations in the cavities between the ICs, interacting with conduction in the solid parts is not easy to simulate. More precisely, it needs very fine modelling with a great number of nodes. Therefore, the computation is to be performed with efficient numerical techniques. First attemps of simulation were performed with 2D models (Habchi and Acharya [1], Afrid and Zebib [2,3], Wang [4]). Although, the influence of the presence of the ICs in a three-dimensional channel on the flow and temperature fields has been demonstrated by Shaw et al. [5], the thermal boundary condition was a simple one, with a given temperature imposed around the obstacle. But a realistic 3D model (3D obstacle, conjugate heat transfer, that means conduction in the IC and in the Printed Circuit Board, power dissipated in a thin layer inside the IC...) needs

E. Beyne et al. (eds.), Thermal Management of Electronic Systems II, 63-72.

about 160000 nodes for a channel with 3 ICs. To perform our simulation, we adapted the multigrid method. After we have shown the conjugate effect of conduction/convection heat transfer in a channel with one and two ICs [6,7], we consider here the case of three ICs and we study in this paper the sensitivity of the heat transfer, to the streamwise spacing between the ICs.

2. The Description of the Problem

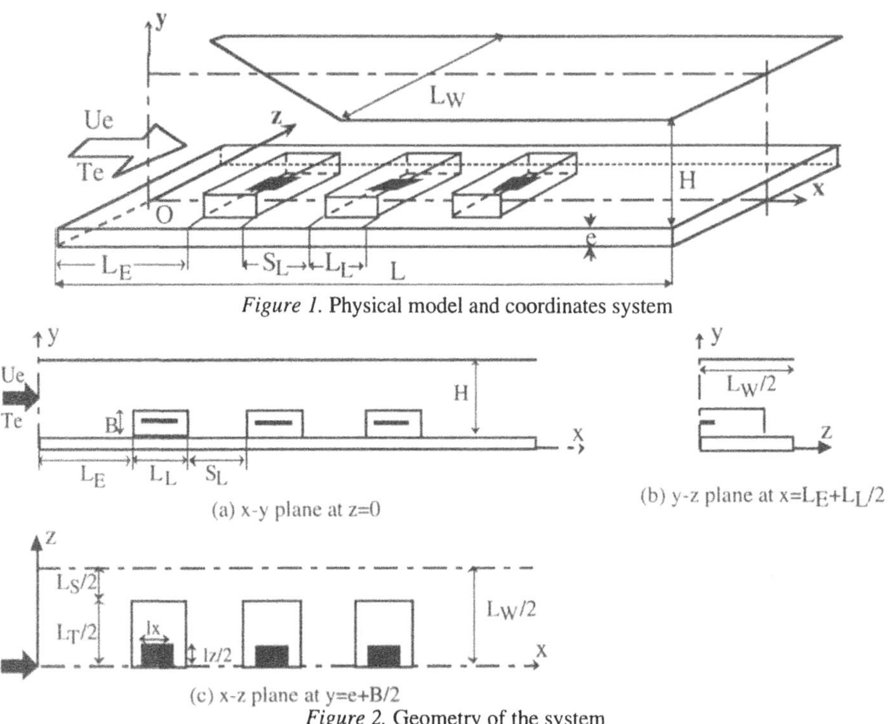

Figure 1. Physical model and coordinates system

Figure 2. Geometry of the system

The problem we present here simulates in a rather realistic way the cooling of an electronic equipment (cf. Fig. 1) in forced convection. One board supports the ICs, and other board delimits a channel in which the cooling air is flowing. We consider the ICs as obstacles, with the presence of the rather small size heated chip (black layer in Fig. 2) within the IC, and we take account of the different thermal conductivities of the materials. Figure 1 shows the physical model to be investigated. Because both geometric properties and physical conditions are symmetric relatively to the surface (z=0), we consider half of the physical domain as our computational domain. The velocity and temperature of the cooling fluid at entrance are Ue=1 m/s and Te=20 °C, respectively. The first IC with length L_L, width L_T, height B and dissipating power Q is located at a distance L_E from the channel entrance (cf. Fig. 2). The ICs are identical and the streamwise spacing between them is S_L. Let us observe that a periodic boundary condition is applied to the surface at z=L_W/2 in order to simulate a series of components located side by side in the z-direction. The planes at y=0 and y=H+e are submitted to cyclic thermal boundary conditions. This means that the thermal condition which rules in the plane y=0 is identical to that in the plane y=H+e. Table 1 summarizes the geometrical dimensions shown in Fig. 1 and Fig. 2, and the physical properties.

Symbol	Value	Symbol	Value	Symbol	Value	Symbol	Value
L (mm)	120	L_T(mm)	20	lz (mm)	5	Q (W)	1
H (mm)	11	L_S(mm)	10	$L_W=L_T+L_S$ (mm)	30	λ_{PCB} (W/m.K)	2
L_E (mm)	20	e (mm)	1.6			λ_{IC} (W/m.K)	0.7
L_L(mm)	9	lx (mm)	5	B (mm)	4	λ_f (W/m.K)	2.5×10^{-2}

Table 1. The geometrical dimensions and physical properties

The effects on the conjugate heat transfer of a variation of the streamwise spacing S_L between the ICs are investigated here. Three configurations are considered, with different values of the spacing range, characterized by $S_L = 5$, 10 and 15 mm.

3. Governing equations and solution procedure

The problem studied is elliptic in nature and all the fluid properties are supposed uniform. The flow is steady, laminar, and the effects of heat radiation and viscous dissipation are neglected. A unique form, used to express the governing equations, is as following:

$$u\frac{\partial \phi}{\partial x} + v\frac{\partial \phi}{\partial y} + w\frac{\partial \phi}{\partial z} = \Gamma\phi\left(\frac{\partial^2 \phi}{\partial x^2} + \frac{\partial^2 \phi}{\partial y^2} + \frac{\partial^2 \phi}{\partial z^2}\right) + S\phi \qquad (1)$$

where ϕ is a general variable, Γ_ϕ is a diffusion coefficient, and S_ϕ is a source term, associated to appropriate boundary conditions.

The finite volume forms of the above differential equations are derived by integrating them over discrete control volumes (Patankar [8]) in the physical domain. A staggered mesh system is used in which the discrete velocities are located on the faces of the finite volume cells, and the discrete pressures and temperatures are situated at the cells centres. The procedure, we have retained in our multigrid technique (Brandt [9]) to solve the equation (1) on each level of grids, is the SIMPLEC algorithm (Doormaal and Raithby [10]). The restriction to the coarse grid velocities is defined as the arithmetic mean of four variables. The prolongation operator is derived in all cases by using bilinear interpolation. The full multigrid procedure FAS-FMG (Vanka [11]), which starts at the coarsest grid and progresses in an adaptive manner to the finest grid, is appropriate.

4. Results and discussion

The influence of the streamwise spacing S_L between the ICs on the velocity, the temperature, the Nusselt number and heat flux fields will be discussed. We also analyze the global heat flux convected by the different surfaces of the ICs and conducted to the board.

4.1. DISTRIBUTIONS OF THE VELOCITY FIELD

The vertical u-velocity profiles across the channel at z=0 just behind the first and the second IC are plotted in Fig. 3. Even if they look rather similar, for a given value of S_L, there are still small local variations between the profiles downstream the first and the second IC. This can provide an idea of the validity of the hypothesis of periodical velocity distribution in the main flow direction, that was classically used in the past. We observe that the intensity of the recirculation flow is quite high across the section just downstream of the first IC (cf. Fig. 3a) for the highest S_L value. We also remark that the flow is very weak in the "open cavity" situated between the components, for S_L=5 mm. However, downstream the third IC, the influence of the streamwise spacing S_L on the u-velocity is not important (cf. Fig. 5).

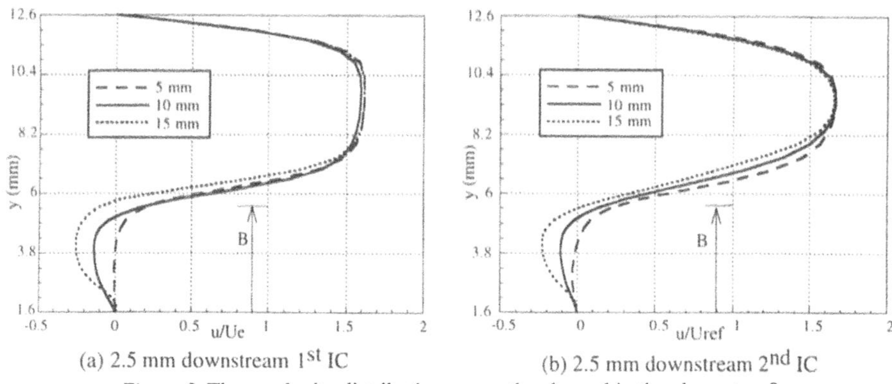

(a) 2.5 mm downstream 1st IC (b) 2.5 mm downstream 2nd IC

Figure 3. The u-velocity distribution across the channel in the plane at z=0

The u-velocity profile in the transverse z-direction, near the leading edge of the second IC is presented in Fig. 4a : this profile still looks similar to the corresponding one, near the leading edge of the third IC (cf. Fig. 4b). In each one, the values of the axial velocity increases with S_L, which is certainly connected to the development of the wake behind the first component. Figure 6 shows the evolution of the u-velocity in the symmetry plane (z=0) just above the second IC (y=e+B+0.2 mm). This longitudinal component u decreases in general for an increase of S_L=5 mm up to S_L=15 mm : the development of the wake, which has not yet involved a reattachment point, leads to a deficit of the flow in plane of symmetry of the second IC. Besides, this deficit is also confirmed by Fig. 4, which shows that the u-velocity values at z=0 are always lower than anywhere else. The evolution of the u-velocity above the third IC is much about the same as the one above the second IC.

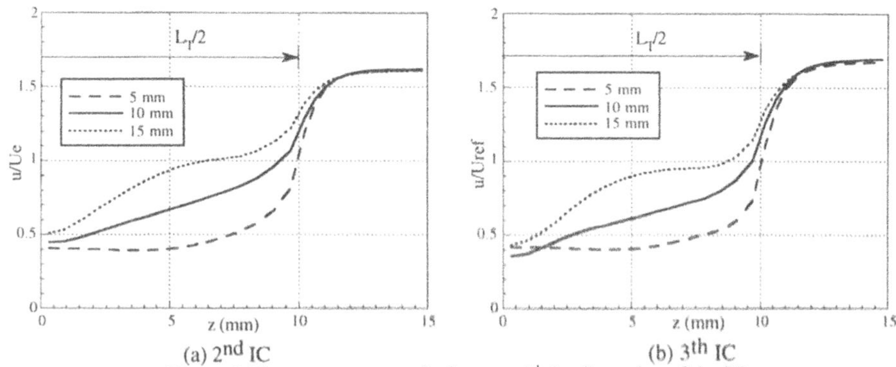

(a) 2nd IC (b) 3th IC

Figure 4. The transverse u-velocity near the leading edge of the ICs

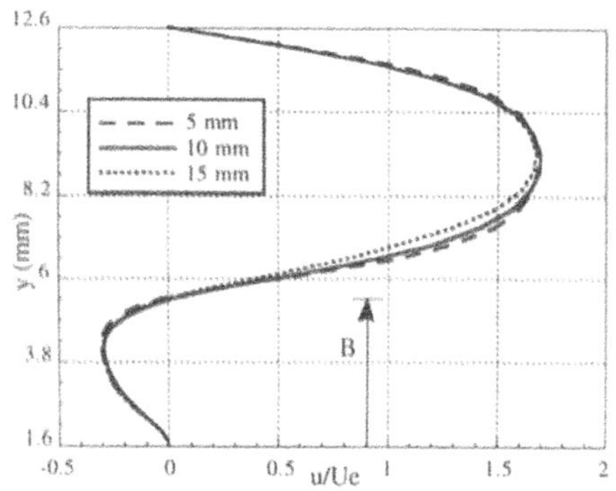

Figure 5. The u-velocity distribution across the channel in the plane at z=0 downstream 3th IC

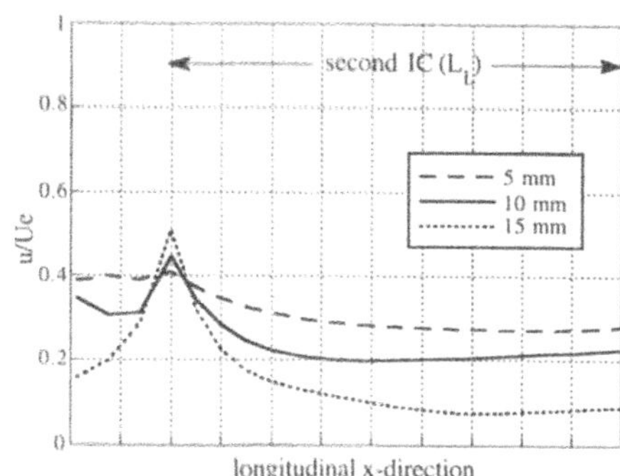

Figure 6. The longitudinal u-velocity in the symmetric plane z=0

4.2. DISTRIBUTION OF THE CHIP TEMPERATURE

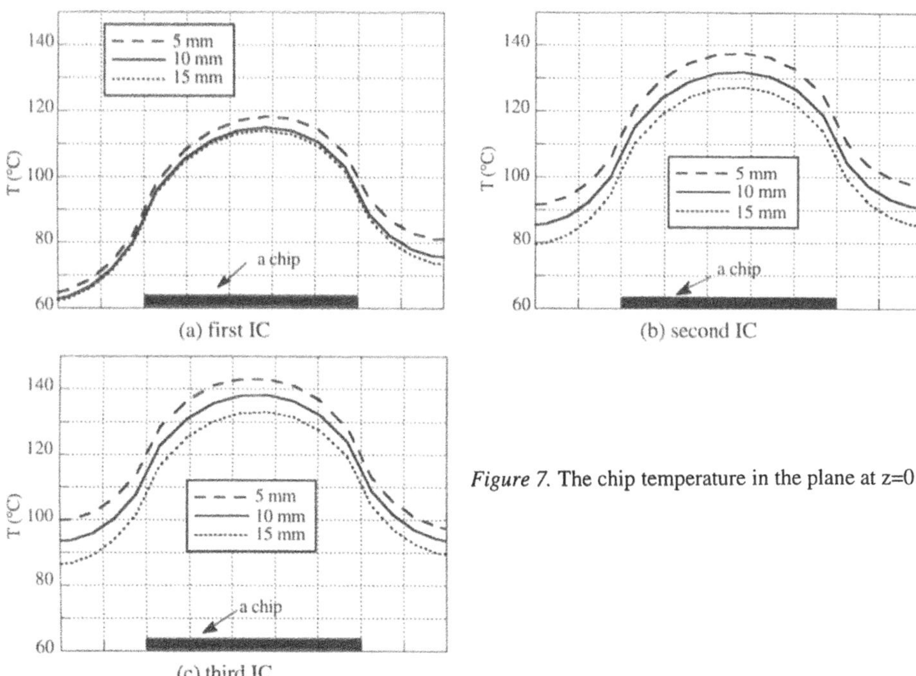

Figure 7. The chip temperature in the plane at z=0

The influence of the streamwise spacing S_L on the temperature distribution along the line passing through the chip of the ICs in the symmetry plane can be observed in Fig. 7. We remark that the third IC is always warmer than the first and the second ones. But the increase of temperature is important between the first and the second ICs (# 20°C for S_L=5 mm) whereas is it quite smaller between the second and the last one (5 °C for S_L=5 mm). This can be explained by an efficient convective heat transfer for the first IC

($\overline{Nu}$ #16), but a rather smaller value, for the two last ones ($\overline{Nu}$ #9 to 7). Besides (see § 4.3.) the part of the PCB attached to the third one is always greater than for the second one: the fin effect developed in the PCB can also explain the small increase between second and third IC. Moreover, the increase of S_L induces a decrease of the maximum temperature of the chip: by 4°C for the first IC and 10°C for the second and the last ones.

4.3. DISTRIBUTION OF THE NUSSELT NUMBER AND THE HEAT FLUX

The local (Nu_x) and average ($\overline{Nu_x}$) Nusselt number at the points located on the surface normal to the x-axis are defined by:

$$Nu_x=\frac{h_x\ H}{\lambda_f} \quad \text{where} \quad h_x=\frac{-\lambda_f(\frac{\partial T}{\partial x})_p}{T_p-T_e} \quad \text{and} \quad \overline{Nu_x}=\frac{\overline{h_x}\ H}{\lambda_f} \quad \text{where} \quad \overline{h_x}=\frac{1}{S}\int_S h_x\ dS$$

S represents the area of surfaces of the IC. Nu_y and Nu_z are defined in a similar way.

For the first IC we observe (cf. Fig. 8) that in the front, upper and lateral surfaces thc local Nusselt number is insensitive to the variation of S_L. But, at the rear surface, as S_L increases the local Nusselt number also increases. It is due to the fact that an increase of S_L induces a higher reverse circulating flow towards the IC, which in the symmetry plane, acts as an impinging jet, and reinforces the local heat transfer.

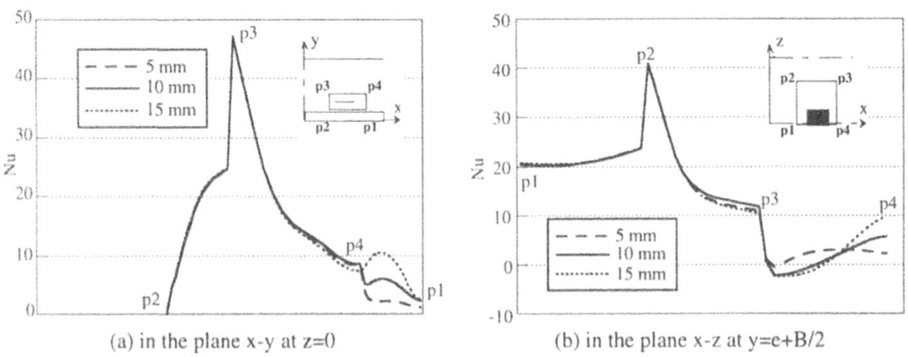

(a) in the plane x-y at z=0 (b) in the plane x-z at y=e+B/2

Figure 8. The local Nusselt number around the first IC

Figure 9 shows the evolution of the local Nusselt number around the second IC. An increase of the S_L value clearly induces a relatively higher value of the local Nusselt number in the front surface (p2p3 on Fig. 9a and 9d and p1p2 on Fig. 9b). It is due to the acceleration of the flow in the upstream of the second IC. Moreover, on the lateral surface, the local Nusselt number is a slightly increasing function of the S_L value (p2p3 on Figs. 9b and 9c). On the upper surface (p3p4 on Fig. 9a), the local Nusselt number in the axial symmetry plane (z=0) evolves as the u-velocity (cf. Fig. 6) and decreases when increasing S_L. But, in the upstream zone of this upper surface when leaving the vicinity of z=0, and scanning the surface (p3p4 on Fig. 9c), the local Nusselt number then increases greatly with S_L. It can be explained by the corresponding evolution of the u-velocity (Fig. 4a). This general trend is confirmed by Fig. 9d, which shows that the behaviour in the symmetry plane z=0 is a singular one both for velocity and Nusselt number. Finally, we observe at the rear surface that the local Nusselt number increases with S_L.

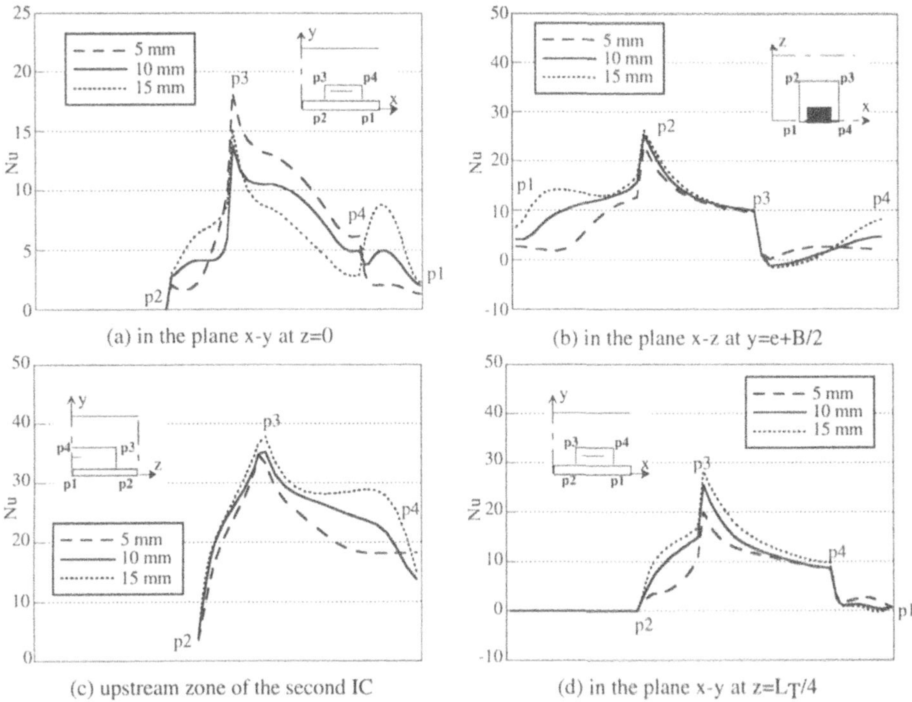

Figure 9. The local Nusselt number around the second IC

Concerning the last IC, we observe that the local Nusselt number values (cf. Fig. 10) are almost always lower than for the second IC. Moreover, on the front, upper and lateral surfaces, the local Nusselt number evolutions are similar to those corresponding to the second IC. Let us comment the particular case of the rear surface: the local Nusselt number is insensitive to the variation of S_L (p4p1 on Fig. 10a and p3p4 on Fig. 10b). Besides, the values of this Nusselt number are rather high, compared to the upper surface (see Fig. 10a), which is due to the effect of the recirculation of flow which at the position is not constrained by a downstream IC.

Let us now discuss the evolution of the average Nusselt number of the different surfaces (tables 2, 3 and 4) and the percentages of heat flux convected and conducted around the ICs (tables 5, 6 and 7). The average Nusselt numbers of the first IC are almost insensitive to S_L. But, we observe that for the second and third ICs, these numbers are monotonously increasing with S_L for the front, upper and lateral surfaces: this confirms that the singular behaviour we discussed in the symmetry plane z=0, has just a small and local influence. The global values (mean value, weighted by the surface areas) are regularly decreasing from the first to the third IC. Of course, this global value is insensitive to S_L for the first IC, and it increases with S_L for the two other ones, in monotonous way.
Concerning the fluxes, an increase of S_L accentuates the conduction from the first IC to the board: the fin effect of the board becomes locally more efficient because the surface of

board involved in this fin effect increases. For the second IC, there is a kind of trade off between the increase of this fin effect, and the modification induced by S_L in the convection: the results is that there is a slight increase of the global convected flux. Finally, this increase of the convected flux is more important for the third and last IC.

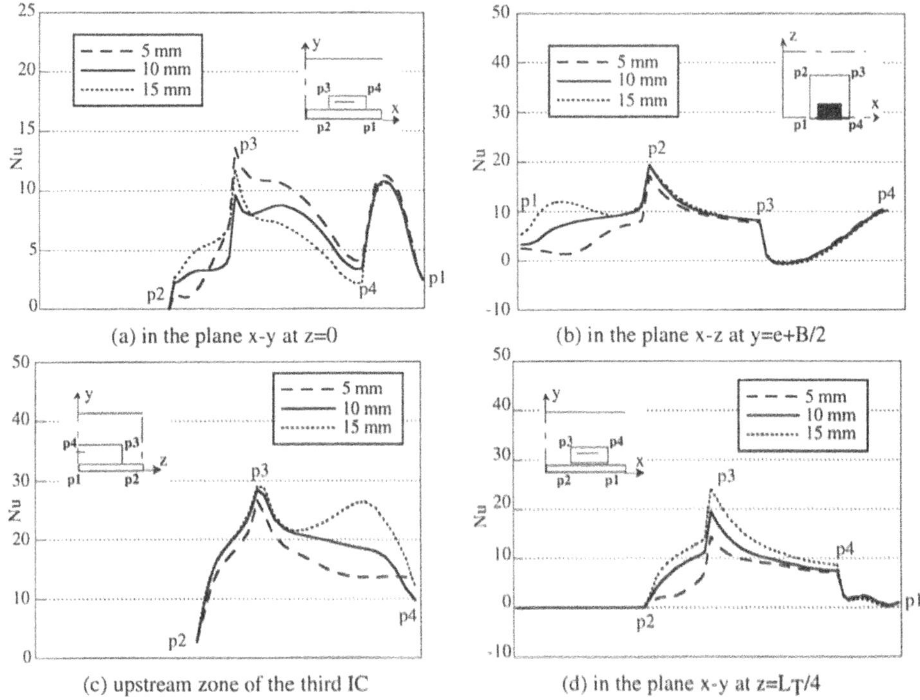

(a) in the plane x-y at z=0

(b) in the plane x-z at y=e+B/2

(c) upstream zone of the third IC

(d) in the plane x-y at $z=L_T/4$

Figure 10. The local Nusselt number around the third IC

S_L (mm)	upper surface	rear surface	front surface	lateral surface	global
5	20.33	1.69	18.57	16.64	15.72
10	20.61	1.37	18.44	16.95	15.81
15	20.03	1.83	18.64	16.47	15.60

Table 2. Average Nusselt number of the first IC

S_L (mm)	upper surface	rear surface	front surface	lateral surface	global
5	12.75	1.67	6.17	12.67	9.31
10	13.33	1.46	9.36	13.24	10.24
15	14.84	1.91	11.85	13.72	11.55

Table 3. Average Nusselt number of the second IC

S_L (mm)	upper surface	rear surface	front surface	lateral surface	global
5	9.94	3.03	4.08	9.83	7.44
10	10.75	2.82	6.93	10.55	8.43
15	12.56	2.63	9.08	10.88	9.66

Table 4. Average Nusselt number of the third IC

S_L(mm)	interface (conduction)	upper surface	rear surface	front surface	lateral surface
5	54.35	29.26	1.43	9.45	5.51
10	56.91	27.86	1.33	8.82	5.08
15	58.24	26.44	1.79	8.77	4.77

Table 5. Percentage of the heat flux repartition around the first IC

S_L(mm)	interface (conduction)	upper surface	rear surface	front surface	lateral surface
5	57.61	28.23	1.84	4.98	7.34
10	58.38	26.27	1.69	6.90	6.76
15	56.66	26.72	2.12	8.22	6.29

Table 6. Percentage of the heat flux repartition around the second IC

S_L(mm)	interface (conduction)	upper surface	rear surface	front surface	lateral surface
5	60.32	24.93	3.90	4.02	6.83
10	59.40	24.39	3.45	6.20	6.57
15	56.91	26.30	3.04	7.60	6.15

Table 7. Percentage of the heat flux repartition around the third IC

4.4. SOME PARTICULAR COMMENTS CONCERNING THE SECOND IC

Let us now compare the results concerning the second IC as the last one (PCB with 2 ICs: [7]) and when it is situated between two ICs (present paper). Table 8 shows that when putting another IC after the second one, the Nusselt number increases and the convection becomes more efficient around this IC. On the contrary, it is the conduction that is more efficient when the second IC is the last one because the fin effect of the PCB is important (cf. Table 9). Finally, relatively to the second IC, the addition of a third one, decreases of course the fin effect created by the board, reinforces locally the convection, but leads to an increase of the chip temperature (see Table 10), which in case of small spacing (S_L=5 mm) is significant (+ 6°C).

	upper surface		rear surface		front surface		lateral surface		global	
S_L(mm)	middle	**last**	middle	**last**	middle	**last**	middle	**last**	middle	**last**
5	12.75	**12.38**	1.67	**3.17**	6.17	**5.78**	12.67	**12.18**	9.31	**9.28**
10	13.33	**13.08**	1.46	**3.10**	9.36	**9.15**	13.24	**12.98**	10.24	**10.36**
15	14.84	**14.57**	1.91	**3.10**	11.85	**11.43**	13.72	**13.30**	11.55	**11.51**

Table 8 Average Nusselt number of the second IC

	conducted		convected	
S_L(mm)	middle	**last**	middle	**last**
5	57.61	**60.74**	42.39	**39.26**
10	58.38	**59.67**	41.62	**40.33**
15	56.66	**57.60**	43.34	**42.40**

Table 9. Percentage of the heat flux repartition around the second IC

S_L(mm)	middle	**last**
5	137.6	**131.2**
10	131.8	**128.5**
15	127.2	**125.5**

Table 10. The maximum chip temperature of the second IC

3. Conclusion

We have shown that we are able to predict in a rather detailed way the temperature, velocity and Nusselt number fields and of course the local value of the chip temperature and the Nusselt number mean values.
The effects on the conjugate heat transfer of a variation of the streamwise spacing S_L between the ICs are investigated here. The increase of the S_L value induces a monotonous evolution of the chip temperature. The maximum chip temperature decreases by about 4°C in the first IC and 10°C in the second and the third ones in our operating conditions.
To summarize when increasing the streamwise spacing between the ICs, we observed two kinds of effects on the first IC:
* the convection around the IC, excepting the rear face, is not affected because relevant of a parabolic flow. Concerning the rear surface, the heat transfer is modified thanks to an elliptic effect, induced by the flow in the opened cavity
* the conduction to the board is reinforced, which is still an elliptic effect connected to the displacement of the second IC.

Concerning the second and the third IC, the convection is rather reinforced when increasing the spacing. But, this effect is conjugated with the conductive barrier through the IC it self and may be limited. We have seen a singular behaviour in the plane of symmetry of the upper surface: the Nusselt number is locally decreasing, at the very place where the temperature is high (vicinity of the chip). On the contrary, when leaving this symmetry plane, the Nusselt number increases, but in zones where the temperature becomes lower, because precisely of a bad conductivity of the IC. As a conclusion the positive effect of increasing S_L would be globally more efficient for ICs of higher conductivities.

References

[1] Habchi, S., and Acharya, S. (1986) Laminar mixed convection in a partially blocked, vertical channel, Int. J. Heat Mass Transfer, 29 (11), 1711-1722.
[2] Afrid, M., and Zebib, A. (1989) Natural convection air cooling of heated components mounted on a vertical wall", Numerical Heat Transfer, Numerical Heat Transfer (A), 15, 243-259.
[3] Afrid, M., and Zebib, A. (1989) Three-dimensional laminar and turbulent natural convection cooling of heated blocks Numerical Heat Transfer (A), 19, 405-424.
[4] Wang, H.Y. (1991) Simulation Numérique par Multigrilles des Transferts Conjugués dans les Cartes de Composants Electroniques, Thèse N° d'ordre 405, Université de Poitiers.
[5] Shaw, H.J., Chen, W.L., and Chen, C.K.(1991) Study on the laminar mixed convective heat transfer in three-dimensional channel with a thermal source, J. of Electronic Packaging, 113, 40-49.
[6] Saulnier, J.B., Wang, H.Y., and Fourka, B. (1993) Differences between 2D and 3D models predictions in a channel with dissipating obstacle, Proceedings of Eurotherm Seminar 29, Thermal Management of Electronic Systems, Delft, 107-116.
[7] Fourka, B., and Saulnier, J.B. (1995) Analysis of 3D conjugate heat transfer around two dissipating obstacles locaed in a channel, Proceedings of International Thermal Energy Congress (ITEC95), Agadir, 2, 493-498.
[8] Patankar, S.V. (1980) Numerical Heat Transfer and Fluid Flow, McGraw-Hill Book Company.
[9] Brandt, A. (1977) Multi-Level Adaptive Solution to Boundary- Value Problems, Mathematics of computation, 31, (138), 333-390.
[10] Doormaal, J.P.V., and Raithby, G.D. (1984) Enhancements of the SIMPLE method for predicting incompressible fluid flow, Numerical Heat Transfer, 7, 147-163.
[11] Vanka, S.P.(1986) Block-Implict Multigrid Solution of Navier-Stokes Equations in Primitive Variables, J. of Computational Physics, 65, 138-158

HEAT TRANSFER IN A VERTICAL STACKING OF PLASTIC PACKAGES

Y.SCUDELLER (*), C.VAL(**)
(*) Laboratoire de Thermocinétique, Equipe de Thermique des Interfaces et de Microthermique, URA CNRS 869, ISITEM, Nantes, FRANCE
(**) THOMSON-CSF, Division Outils Informatiques, Colombes, FRANCE

abstract-This paper presents a study of heat transfer in a vertical stacking of plastic packages. The work is based on experiments and simulations. An experimental module, integrating a great number of temperature sensors, has been set up. Temperature profiles, inside the module, have been measured and correlations with natural convection have been identified. An original compact model has been developed.

1. Introduction

Tridimensionnal stackings of plastic packages increase densification of mass memories but heat dissipations constitute a strong limitation [1][2][3]. Temperature levels require an optimal design.

This paper presents an analysis of heat transfer in a vertical stacking of plastic packages. The work is simultaneously based on experiments and simulations. An experimental module, integrating a great number of temperature sensors, has been set up. Temperature profiles have been measured, inside the module, and correlations have been identified with natural convection. Moreover, an original compact model has been developed.

2. Description of the experimental module

The experimental module is represented in figure 1. It is a vertical stacking of 8 thin plastic packages. Dimensions of the module are about 11.2 mm (x), 9.6 mm (y), 19.5 mm (z). Dimensions of a package are 1 mm (x), 10 mm(y), 15 mm (z) each. Each chip dissipates up to 0.7 W. Packages are stacked in the direction of x axis, with 8 copper sheets (thickness of 0.15 mm each). All sheets are connected with the face upper of the module (side 1). y axis is the direction of sheet lengths. Sides 2 and 3 are the lateral faces of the module, perpendicular to the x axis.

33 microthermocouples (25 μm wires) have been integrated during the manufacturing process of the module. Lead Frames (FeNi) and Copper sheets are used as thermoelectric elements to measure the temperature. In this way, two temperature measurements require only three wires (semi-intrinsic technique). In the whole, 48 wires are incorporated (see figure 1) at:

E. Beyne et al. (eds.), Thermal Management of Electronic Systems II, 73-82.

-copper sheet- metallization interfaces (side1)
-copper sheet- plastic package interfaces,
- center and tip of each Lead Frame.

Each thermoelectric wire is attached on side 4 (bottom face) and connected with NiAu Layers. A Nd:YAG laser is used to insulate, by ablation, each electrical contact (area of 100 µm square). Thermocouples wires of 80µm are then soldered on these layers

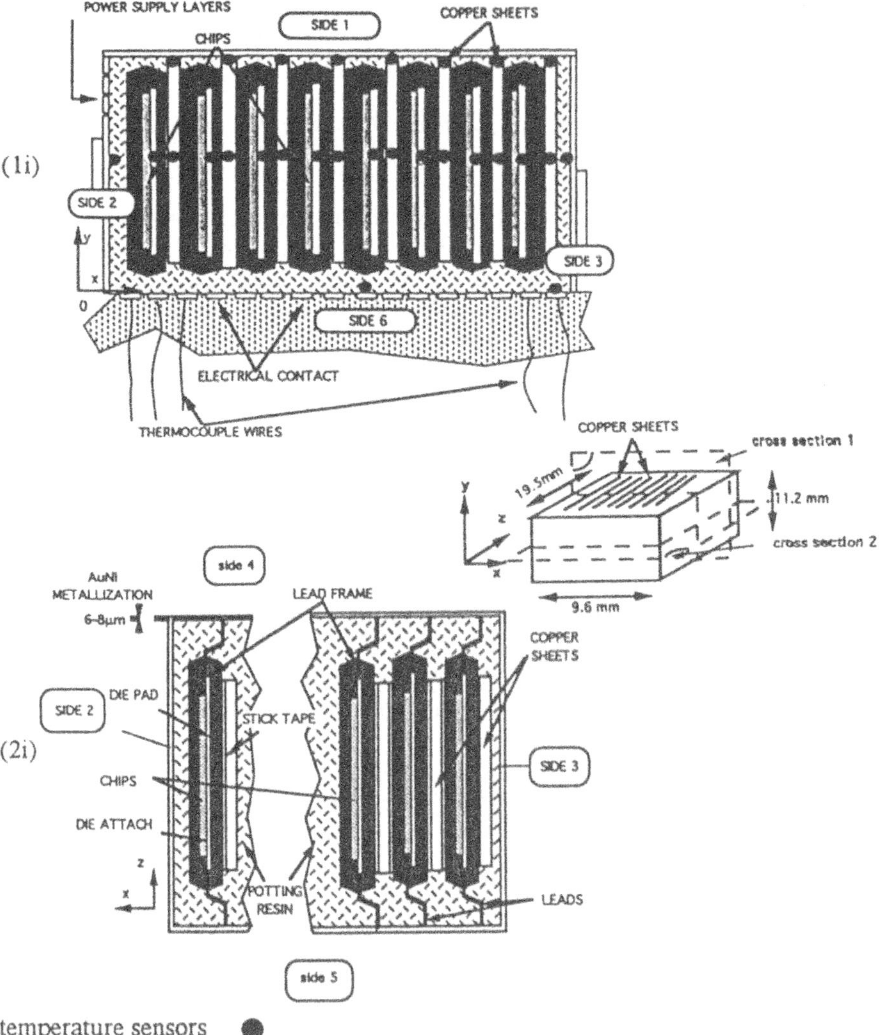

Figure 1: Experimental module (1i) cross-section 1 (2i)cross section 2

3. Experimental arrangement

The experimental arrangement is shown in figure 2. The module (I) is set inside an isothermal enclosure (II). Temperature is controlled, from -40°C to +140°C, by means of a thermostated fluid flow and measured by a thermocouple probe (VI) shielded from thermal radiation. The module is hold on an hollow tube (III) fixed itself on an Aluminium block.(V). This one contains the cold junctions of each thermocouple. Power supply is composed of a tension source 0-6V and a switchboard unit which distribute the current among the chips. Data recording and processing system is composed of a scanning unit and a Digital Microvolmeter

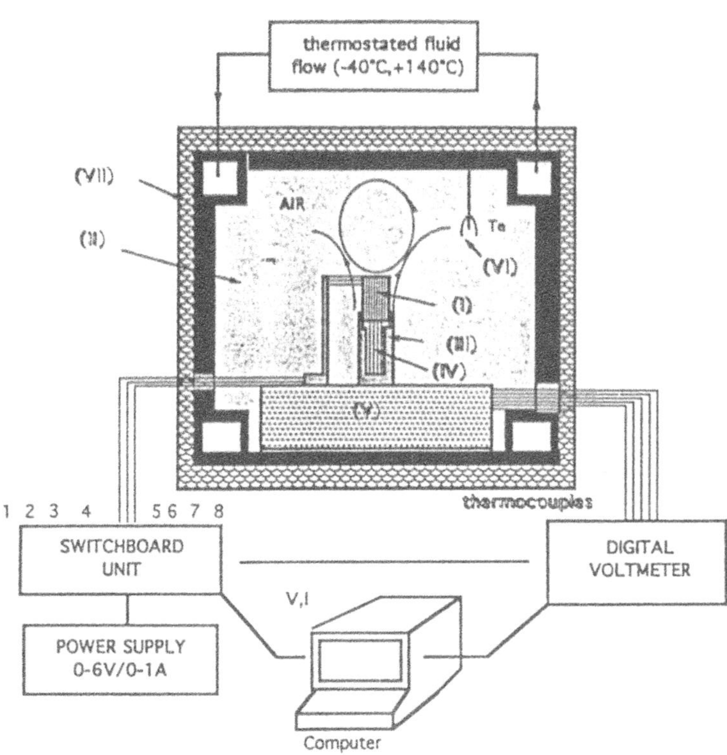

(I)Experimental module
(II)Enclosure
(III)Holder
(IV)Thermocouple wires
(V)Aluminium block (cold juntions)
(VI)Temperature probe
(VII)Insulating foam

Figure 2: Experimental arrangement

4. Experimental results

Typiccal temperature distributions, measured inside the module, are in figure 3. They depend on x and y directions only, because vertical conducting sheets are not in contact with the two lateral sides 4 and 5. Clearly, the two directions, x and y, participate to heat transfer. Surface area of sides 2,3 is approximately equivalent to the one of side 1. Temperature rises on x and y axes, are in the same order of magnitude. Small sections of packages induce a poor thermal resistance in y direction. Temperature rises, inside the module, are about 2-3K/W if all chips work. Important temperature rises occur on x axis if a single chip work (see figure 3). This state bring to the fore that the conducting sheets close to the dissipating element are the most efficient Experimental correlations have been identified. Average temperature T for all dissipating chips may be given by the following relation:

$$P = 0.0080(T - T_e)^{1.16} \tag{1}$$

where P is the total dissipated power and Te, the temperature of the surrounding ambient air. The average surface temperature Ts, for two vertical and one horizontal surfaces (sides 1,2,3, area of 6.2 10^{-4} m^2) is given by:

$$P = 0.00812(T_s - T_e)^{1.15} \tag{2}$$

It follows that the average natural convection coefficient has the expression:

$$h = 13.1(T_s - T_e)^{0.15} \tag{3}$$

It may be observed that the exponent is differente from the classical value (0.25) because the characteristic length of the surfaces is very small - Rayleigh Number is approximatly 100-.

Conduction thermal resistance (from the dissipating chips to the surface) has been founded equal to 2.7 K/W.

Correlations for a single dissipating chip are :

$$P = K(T - T_e)^m \tag{4}$$

It has been identified (m=1.14,K=0.00754 W.$K^{-1.14}$) for a dissipating chip located at one edge of the module and (m=1.15,K=0.0080 W.$K^{-1.15}$) for a center chip. Conduction thermal resistances, corresponding to these configurations, are 10 K/W and 3 K/W.

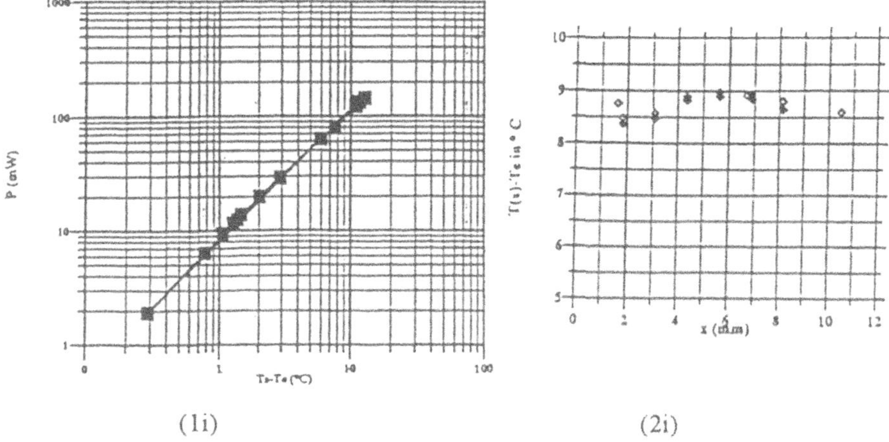

(1i) (2i)

Figure 3: results with 8dissipating chips (1i) average surface temperature in fonction of total dissipated power P (2i)Temperature distribution T_e=20.23°C, P=96.45 mW

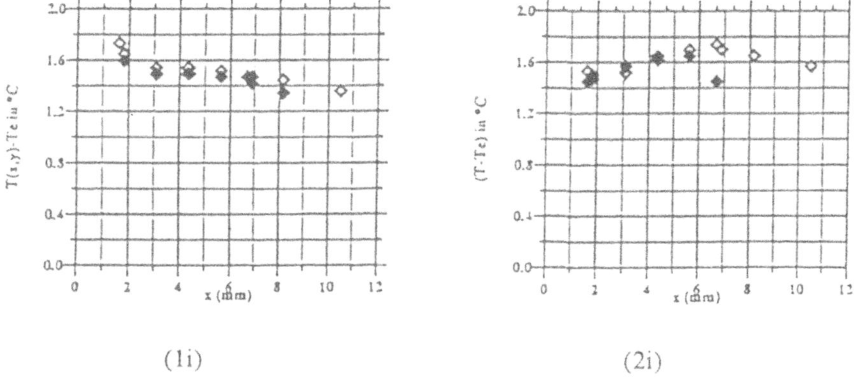

(1i) (2i)

Figure 4:Temperature distributions (1i) dissipating chip located at one end of the module T_e=17.2°C, P=14.26 mW (2i) dissipating chip located at the center of the module T_e=16.1°C, P=15.44mW

5. Theoritical model

A physical model has been set up to analyse the behavior of any stacking during various conditions. Any heat source distributions as well as boundary conditions (forced or natural convection, conduction) may be considered. The model take into account the multidimensionnal feature of temperature profiles but constitutes basically a compact model. A very good agreement is observed with tridimensional simulations[4].

1. principle of the model

The model may be simplified by breaking up the module into several elements where the conduction heat flow is assumed occuring in one major direction.. Temperature distribution may be supposed depending on x, inside each plastic package and metallization, and y, inside each copper plate. Each chip is taken isothermal. All of these elements are interconnected. In the methodology of interconnection, each conducting sheet and metallisation piece may be view as fins. Boundary condititions are characterized by heat transfer coefficients (h_1,h_2.....h_6).

Considering a thin conducting sheet in contact with each two packages (figure5). Each of two sides is connected with a chip (temperature θ_1 ,θ_2) through a thin layer of plastic (conductances K_1 , K_2). The conduction heat flux penetrating at a point y of a lateral side i (i=1 or 2) of the conducting sheet may be written as

$$\varphi_i(y) = K_i(\theta_i - T(y)) \tag{5}$$

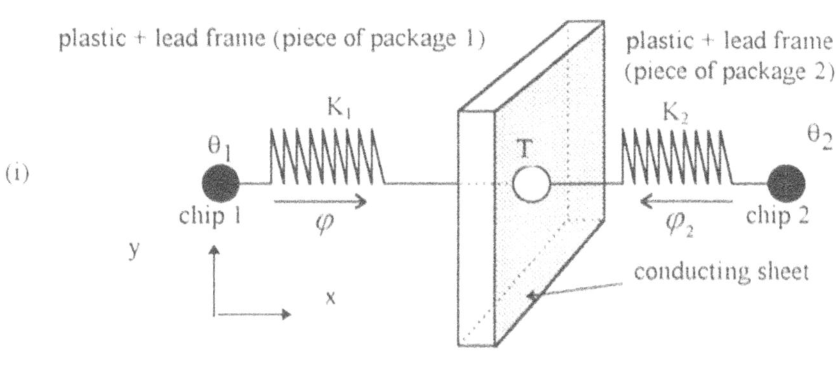

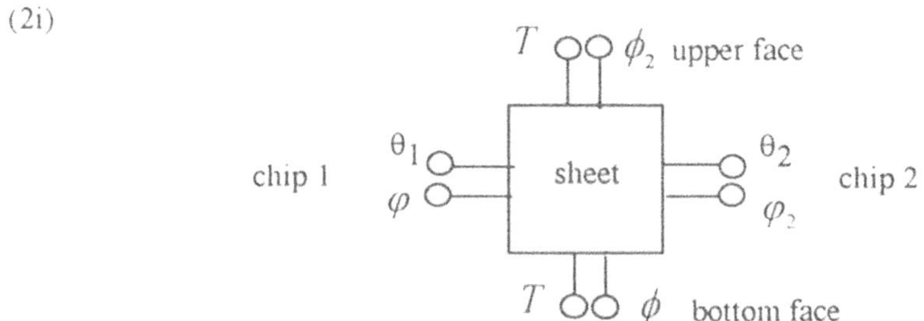

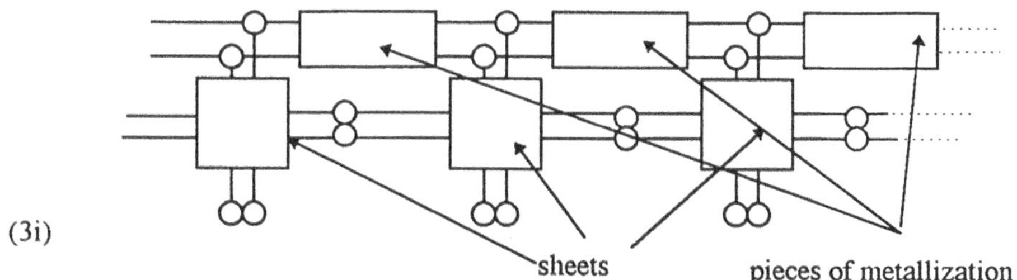

Figure 5: (i) physical representation of the unit element (2i) mathematical representation of the unit element (3i) network equivalent to the stacking

The solution of the one-dimensional steady state energy equation for fins of uniform cross-section (cross sectional area S, length L) gives a vector (temperature, heat flux) defined at one end (T_2, ϕ_2) in function of the one at the other end (T_1 , ϕ_1). The relations depends of the temperature θ_1, θ_2 of the two chips (fig.5). Matricial formulation make appear a first matrix **[m]** and also a second **[m']**. The vector (T_2, ϕ_2) can be written as :

$$\begin{pmatrix} T_2 \\ \phi_2 \end{pmatrix} = \begin{bmatrix} m_{11} & m_{12} \\ m_{21} & m_{22} \end{bmatrix} \begin{pmatrix} T_1 \\ \phi_1 \end{pmatrix} + \begin{bmatrix} m'_{11} & m'_{12} \\ m'_{21} & m'_{22} \end{bmatrix} \begin{pmatrix} \theta_1 \\ \theta_2 \end{pmatrix} \tag{6}$$

with

$$m_{11} = ch(\gamma L); m_{12} = -\frac{sh(\gamma L)}{\lambda\gamma S}; m_{21} = -\lambda\gamma S sh(\gamma L); m_{22} = ch(\gamma L)$$

$$m'_{11} = \alpha_1 (1 - ch(\gamma L)); m'_{12} = \alpha_2 (1 - ch(\gamma L)); m'_{21} = \alpha_1 \lambda\gamma S sh(\gamma L); m'_{22} = \alpha_2 \lambda\gamma S sh(\gamma L)$$

$$\gamma = \sqrt{\frac{K_1 + K_2}{\lambda . e}}; \alpha_1 = \frac{K_1}{K_1 + K_2}; \alpha_2 = \frac{K_2}{K_1 + K_2}$$

where λ denotes the thermal conductivity of the fin.

The expression of heat conduction flow, φ_1, φ_2 , from each chip, in and out of the conducting sheet may be deduced from integration of the temperature distribution T(y):

$$\varphi_1 = \int_0^L K_1 p_1 (\theta_1 - T(y)) dy$$
$$\varphi_2 = \int_0^L K_2 p_2 (T(y) - \theta_2) dy \tag{7}$$

p_1 ,p_2 are wides of each side of the conducting sheet. It leads to the following expressions:

$$\varphi_1 = \frac{K_1 \cdot p_1}{\gamma} \left\{ -sh(\gamma L) T_1 + \frac{ch(\gamma L) - 1}{\lambda\gamma S} \phi_1 + (\gamma - \gamma\alpha_1 + \alpha_1 sh(\gamma L)) \theta_1 + (-\gamma\alpha_2 + \alpha_2 sh(\gamma L)) \theta_2 \right\}$$
$$\varphi_2 = \frac{K_2 \cdot p_2}{\gamma} \left\{ sh(\gamma L) T_1 - \frac{ch(\gamma L) - 1}{\lambda\gamma S} \phi_1 + (\gamma\alpha_1 - \alpha_1 sh(\gamma L)) \theta_1 + (-\gamma + \gamma\alpha_2 - \alpha_2 sh(\gamma L)) \theta_2 \right\} \tag{8}$$

If relations (5) and (6) are gathered together, then a matrix of order 4 allows to link the vectors (temperature, heat flux) at the ends of a group of elements (plastic package and conducting sheet). It can be written:

$$\begin{pmatrix} T_2 \\ \phi_2 \\ \theta_2 \\ \varphi_2 \end{pmatrix} = [M] \begin{pmatrix} T_1 \\ \phi_1 \\ \theta_1 \\ \varphi_1 \end{pmatrix} \qquad (9)$$

A Stacking is represented as a network of such Matrix (figure5). N packages are finally characterized by a matrix of order (3N+2). Inlet parameters are power dissipated by each chip and the surrounding temperature T_e. Outlet pameters are temperature and flux at each node.

2. Results

A very good agreement is observed with experimental datas (figure 6) and tridimensionnal simulations. Copper sheets appear very efficient. Heat flux that can be removed from each face and each sheet is given in figure 7 as a fraction of total dissipated power P. Distribution is uniform on side 1 if all chips dissipate. Nearest sheets are more efficient if a single chip dissipate.

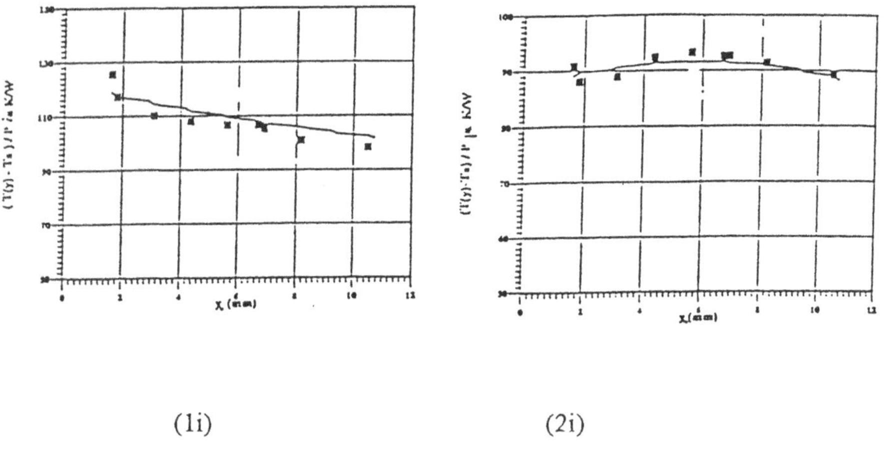

(1i) (2i)

Figure 6: comparaison of experimental (■)and theorical (-----)temperature distributions with natural convection (1i) one dissipating chip (2i) 8 dissipating chips

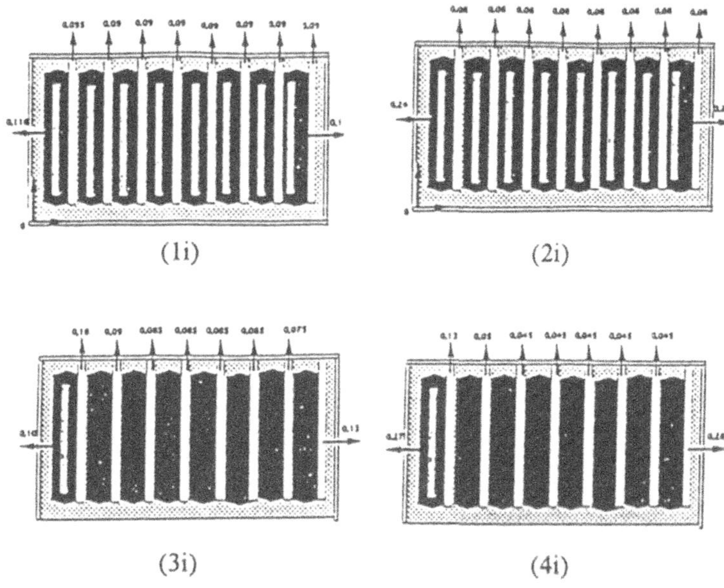

Figure 7: Heat flux distributions (1i) 8 dissipating chips $h_2=h_3=5W/m^2.K$ (2i) 8 dissipating chips $h_2=h_3=20W/m^2.K$ (3i) one dissipating chip $h_2=h_3=5W/m^2.K$ (4i) one dissipating chip $h_2=h_3=20W/m^2.K$

Temperature is reduce from a fraction 5 to 10 with forced convection ($h_1 \approx 100W/m^2K$) in comparaison with natural convection ($h_1 \approx 20W/m^2K$). A velocity of 5 m/s gives an average temperature rises of about 20 K/W.

A good way is to extend the metal sheets in surrounding air. Average temperature rise is reduce with natural convection, from 120 K/W to 40 K/W (case of copper sheets of 5 mm height from the side 1). Fins can be also fixed on sides 2,3.

Basically, temperature distributions appear sensitive to the thermal conductivity of plastic (figure 8) and to the number of package stacked (Figure 9)

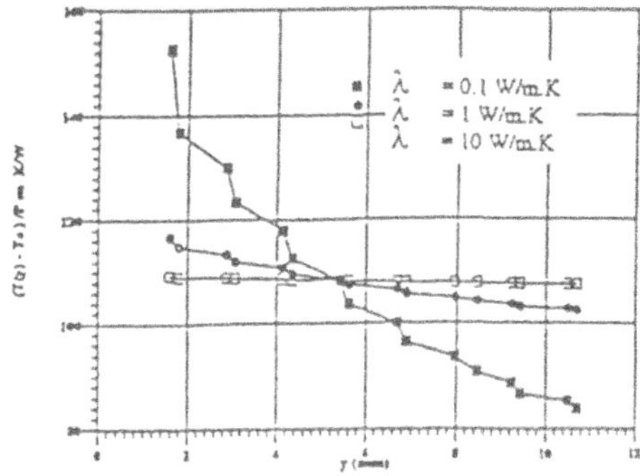

Figure 8: Effect of the thermal conductivity of the package on temperature distribution

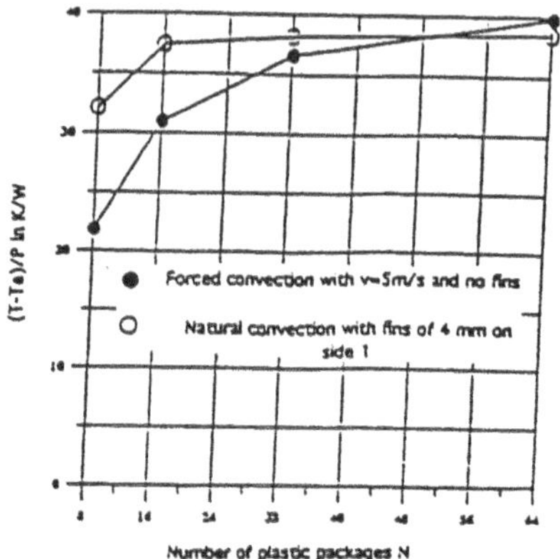

Figure9: Effect of the number of package on temperature

6. CONCLUSION

A comprehensive heat transfer study on vertical stackings of plastic packages based on original experiments and modelling has been performed.

An experimental module integrating many temperature sensors has been set up. Experiments provide correlations given thermal resistances and average natural convection coefficients. Measurements demonstrate the efficiency of the vertical conducting sheets.

A compact model, based on an original concept and taking into account the multidimensionnal feature of heat transfer, has been developed. Analysis has lead to an optimal design.

REFERENCES

1. M. YAMADA, T.KUDAISHI, 9th European Hybrid Microelectronics Conference,Nice, 1993
2. C. VAL&al, l'Onde Electrique (in french), vol.74, n°2, 1994
3.C. VAL, 9th European Hybrid Microelectronics Conference,Nice, 1993
4. Y.SCUDELLER- to be published
5.Allan D. KRAUS,in "advances in thermal modeling of electronic components and systems",vol.2, pp41-106, 1990
6.Y. SCUDELLER,Société Française des Thermiciens, Paris, 1994

ELECTROTHERMAL SIMULATION OF ANALOGUE INTEGRATED CIRCUITS WITH ETS

W. VAN PETEGEM, B. GEERAERTS AND W. SANSEN
K.U. Leuven, ESAT-MICAS,
Kardinaal Mercierlaan 94, B-3001 Heverlee (BELGIUM)

1. Introduction

Nowadays, high-power devices are integrated together with complex digital and high performance analogue circuits on the same chip. Inappropriate design and/or layout cause major problems with maximum temperature and temperature gradients between components. An accurate prediction of the electrothermal interaction is required to design this type of circuits more efficiently (Poulton *et al.*, 1992). A new electrothermal simulation tool, called ETS, has been developed (Van Petegem *et al.*, 1990; Van Petegem, 1993; Van Petegem *et al.*, 1994). It allows static and transient simulations of circuits where electrical behaviour is seriously affected by thermal effects on chip.

2. Software Description

2.1. GENERAL CONSIDERATIONS

Two major approaches are described in literature to analyse both electrical and thermal behaviour of a given circuit: a coupled set of electrothermal equations, by introducing special (electro)thermal entries in the admittance matrix (Fukahori *et al.*, 1976), or a relaxation procedure, in which the electrical and thermal equations are solved separately (Poulton *et al.*, 1992).

The latter solution has some major advantages. Existing software, dedicated to two subproblems, may be used, with no need to change the source code. Also, the construction of the electrical and the thermal model of a specific circuit can be done separately. This results in the most appropriate models for both subproblems: a typical SPICE-like electrical circuit description and a thermal model, based on finite elements. With this complete physical thermal model, no approximation of reality is made to equivalent

E. Beyne et al. (eds.), Thermal Management of Electronic Systems II, 83-92.

lumped components, associated with heating circuit elements. This model accounts better for thermal gradients on chip.

The above considerations are in favour of using a relaxation procedure as basic technique to develop a new electrothermal simulator, called ETS.

2.2. INSIDE ETS

The approach taken in ETS is based on a blockwise Gauss-Seidel waveform relaxation method with time windowing (Van Petegem *et al.*, 1990; Van Petegem, 1993; Van Petegem *et al.*, 1994). Electrical and thermal equations are decoupled and solved separately The electrical analysis is performed by a conventional circuit analysis program, SPICE, and the thermal equations are solved with a commercial finite element simulator, SYSTUS.

The flowchart of ETS is given in Fig. 1.

A transient analysis starts with a static electrothermal simulation to determine the electrical and thermal operating point parameters. With this initial condition, an electrical transient simulation over the whole time interval is performed (at a constant temperature). From this result the time dependent temperature sources are derived. A thermal transient simulation follows next. Careful analysis of the temperature profiles on chip allows to subdivide the whole time interval into smaller *critical time intervals*, in which temperature is almost constant. This is possible because thermal phenomena generally occur much slower than electrical events.

In each critical time interval, an electrical transient analysis is performed at the mean temperature within this interval. The time dependent power dissipation for each circuit component is then taken as a thermal source for a transient thermal analysis, in which only small variations in the temperature profiles are expected.

This iterative process is repeated until simulation results converge within specified accuracy limits.

Convergence of ETS is guaranteed. Applying Lipschitz's criterion to the system of partial differential equations governing the coupled electrothermal problem, yields sufficient conditions for the proposed method to converge. These conditions are met in all practical circuits. Due to the blockwise integration (separate electrical and thermal analysis) and the time windowing (critical time intervals), convergence will even be enhanced (Dumlugöl, 1986; Van Petegem, 1993; Van Petegem *et al.*, 1994). Problems may occur in unstable situations, like secondary breakdown, but this is beyond the application scope of ETS.

Accuracy is attained by using a fully realistic thermal model instead of some equivalent lumped components. In each iteration a finite element thermal analysis is performed on a complete three-dimensional model of the

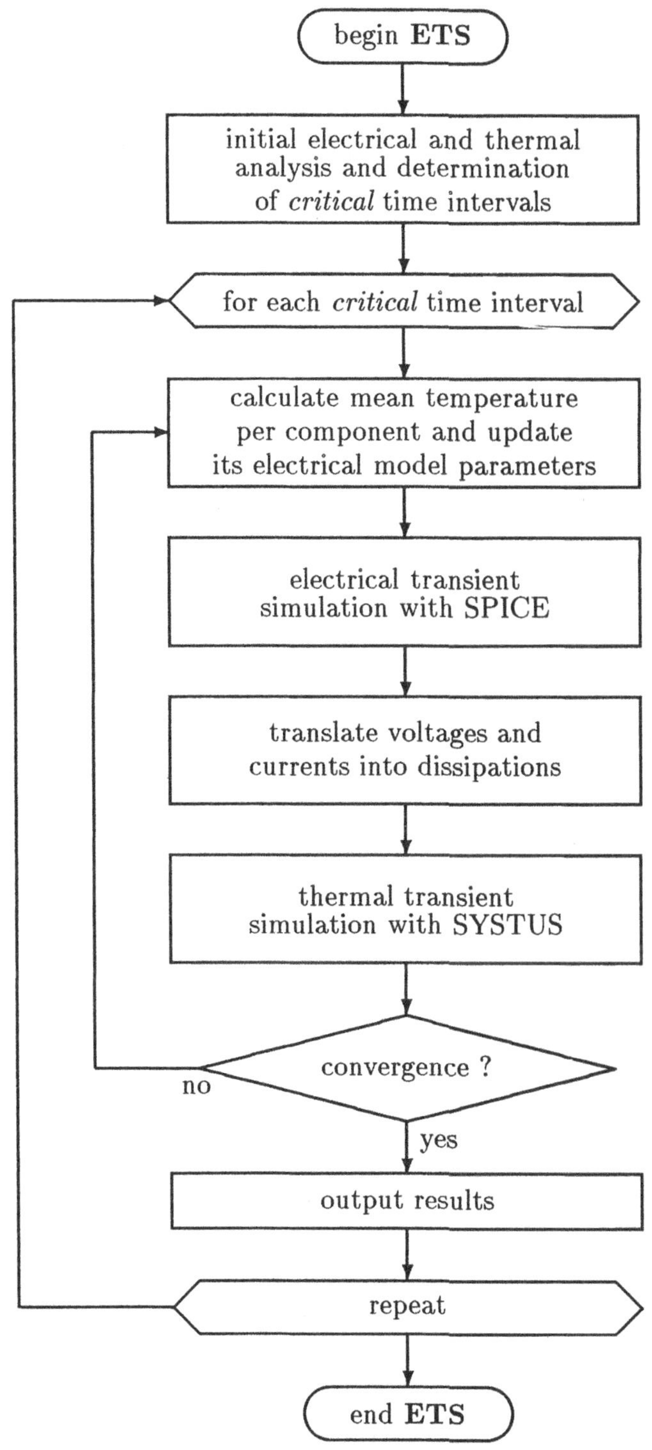

Figure 1. ETS flowchart

chip. This takes about 80% of the overall CPU time for one simulation with ETS. Fortunately, in typical examples only a few iterations are necessary. As a result, high accuracy is achieved in a moderate execution time (Van Petegem *et al.*, 1990; Van Petegem, 1993; Van Petegem *et al.*, 1994).

2.3. PRACTICAL SIMULATION EXAMPLE

Several examples are simulated with ETS. A characteristic circuit is the SBIMOS [1] driver of Fig. 2. It is a general purpose, high current driver from MIETEC-ALCATEL (Belgium). It is a good example to simulate with ETS illustrating the mutual effects of electrical behaviour and thermal gradients on chip.

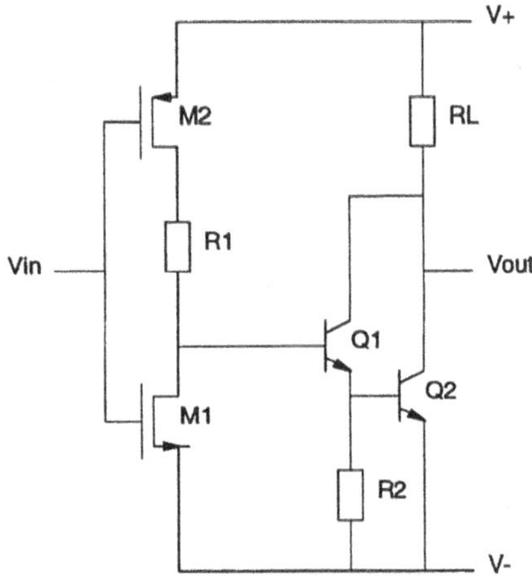

Figure 2. SBIMOS driver circuit - electrical schematic

The NPN driver transistor has four ballasted emitter stripes, denominated A to D, from left to right position on the chip surface. In the finite element model, the interdigitated structure of these emitter stripes is simplified to one large rectangular heat source.

A three-dimensional two-layer model is used for the thermal simulations. One layer represents the silicon wafer, with a thermal conductivity of $150Wm^{-1}K^{-1}$, a specific heat capacity of $700Jkg^{-1}$ and a mass density of $2330kgm^{-3}$. The additional layer models the imperfect contact between the smooth silicon wafer and the environment. Material properties for this layer are chosen as: a thermal conductivity of $8Wm^{-1}K^{-1}$, a specific heat

[1] SBIMOS is a high voltage BiCMOS technology of MIETEC-ALCATEL (Belgium) (Graindourze *et al.*, 1992)

capacity of $990Jkg^{-1}$ and a mass density of $120kgm^{-3}$. These values are reasonable for a layer of about 90% silicon and 10% air or other material. They also give good agreement with measurements.

The thermal boundary conditions are: an isothermal bottom surface, adiabatic side walls and natural convection at the top surface with a heat transfer coefficient of $15Wm^{-2}K^{-1}$.

Table 1 shows the static simulation results for the driver circuit. An input voltage of $0V$ is applied, between supply voltages $V_- = 0V$ and $V_+ = 18V$, with a load $R_L = 20\Omega$.

TABLE 1. Static Currents and Temperatures in Output Transistor of SBIMOS Driver Circuit

current I_C $[A]$				
Part	Iter. 1	Iter. 2	Iter. 3	Meas.
Q2A	0.2196	0.2170	0.2169	0.215
Q2B	0.2196	0.2228	0.2230	0.227
Q2C	0.2196	0.2228	0.2230	0.224
Q2D	0.2196	0.2171	0.2169	0.216
voltage V_{CE} $[V]$				
Part	Iter. 1	Iter. 2	Iter. 3	
Q2A,B,C,D	1.479	1.4571	1.4574	
junction temperature $[K]$				
Part	Iter. 1	Iter. 2	Iter. 3	
Q2A	311.2	311.0	311.0	
Q2B	312.3	312.2	312.2	
Q2C	312.3	312.2	312.2	
Q2D	311.2	311.0	311.0	

Remarkable in these results is the different electrical behaviour of the several emittor parts of the output transistor Q_2, due to the temperature gradient over the chip, which can never be simulated with an electrical simulator as such. The results become even more obvious, when a load $R_L = 10\Omega$ is considered. The temperature increases to $60K$ above ambient temperature in parts A and D of Q_2 and to $69K$ in the middle stripes B and C. This temperature difference in turn causes a difference of $40mA$ between the respective emitter parts.

A special version of the SBIMOS driver has been designed for measurement purposes. Because the junction voltage over a diode is used as temperature sensor, some extra components have been provided. Measurements of the currents in the different parts of Q_2 are included in Table 1 as well. They were performed on a chip in its package, which was cooled with a fan. After temperature stabilization, the current agrees remarkably well with the simulation results. Temperature itself can be measured too, as is extensively explained in (Geeraerts *et al.*, 1993; Geeraerts *et al.*, 1993).

The transient behaviour of this driver circuit is also analysed. The step response output parameters are given in Fig. 3. A high voltage ($V_{in} = V_+ = 18V$) is applied at the input of the invertor, to calculate the steady-state solution of this circuit (with $R_L = 20\Omega$). Then, at $t = 0ms$, an abrupt step to $V_{in} = V_- = 0V$ is applied. A period of $50\mu s$ is chosen for the critical time intervals.

The electrical outputs change within less than $1\mu s$ and the typical behaviour of the four emittor parts of Q_2 becomes visible. The temperature however changes much slower (in the range of ms). During this temperature increase, the voltage and currents do no longer change considerably. This behaviour confirms the basic hypothesis for the implementation of the transient analysis module in ETS, i.e. the electrical time constant differs some orders of magnitude from the thermal one. Consequently, when the simulation interval is taken large enough (about $100ms$), the transient simulation results converge to the static ones, presented in Table 1.

It is also worth to note that the number of iterations to converge a simulation with ETS is quite small (in the order of 3 to 5). In one iteration, the most CPU-time consuming part is the thermal finite element simulation. This depends of course on the number of nodes in the mesh. A good compromise between accuracy and calculation time for the SBIMOS driver means some 2000 nodes. This takes about half an hour on a $16Mips$ machine with $24MByte$ memory. With the number of iterations mentioned, it is clear that the total simulation time with ETS is in the order of one hour and a half. A transient simulation has to be done overnight.

3. Electrothermal Design Environment

Nowadays, extensive research is going on to automate mixed analogue / digital ASIC design, from performance up to layout. At ESAT-MICAS, a complete environment for synthesis of these circuits is under investigation (Gielen *et al.*, 1995). A framework is built around certain in-house developed CAD-programs. It combines artificial intelligence with advanced

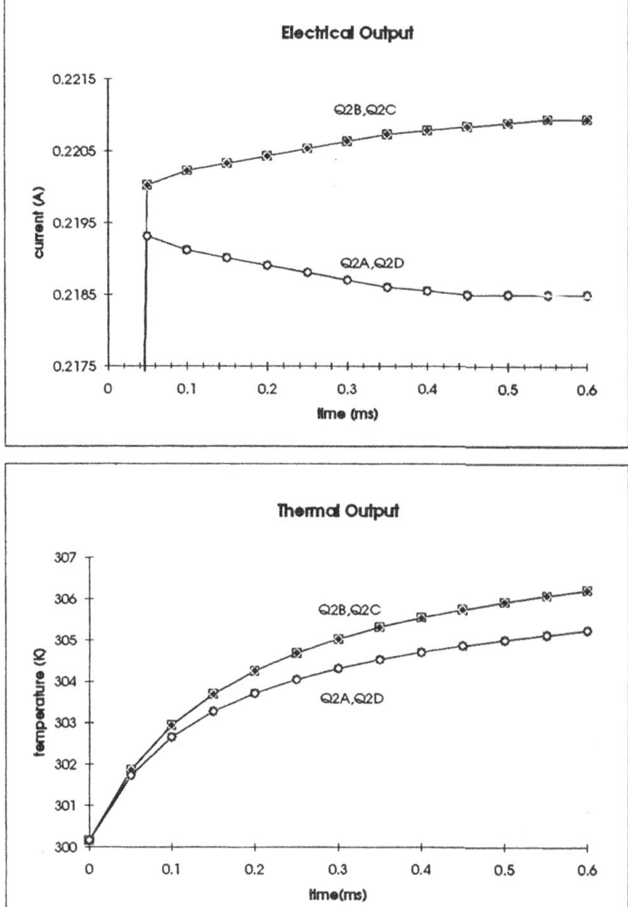

Figure 3. Step response output parameters of SBIMOS driver circuit. (a) Electrical output. (b) Thermal output.

symbolic and numerical analysis techniques. Due to its open structure, it is easy to include new components, like e.g. ETS.

Indeed, the value of ETS is enlarged when integrated in a complete silicon compilation framework. One possible scheme of a design strategy is given in Fig. 4. It is based on ARIADNE, (Advanced Routines for Interactive or Automated Design of Analogue Transistor Networks) (Swings *et al.*, 1993).

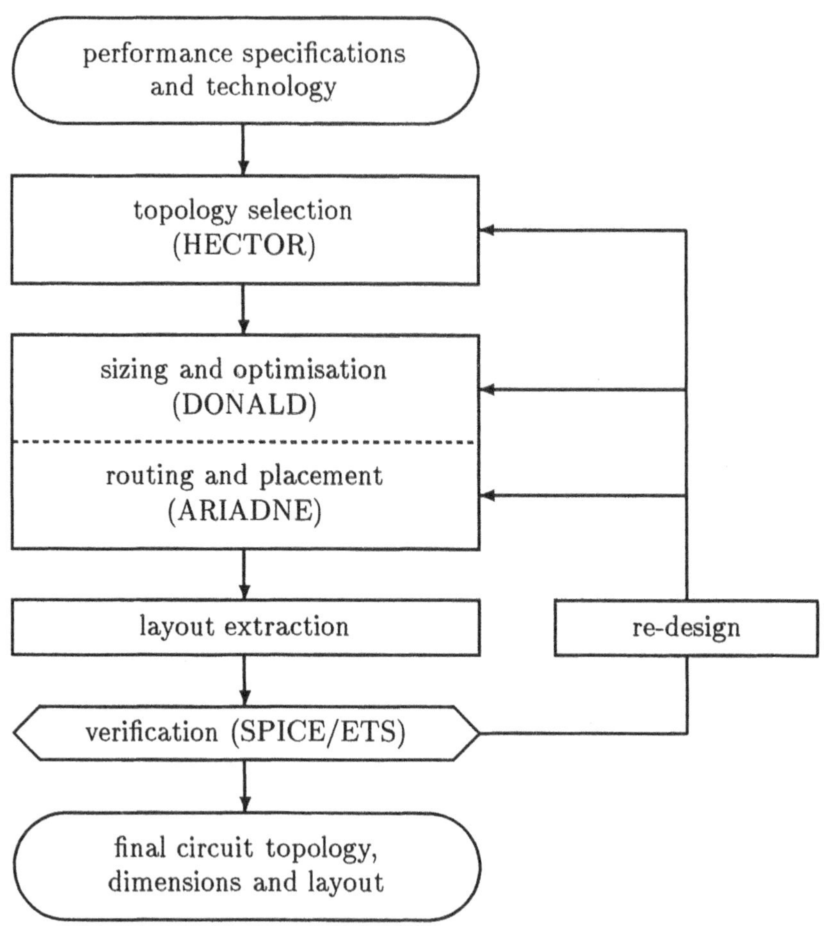

Figure 4. Integration of ETS in an ASIC Design Environment

ETS is a perfect verification tool to analyse a potential design for a circuit in which temperature distributions are expected to be important. The design phase starts from a set of specifications for the circuit (like accuracy, noise sensitivity, frequency range, temperature range) and the technology (e.g. bipolar, CMOS, SBIMOS) in which it has to be realised. The next step is the selection of an appropriate circuit topology, an unsized scheme, with a set of equations describing the circuit. This circuit is then dimensioned and further optimised. If a satisfactory solution is found, it has to be layout.

Finally, a verification is performed on the completely designed circuit to check if the possible candidate meets the predefined specifications. At this point, ETS comes in, especially when it concerns a power integrated circuit or another circuit in which temperature distributions play a substantial role. If the ETS results are not satisfactory, a redesign procedure is started, presumably on several levels in the design strategy (layout, sizing or even topology selection). The whole process is repeated until a suitable design is found. In this way, it is perfectly possible to yield a fully justified electrothermal design, meeting all required specifications, in a reasonable time and with minimum designer effort.

4. Conclusions

This paper presents a new tool, ETS, for simulation of the electrothermal behaviour of integrated circuits. It is a fully automatic iterative relaxation procedure to preduct quantitatively the combined elctrical and thermal performance of a given (critical) circuit. Its significance compared with existing simulators is demonstrated on a typical test device and is verified with measurements. It is also explained how ETS is a valuable component in a general analogue module generator for design of ASIC's in which a mutual influence between thermal and electrical behaviour is expected. With ETS, the complete design phase of circuits with electrothermal coupling, intentional or not, can be optimised in a new way, that was not possible before.

Acknowledgment

The authors wish to thank B. Graindourze, H. Casier, G. Schols and S. Blieck (with MIETEC-ALCATEL) and C. Van Grieken (K.U. Leuven, ESAT-MICAS).

This work was supported by IWONL Brussels.

References

Dumlugöl, D. (1986), The segmented waveform relaxation method for mixed-mode simulation of digital MOS VLSI circuits, *Ph.D. dissertation, Katholieke Universiteit Leuven, Belgium.*

Fukahori, K. and Gray, P. (1976), Computer simulation of integrated circuits in the presence of electrothermal interaction, *IEEE Journal of Solid-State Circuits* **Vol. 11, No. 6**, pp. 834–846.

Geeraerts, B., Van Petegem, W. and Sansen, W. (1993), A SBIMOS diode matrix for the characterisation of static and transient thermal phenomena on silicon, *Proc. 9th IEEE SEMI-THERM Symp. (Austin, TX)*, pp. 108–111.

Geeraerts, B., Van Petegem, W. and Sansen, W. (1993), Thermal measurements by use of a SBIMOS diode matrix, *Proc. of the IEEE International Conference on Microelectronic Test Structures 1993 (Barcelona, Spain)*, pp. 183–187.

Gielen, G., Debyser, G., Lampaert, K., Leyn, F., Swings, K., Van der Plas, G., Sansen, W., Leenaerts, D., Veselinovic, P. and van Bokhoven, W. (1995), An Analog Module Generator for Mixed Analog/Digital ASIC Design, accepted in 1995 for publication in *International Journal for Circuit Theory and Applications.*

Graindourze, B., Blieck, S., Casier, H. and Bardyn, J.-P. (1992), High performance op amps in a 40V BiCMOS process and their applications, *Analog Integrated Circuits and Signal Processing*, Kluwer Academic Publishers, pp. 33–42.

Poulton, K., Knudsen, K.L., Corcoran, J.J., Wang, K.-C., Pierson, R.L., Nubling, R.B. and Chang, M.-C. (1992), Thermal design and simulation of bipolar integrated circuits, *IEEE Journal of Solid-State Circuits* **Vol. 27, No. 10**, pp. 1379–1386.

Swings, K. e.a. (1993), ARIADNE: A constraint-based approach to computer-aided synthesis and modeling of analog integrated circuits, *Analog Integrated Circuits and Signal Processing*, Kluwer Academic Publishers, pp. 197–215.

Van Petegem, W., De Wachter, D. and Sansen, W. (1990), Electrothermal simulation of integrated circuits, *Proc. 6th IEEE SEMI-THERM Symp. (Phoenix, AZ)*, pp. 70–73.

Van Petegem, W. (1993), Iterative solutions to electronic problems, based on finite elements: electrothermal simulation and electrical impedance tomography, *Ph.D. dissertation, Katholieke Universiteit Leuven, Belgium.*

Van Petegem, W., Geeraerts, B., Sansen, W. and Graindourze, B. (1994), Electrothermal simulation and design of integrated circuits, *IEEE Journal of Solid-State Circuits* **Vol. 29, No. 2**, pp. 143–146.

CALCULATIONS OF THE TEMPERATURE DISTRIBUTION THROUGHOUT ELECTRONIC EQUIPMENTS BY THE BOUNDARY ELEMENT METHOD WITH THE REBECA-3D SOFTWARE

by J.P. FRADIN (*) and B. DESAUNETTES (*)

(*) EPSILON INGENIERIE,
California - Voie 5 - B.P. : 653 - 31319 Labège Cedex - France

Abstract
In order to allow accurate calculations of the temperature distribution throughout electronic packages, the REBECA-3D Software, based on the Boundary Element Method, has been developed. The principles of this method for steady-state three-dimensional problems are given. It appears to be very well-adapted for parametric studies then for conception phase. It is finally applied to two cases of interest for engineers.

Nomenclature

$\widetilde{A}, \widetilde{B}, \widetilde{C}$...:	matrices	R :	contact resistance
K :	conductivity	h :	convective coefficient
Bi, Gij, Hij :	matrix coefficients	φ :	flux surface density
n_d :	number of domains	Γ,Ω :	boundary, domain
p,Q :	points		
Q' :	heat generation	* :	fundamental functions
T :	temperature	_ :	prescribed values
T_c:	convective temperature	~ :	matrices

1. Introduction

In order to improve reliability and performances, one of the objectives of electronic companies is to predict the operating temperatures of critical electronic parts. This problem becomes more and more difficult to solve because many industrial applications require high power range for a minimum occupied volume.

Classically, the Boundary Element Method (B.E.M.) allows to solve the conductive problems according to boundary conditions, thermal properties and geometrical characteristics. It calculates boundary temperatures and boundary fluxes without the discretization of the domain. It is very well-adapted to the class of potential problems expressed by electronics. A 3D Boundary Element Software named REBECA-3D (REliability Boundary Element Conductive Analysis in Three-Dimension) has then been developed.

E. Beyne et al. (eds.), Thermal Management of Electronic Systems II, 93-102.

This communication proposes the use of this software in the particular case of the electronic packages. First, the Boundary Element Method is presented in emphasizing on the capacity to make parametric studies. Secondly we describe rapidly the REBECA-3D Software. Finally, we use it to determine the three-dimensional temperature distributions corresponding to classical examples.

2. Bases of the B.E.M. for steady-state three-dimensional problems

The direct formulation of the Boundary Element Method is based on the use of weighted residual techniques [1] [2]. Classically, this method allows to calculate boundary temperatures and boundary fluxes without the discretization of the domain. With special developments, it is possible to reach the same results for mean temperatures [3].

2.1. DEVELOPMENT OF THE METHOD FOR LINEAR PROBLEMS

Our presentation is limited to an isotropic and linear problem (i.e. to a heterogeneous medium with boundary conditions of Dirichlet, Neumann or mixed type).

2.1.1. *Thermal equations*

According to the generalized Fourier theory, in steady-state, the equation of heat conduction for an homogeneous medium Ω with a boundary Γ, in which a heat rate per unit volume Q' is generated, is formulated as follows :

$$K\,\nabla T + Q' = 0 \qquad \text{EQ(1)}$$

The boundary Γ is divided into three distinct parts $\left(\Gamma_i\right)_{i=1,3}$ such as $\Gamma = \Gamma_1 \cup \Gamma_2 \cup \Gamma_3$. Each part corresponds to a different type of boundary condition applied to the boundary :

$\Rightarrow$ on Γ_1 : boundary condition of Dirichlet type : $T = \overline{T}$ ($\overline{T}$ prescribed value of T)

$\Rightarrow$ on Γ_2 : boundary condition of Neumann type : $\varphi = K\dfrac{\partial T}{\partial n} = \overline{\varphi}$ (φ, the inward normal flux surface density defined by $\varphi = \vec{\varphi}\,.\vec{n}_{inward} = \vec{\varphi}\,.(-\vec{n}_{outward}) = -\vec{\varphi}\,.\vec{n}$ where $\vec{\varphi}$ is expressed according to the generalized Fourier theory).

$\Rightarrow$ on Γ_3 : boundary condition of mixed type : $\varphi = h(T_c - T) = \overline{\varphi} - hT$.

2.1.2. *Boundary integral equation of the linear homogeneous problem*

In the case of a constant heat generation rate Q', the integral formulation, based on the use of Green's third identity, gives for a point p the following relation :

$$c(p)KT(p) + K\int_\Gamma \frac{\partial T^*(p,Q)}{\partial n(Q)} T(Q)d\Gamma(Q) + \int_\Gamma hT^*(p,Q)T(Q)d\Gamma(Q)$$
$$= \int_\Gamma T^*(p,Q)\varphi(Q)d\Gamma(Q) + \frac{Q'}{K}\int_\Gamma \frac{\partial v^*(p,Q)}{\partial n(Q)} d\Gamma(Q) \qquad \text{EQ(2)}$$

with $\Rightarrow$ the value of the coefficient c(p) depends on the position of the collocation point p (for p located inside the domain c(p)=1, for p located outside the domain c(p)=0, for p located at the boundary, for constant element where the boundary is "smooth", c(p)=0.5).

⇒ T^* denotes the fundamental Green's function, the basic solution of the Laplace's differential equation having a source located at a source point p. The function v* verifies $\nabla v^* = T^*$.

2.1.3. *Discretization and expressing in matrix form*

For our application, the boundary is discretized into a series of n geometric elements (triangles and quadrangles in a first time) over which the unknowns are assumed to vary according to constant interpolation functions. The nodes are taken in the centroid. The discretization of the integral equation EQ(2) for a point I at the boundary gives a relation coupling the n values of the temperatures and the n values of the fluxes :

$$K\sum_{j=1}^{n} H_{ij}T(J) + \sum_{j=1}^{n} G_{ij}h(J)T(J) + \frac{Q'}{K}B_i = \sum_{j=1}^{n} G_{ij}\varphi(J) \qquad \text{with}$$

$$H_{ij} = \int_{\Gamma_j} \frac{\partial T^*}{\partial n}\, d\Gamma_j + \frac{1}{2}\delta_{ij}\ ;\ G_{ij} = \int_{\Gamma_j} T^* d\Gamma_j\ ;\ B_i = -\sum_{j=1}^{n} \int_{\Gamma_j} \frac{\partial \varphi(v^*)}{\partial n}\, d\Gamma_j \qquad \text{EQ(3)}$$

If the discretized relation is written for the n nodes, a system of n equations (3) with n unknowns is obtained because a set of n values is known. If we note k the studied domain, the whole set of equations can also be expressed in matrix form as :

$$\left(K_k\widetilde{H}_k + \widetilde{h}_k\widetilde{G}_k\right)\widetilde{T}_k + \frac{Q'_k}{K_k}\widetilde{B}_k = \widetilde{G}_k\widetilde{\varphi}_k \qquad \text{EQ(4)}$$

2.1.4. *Application for heterogeneous problems*

In the case of heterogeneous applications, the previous numerical procedure can be applied to each homogeneous subregion as they were separated from the others. The final system of equations for the whole region is obtained by adding the set of equations together with compatibility and equilibrium conditions between their interfaces.

If we note kl the interface between the domain k and a neighbouring domain l, the previous matrices $\widetilde{G}_k$ and $\widetilde{H}_k$ relative to the domain k may be considered as the association of rectangular blocks matrices (dimension : number of unknowns of the k domain on the exterior boundary x number of unknowns on the boundary kl) :

$$\widetilde{G}_k = \left[\widetilde{G}_{k1} \quad \cdots \quad \widetilde{G}_{kl} \quad \cdots \quad \widetilde{G}_{kn_d}\right] \text{ and } \widetilde{H}_k = \left[\widetilde{H}_{kl} \quad \cdots \quad \widetilde{H}_{kl} \quad \cdots \quad \widetilde{H}_{kn_d}\right]$$

The final system is written with all the unknowns on the left hand side and with a right hand side obtained with the known values :

$$\widetilde{A}\widetilde{X} = \widetilde{C}\widetilde{D} - \widetilde{P} \qquad \text{EQ(5)}$$

with :

⇒ $\widetilde{A}$ the characteristic matrix of the system expressing couplings between the unknowns and consisting of blocks of the form :

$$-\widetilde{G}_{kl}\ ;K_k\widetilde{H}_{kl}\ ;-(\widetilde{G}_{kl} + K_k\widetilde{R}_{kl}\widetilde{H}_{kl})\ ;(K_k\widetilde{H}_{kl} + \widetilde{h}_{kl}\widetilde{G}_{kl})$$

$\Rightarrow \widetilde{P}$ the vector including terms connected with heat generation consisting of terms of the form : $\widetilde{P}_k = \frac{Q'_k}{K_k}\widetilde{B}_k$

$\Rightarrow \widetilde{C}$ the matrix expressing the couplings with boundary conditions consisting of terms of the form : $\widetilde{G}_{kk}\ ; K_k \widetilde{H}_{kk}$

$\Rightarrow \widetilde{D}$ the vector assembling exterior boundary conditions

$\Rightarrow \widetilde{X}$ the vector assembling unknowns.

2.1.5. *Determining of the mean temperature*

Classically, to calculate the mean temperature of a medium, this one is divided into a series of cells on which the temperature is supposed to be isothermal. The mean temperature is then deduced from the arithmetic average of the temperature of these cells balanced with their area. Then, accuracy of the value of the mean temperature depends on the number of theses cells.

As the Boundary Element Method allows to express the temperature of any point of the domain according to fluxes and temperatures at the boundary, it is interesting to calculate the mean temperature according to the same parameters.

To realize this, an original method based on the definition of the mean temperature (surface integration of the field of internal temperatures given by EQ(3)) and on special formulae allowing to transform domain integral into boundary integral has been developed [3]. Both performance of the method and accuracy of the results have been proved. Such a process improves the quickness of parametric studies while reducing calculation of the mean temperature to a series of multiplications.

2.2. ADVANTAGES OF THE BOUNDARY ELEMENT METHOD

2.2.1. *Particular advantages of the method*

We give below advantages of the Boundary Element Method which are directly linked with our electronic application.

$\Rightarrow$ **Reduction in dimension**. For constant powers, the procedure provides a reduction in dimension, so that three-dimensional problems are reduced to a sequence of two dimensional problems, involving only surface integrations. No internal unknown and no internal mesh is necessary.

$\Rightarrow$ **Accuracy and reliability**. The performances relative to the reduction of the dimension of the problem should not come at the expense of accuracy and reliability. In addition, this method generally gives better results than domain type techniques (Finite Element Method and Finite Difference), especially for regions with high gradients.

$\Rightarrow$ **Treatment of singularities**. The Boundary Element Method allows the treatment of power dissipated in a point.

$\Rightarrow$ **Parametric studies**. The Boundary Element Method is a very powerful tool to carry out a great number of parametric studies with very few calculations. The conception of the software was brought out in accordance with this requirement. In order to understand the parametric methodology, we preliminary remembered that there is in fact three distinct steps as far as the matrix calculations are concerned :

➲ first step : formation of the $\widetilde{B}_k$, $\widetilde{G}_k$ and $\widetilde{H}_k$ matrices with regard to each k domain. The coefficients of these matrices only depend on geometric properties of each domain.

➲ second step : assembling of the whole matrices $\widetilde{G}_k$ and $\widetilde{H}_k$ to form the matrices $\widetilde{A}$ and $\widetilde{C}$ and of $\widetilde{B}_k$ to form the matrix $\widetilde{P}$

➲ third step: solution of the global system. For each parameter study, it is obvious that this step must be done again.

We are going to review the modifications influenced by variations of several geometric and thermal parameters relative to a domain k. We restrict our study to modifications not meaning new integral calculations.

➲ power Q'_k: only modification of the value concerning this domain, so the sub-vertor of $\widetilde{P}$ relative to this domain is changed : $\widetilde{P}_k = \dfrac{Q'_k}{K_k}\widetilde{B}_k$.

➲ modification of the exterior boundary conditions : for Dirichlet, Neumann and mixed conditions (if only T_c modified), only modification of the value of the $\widetilde{D}$ vector.

➲ modification of the interior boundary conditions between two domains (or of the convective coefficient in the case of mixed boundary conditions) : only modification in the assembling of the $\widetilde{A}$ matrix of terms where the contact resistance between both considered domains intervene : $-(\widetilde{G}_{kl} + K_k\widetilde{R}_{kl}\widetilde{H}_{kl})$ (respectively where the convective coefficient intervene : $-(K_k\widetilde{H}_{kl} + \widetilde{h}_{kl}\widetilde{G}_{kl})$)

➲ modification of the type of boundary conditions : modification in the assembling of the matrices $\widetilde{A}$ et $\widetilde{C}$.

➲ modification of the values of the conductivities : modification in the assembling of the matrices $\widetilde{A}$ et $\widetilde{C}$ and of the vector $\widetilde{P}$ of terms making K_k intervene.

➲ for general geometric modifications, that is to say for modifications of the relative position of two domains or of the position of the boundary with regard to one domain, the problems are more complicated because of the difficulties given by the similarity concerning the mesh on each side of the boundary.

Such advantages make the software being a powerful tool for the conception of an electronic package. In a general case, the preliminary calculations are made only one time and only the solution of the system must be repeated.

(little system because of the reduction of the unknowns on the boundary)

⇒ **Reduction of CPU time consumption** : The consequence of the previous aspects is the time saving produced by the Boundary Element Method.

2.2.2. Advantages of the Boundary Element Method in comparison with other methods

Semi-analytical methods. These methods are very efficient on a CPU time point of view because the mathematical calculations are very developed. Their main disadvantages are linked to the hypotheses that have to be respected for their application. Even for the most high-performance ones [4], the variety of geometrical shape and boundary conditions which can be considered is inevitably and significantly restricted. Heat sinking and arbitrary heat sources are assumed to be provided at the opposite limiting planes of the the system.

Numerical methods. These methods are very spread and very precise but their use is difficult to bring into play and the calculations are too long. So, their application in a conception phase is not very interesting.

The Boundary Element Method appears as a method which has all the previously described advantages without collecting all their disadvantages.

3. Description of the software

The current version takes into account the following hypotheses. A steady-state temperature field in a three-dimensional heterogeneous system bounded by a closed surface is considered : the developed technique is valid for systems consisting of arbitrary numbers of homogeneous isotropic domains for which the conductivity may be a function dependent on the temperature. In the general case, for potential problems, boundary conditions of Dirichlet, Neumann and mixed type are prescribed at the external surface of the body. Both imperfect and ideal contact boundary conditions can occur at the interface of both joined domains. In a first time, to avoid the calculations linked to an internal mesh, the heating dissipation of each domain is limited to a constant function.

The final version will also take into account anisotropic materials (development in course in a CNRS laboratory). To simulate radiation, two possibilities will be offered. In the first way, the autonomous software will allow to modelize radiative transfers towards constant temperatures. In a second way it will be possible to modelize multireflexion between surfaces thanks the connection of Monte-Carlo radiative software [5]. In the same way, to modelize the convective transfers, it is possible to use software given exchange coefficients according to temperature, geometry, flow...

In fact, the Boundary Element Method Software is developed in such a way that it is an open and evolutive system. So, it may be connected to many software analysis tools used for thermal simulations or for other associated simulations (electricity, ...) in order to give the boundary conditions for the REBECA-3D Software. The procedure is then easy to integrate in a design environment.

To reach these purposes and efficiently realize a software that satisfies the goals of modularity, reusability, extensibility, reliability, portability..., we chose the Object Oriented Conception. The evolutivity of the software is then certain. To implement the various objects of the application, the C^{++} language has been chosen. C^{++} is a high level standard language which offers all the proprieties of the object paradigm. Moreover, C^{++} is available on many platforms.

Thanks to all these advantages, the software is easy to use even for a non thermal specialist.

In order to characterize the response of any electronic system, in immediate future, the diffusion problems will be taken into account with hypotheses in agreement with those previously presented.

4. Examples

We study two simple examples to show some aspects of the use of the REBECA-3D Software to predict the temperatures of an electronic package. The first model is essentially a homogeneous block of GaAs with a small uniform heat source located on the top surface. The second model consists in an hybrid circuit.

4.1. MMIC APPLICATION

The Boundary Element Technique is examined for predicted MESFET channel temperature on a GaAs MMIC (Monolithic Microwave Integrated Circuit) [6].

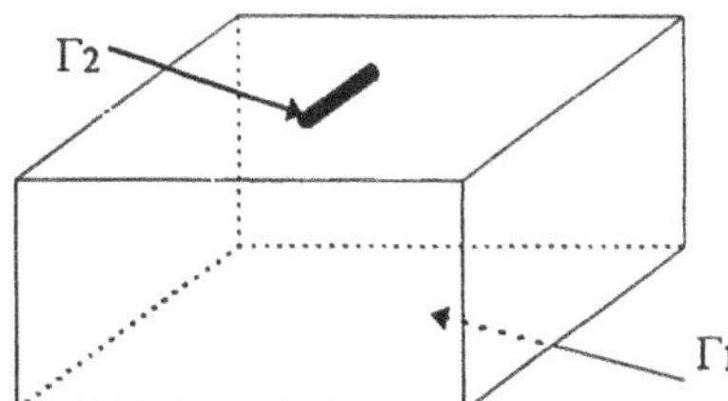

fig.1 : Schematic representation of the MMIC application

The heat sources on a MESFET are located in a very thin layer at the top surface of the GaAs chip. For the mathematical model, the chip is approximated as a homogeneous layer of pure GaAs (400μm*400μm*95μm). A constant GaAs thermal conductivity (K=41.8W/mK) was used during the development.
A power input of 0.1W is uniformly applied on the center top surface of the chip in the 0.5μm*65μm area defined by a single transistor gate. The base of the carrier plate is assumed to be isothermal (T=50°C). For simplicity, all other surfaces are considered adiabatic : natural convection and radiation heat transfer from the chip are assumed to be negligible.
On the following figure, we represent the 2D temperature distribution on the top surface :

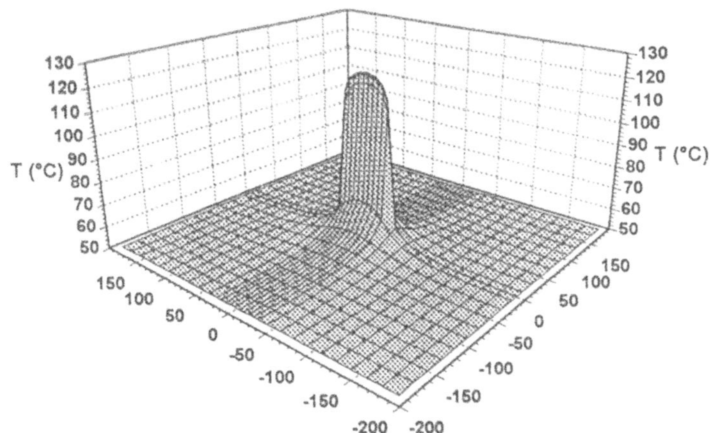

fig. 2 : Temperature distribution on the top surface of the MMIC

In this type of application, the difficulties lie in the difference of scale between the heat source which is the point to focus and the dimension of the entire model. It is the limit between finite and semi-infinite medium. Moreover, large temperature gradient around the heat source require smaller mesh spacings to be used near the transistor.

4.2. HYBRID CIRCUIT

We considered an alumina-copper substrate because of its frequent use in the hybrid circuits. The below figure gives the dimension of the substrate and the location of the MOS transistors on its structure. [7]
We relate the power dissipation to 1W on the MOS transistors.

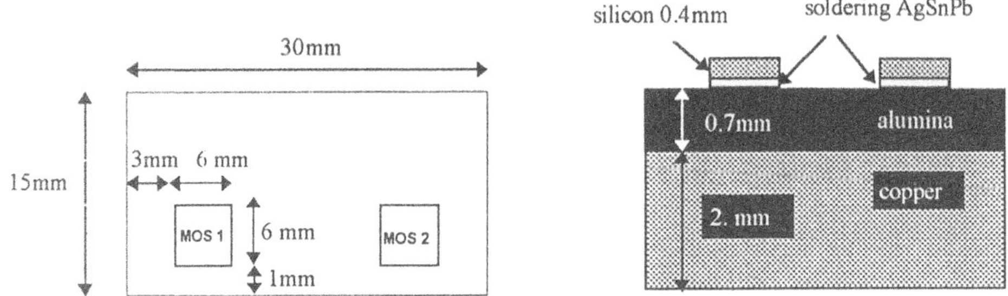

fig 3 : Geometrical data and thermal structure of the hybrid circuit

The previous data are completed with the following hypotheses :

⇒ conductance at the interface between the Alumina and the Copper layer : $C=1.6.\ 10^5 W/m^2K$

⇒ soldering of the chip on the structure $e=75\mu m$ and $K=0.35 W/°C$ cm so $C=4.7.\ 10^5 W/m^2K$

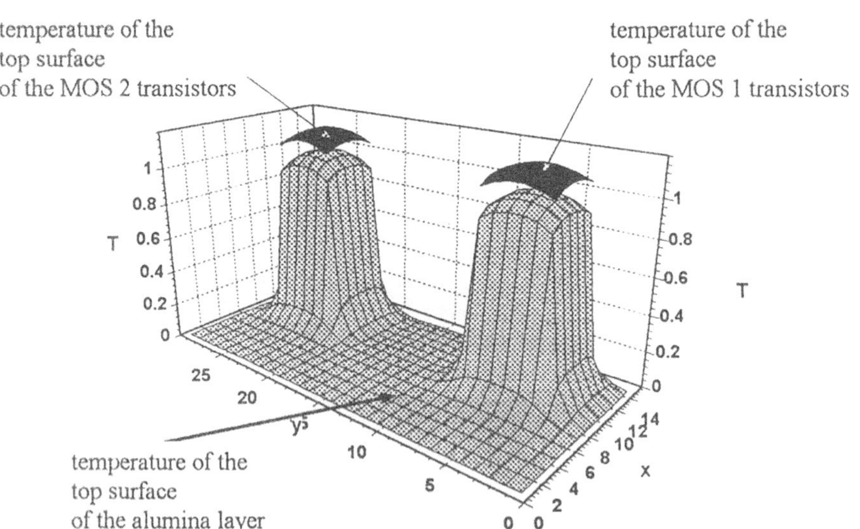

fig. 4 : Temperature distribution on the top surface of the alumina layer and on the top surface of the MOS transistors

The boundary conditions are the following ones :

⇒ on the top surface of the MOS transistors, the power input is uniformly applied on the whole surface.

⇒ on the lower surface, the temperature is supposed to be maintained at 0°C.

⇒ all the other surfaces are supposed to be adiabatic.

On the figure 4, we present the temperature distribution on the top surface of the alumina layer and on the top surface of the MOS transistors.

5. Conclusions

In the previous developments, we have presented the REBECA-3D Software based on the Boundary Element Method allowing an automatic conception for electronic packages.
We demonstrated that this method improves accuracy and reliability. It appears as an attractive alternative compared to the other approximate methods as well as to the semi-analytical ones.

We proved that the Boundary Element Method is particularly well-adapted to these conductive problems because it transforms a three-dimensional boundary problem into a two-dimensional one. Moreover, parametric calculations are very easy to realize with this method. The capacity to modelize each type of boundary conditions makes easier the study of thermal regulation system. So the predesigning of such system (optimisation of the efficiency, mass...) becomes more easy. So, the Boundary Element Method may appear as an help for the development of a model.

On a pratical point of view, for the engineer, the use of such a software means :

⇒ end of the restrictions concerning the variety of geometrical shapes and boundary conditions

⇒ end of the use of empirical procedures for the conception of thermal regulation systems

⇒ great facilities for conception phase

⇒ great facilities for parametric studies

Such advantages provide time consumption saving and accuracy improvement, two generally opposed aspects which are here combined thanks to the effectiveness of the method.

Of course, the software is not limited to electronic applications. It may be applied in every domain whose governing equations are potential.

6. Bibliography

1 Brebbia C.A., Telles J.C.F. & Wrobel L.C., *Potential problems : Boundary Element Techniques : Theory and Applications in Engineering*, Springer-Verlag, chap 2, p47-p107, 1984

2 Brebbia C.A. & Dominguez J, *Potential problems - Boundary Elements , An Introductory Course*, Comp. Mech. Publications, McGraw-Hill Book Company, chap 2, p45-151, 1991

3. Fradin J.P. - *Application de la méthode des éléments de frontière au calcul des conductances dans le cadre du contrôle thermique des engins spatiaux*, Mémoire de thèse, septembre 1994, Poitiers

4 Leturcq Ph, Dorkel J.M, Ratolojanahary F.E. and Tounsi S : A Two-Port Network Formalism for 3D Heat Conduction Analysis in Multilayered media : *Int. J. Heat Mass Transfer*, Vol.36, No 9, pp2317-2326, 1993

5 Fradin J.P. and Desaunettes B., Automatic computation of conductive conductances intervening in the thermal chain, 25th International Conference on Environmental Systems, San Diego, July 10-13, 1995

6 Wright J.L., Marks B.W. & Decker K.D., Modeling of MMIC Device for Determining MMIC Channel Temperatures During Life Tests : *7th IEEE SEMI-THERM Symposium*

7 Tounsi P., *Méthodologie de la Conception Thermique des Circuits Electroniques Hybrides et Problemes Connexes*, Mémoire de Thèse, Décembre 1992, Toulouse

AN INTERACTIVE THERMAL CHARACTERIZATION OF COMPONENT PLACEMENT

J.L. BLANCHARD & S. LOUAGE

Zuken-Redac TAD
Parc Technopolis. Bâtiment Alpha. ZAC de Courtaboeuf
91974 Les Ulis Cédex. France

Abstract

This paper describes a first-order approach for the thermal simulation of electronic boards, intended especially to appreciate at an early stage of a design the effect of data change. This tool is conceived so as to ensure ease of use, interactivity and speed. These features are obtained by using an analytical numerical method allowing the instantaneous update of the board thermal map as the component placement is modified and minimizing the number of control parameters. This paper describes the principle of the method and investigates its accuracy using comparisons with experimental measurements.

1. Introduction

Board thermal analysis is frequently performed using the conventional finite difference (FD) or finite element (FE) methods. These numerical approaches are potentially very accurate, provided that the data is available with a consistent degree of accuracy. These methods may be also viewed as best suited for post-layout analysis since the meshing phase they require must generally be reprocessed each time the geometrical data is modified, although specific numerical procedures may avoid that [1].
By contrast, during the pre-design phase of a project, the data is subject to change. Accordingly, the capability to examine rapidly the consequences of such a change with respect to the board thermal behaviour takes precedence over accuracy. This means for instance that the designer may need to appreciate as rapidly as possible how a board thermal map is affected by a new component placement, a modification of component power dissipation or an update of the operating conditions. [2] or [3] emphasize for instance the need for fast estimation methods.

This paper describes an approach suited to this context. The numerical method and computational process are first briefly reviewed. Then an investigation of the main factors influencing the method accuracy follows, based on comparisons with experimental measurements.

E. Beyne et al. (eds.), Thermal Management of Electronic Systems II, 103-111.

2. Numerical method description

Ease of use, interactivity and speed are the goals of the proposed approach. The last two are best illustrated by the capability to update instantaneously the board thermal map as the component placement is modified. The algorithm speed is actually such that it could even be possible to perform this update in real-time. To achieve this, the FD and FE methods were *a priori* discarded and an analytical approach based on Fourier expansion is used instead. [4] illustrates an application of this method for thermal purposes in electronics.

The main assumptions used are as follows:

i. the temperature distribution $T(x, y)$ within the rectangular board $[0, a] \times [0, b]$ of conductivity K and thickness w is two-dimensional; this last feature is dictated by execution speed considerations;

ii. the convective heat transfer on the upper and lower board faces are characterized by a unique pair (h, T_h) where the heat transfer coefficient h and the fluid temperature T_h are assumed to be uniform;

iii. component influence is modelled by rectangular heat sources with uniform rate Q_k;

iv. steady-state is only considered.

Assumption (ii) is relaxed using an *equivalent pair* obtained through the formulae:

$$T_h = \frac{h_{bot}T_{bot} + h_{top}T_{top}}{h_{bot} + h_{top}} \; ; \; h = h_{bot} + h_{top} \tag{1}$$

The left formula is easily interpreted as a weighted average of the top and bottom fluid temperatures. The pairs (h_{top}, T_{top}) and (h_{bot}, T_{bot}) are themselves either explicitly provided by the user or obtained with the help of a *heat transfer coefficient assisting tool*. If the board edges are further assumed to be adiabatic, the temperature map is such that:

$$-\left(\frac{\partial^2 T}{\partial x^2} + \frac{\partial^2 T}{\partial y^2}\right) + \lambda^2 (T - T_h) = \sum_k \frac{Q_k}{K} \tag{2}$$

with $\lambda^2 = h/(Kw)$. At that stage, the superposition principle [5] is involved and allows the use of the following computational process:

- the *bare board map*, corresponding to the board assumed free of components, is first calculated; for adiabatic edges, this map is actually isothermal;
- the *individual component maps* are then computed successively;
- the *global board map* is finally obtained by adding up the previous ones.

The superposition principle is the key of the algorithm speed, since it allows a component placement modification to be taken into account according to the following procedure:

- the individual map of the moved component is substracted from the global board map;

- the component map is re-calculated at the new position;
- the new component map is added to the global board map.

With $\lambda_i = (i\pi)/a$ and $\mu_j = (j\pi)/b$, thermal maps solution of (2) have the general form:

$$T(x, y) = \sum_{i=0} \sum_{j=0} a_{ij} \cos(\lambda_i x) \cos(\mu_j y) \tag{3}$$

with, for instance, when $i \neq 0$ and $j \neq 0$:

$$a_{ij} = \frac{4Q_c/\lambda_i\mu_j}{ab\left(\lambda^2 + \lambda_i^2 + \mu_j^2\right)} [\sin(\lambda_i x_r) - \sin(\lambda_i x_l)] [\sin(\mu_j y_t) - \sin(\mu_j y_b)] \tag{4}$$

where $[x_l, x_r] \times [y_b, y_t]$ stands for the rectangular source outline modelling the component influence. Although beyond the scope of this paper, it is worth mention that the algorithm optimization for the fast computing of (3) and (4), through their reformulation and implementation, was a major task.

The method is actually not restricted to adiabatic board edges. Any combination of board edges at a given temperature can be processed using an appropriate set of Fourier functions, provided that the prescribed temperature is the same for all edges. In that case, the bare board map is obviously no longer isothermal. The processing of edges at a fixed temperature is a useful functionality to model board rails.

A beneficial side effect of the Fourier approach for the ease of use is that very few parameters control the algorithm behavior. The most important is the grid size, which sets up the point density. For the need of fast graphical processing, the temperature is calculated at the centre of the grid cells. Another parameter allows control on the convergence of the series expansion.

3. Component models

Eq. (3) allows the computation of the board thermal map and indication of the hot spots. However, junction temperature is a relevant information for the designer. To estimate it, the method adopted consists of computing the temperature rise due to the power flow through the junction-to-case resistance. This rise is computed from the board cell temperature closest to the component centre, which means that the component case temperature is assumed to be identical to the board one at that location.

4. Accuracy influence factors

Although maximum accuracy is not the target of the proposed approach, the knowledge at least qualitative of the factors which influence the method precision is crucial.

From the mathematical viewpoint, the algorithm is based on an exact solution of the problem considered, which means in particular that, by contrast to the FD or FE methods, the temperature at a given point is independent of the others. In short, the grid is not a mesh. Fine grids lead to better results in the sense that they provide more detailed maps and so more resolution in graphical outputs.

From the computational viewpoint, the rapid convergence of the series expansion means that the formulation is effectively usable to provide numerical results. However, the computational process may obviously itself induce errors when the partial sum approximating the series expansion has not a sufficient number of terms.

But, *from the application viewpoint*, the most important source of discrepancy stems from the possible violation of the strict method assumptions when processing a real case. The rest of this sections reviews some of the most significant influence factors.

To illustrate those through histograms, a fairly simple statistical approach is used, based on the absolute values $|\delta e_r|$ of the relative errors $\delta e_r = (v_i^* - v_i)/v_i$, where v_i^* and v_i are respectively the reference and approximated (i.e. simulated) values.

The reference values used to build the histogram are junction temperature measurements shown from an experimental device [9] whose brief description follows.

The device is made of a tunnel enclosing three boards equipped with components and two dissipative lateral boards (Fig. 1). The external sides of these latter are insulated. The tunnel dimensions are $82 \times 175.26 \times 500$ mm. The board dimensions are $175.26 \times 233.68 \times 1.6$ mm. The substrate conductivity is $K = 0.9$ Wm^{-1} °C. 60 components of 14 pin plastic dual in line packages are mounted on each single-sided board.

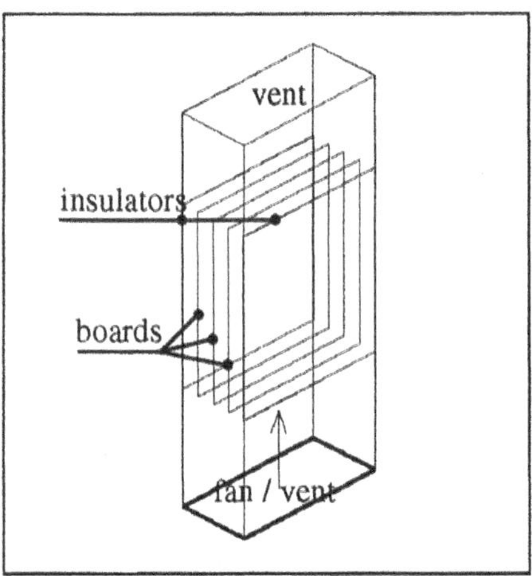

Figure 1. Sketch of the experimental device.

Five arbitrary classes of accuracy, named A to E are introduced as follows:

Class	A	B	C	D	E
$\|\delta e_r\|$ (%)	[0.0-5.0 [	[5.0-10.0 [	[10.0-15.0 [	[15.0-20.0 [	[20.0-∞ [

Table 1. Error class definition

With n_i standing for the number of observations in the ith class and n for the total number of observations, histograms for the relative frequency $f_i = n_i/n$ and the cumulative frequency F_i defined by $F_i = \sum_{j \leq i} f_j$ allow a quick visual appreciation of the result quality[1].
For instance, Fig. 2 shows that there is 5% of components for which $0.05 < |\delta e_r| < 0.10$ (f_2 -class B), while there is almost 50% of them for which $|\delta e_r| < 0.20$ (F_4 -class D).

4.1 THE INFLUENCE OF ENVIRONMENTAL DESCRIPTION

Boards may operate in environmental conditions which are difficult to characterize, especially when the electronic system geometry is complex. Two parameters play a key role in this characterization: the heat transfer coefficient and the fluid temperature.

For a natural convection cooling, Fig. 2 illustrates the accuracy obtained with $h = 6.4\ \mathrm{Wm^{-2}\ °C}$ directly derived from the internal heat transfer coefficient assisting tool. The same case processed with the help of a computational fluid dynamics (CFD) package gives $h = 7.7\ \mathrm{Wm^{-2}\ °C}$ and Fig. 3 illustrates the improvement brought by this second value.

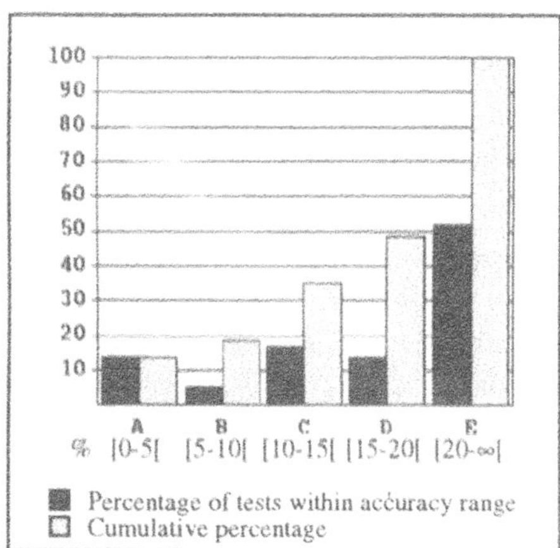

Figure 2. Natural convection ($h = 6.4\ \mathrm{Wm^{-2}\ °C}$)

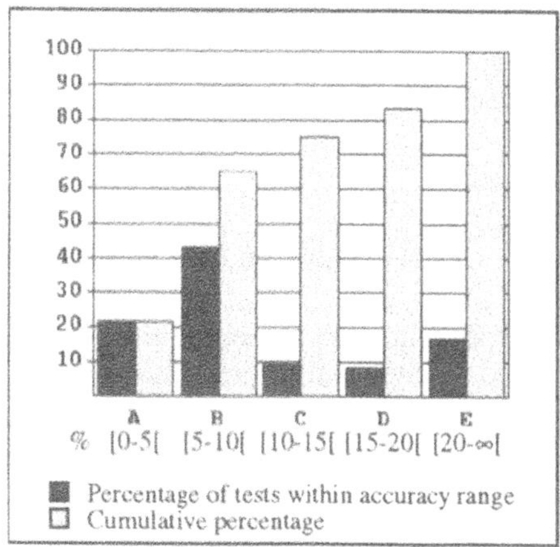

Figure 3. Natural convection ($h = 7.7\ \mathrm{Wm^{-2}\ °C}$)

So, the conventional flat-plate correlations used in the assisting tool are helpful, but they may poorly model real configurations where components induce significant fluid flow disturbances. Actually, even the values derived from CFD techniques cannot be considered as fully reliable. The accurate characterization of heat transfer coefficients remains a difficult and open problem [6].

1. The higher the leftmost bars are, the more accurate the simulation is.

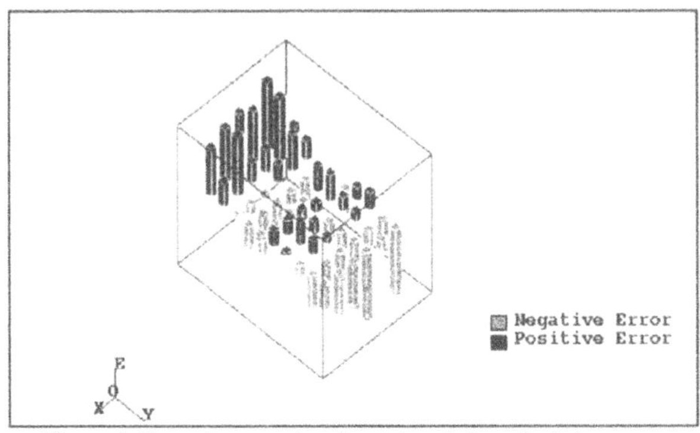

Figure 4. Natural convection. Spatial variation of relative error

The examination of the spatial variation of the relative errors also shows that the positive errors $(v_i^* > v_i)$ are mainly located at the board bottom (small y values) while the negative ones are located at the board top (Fig. 4). This remark may be explained by the variation of the fluid temperature as it flows along the board. Long vertical boards cooled by natural convection may in particular give rise to significant fluid temperature variation. Nevertheless, the visual comparison of isotherm maps as generated by the current approach (Fig. 8) and a FD based simulation (Fig. 9) shows that a satisfactory snapshot of the board hot spot is obtained, which is precisely what this approach is intended to achieve.

4.2 COMPONENT MODELLING INFLUENCE

The approximation modelling the component influence through heat sources holds as long as the temperature difference between the board and the component case is reasonably small. In particular, when the cooling mode is natural convection, heat transfer coefficients have moderate values and the approximation may be viewed as realistic. By contrast, in forced convection, the thermal coupling between the board and the component may be loose, especially when realized for instance by long pins of discrete components. In such a case, a major part of the heat dissipated by the component is directly removed by the ambient fluid. For a forced convection case with $h = 12.8\,\mathrm{Wm^{-2}\,°C}$, this effect is illustrated in Fig. 5.To avoid temperature overestimation, a simple workaround is then to ignore the component. Another less radical workaround consists of decreasing the source rate Q_k by an estimated amount of the heat directly evacuated to the fluid. Already tested, this approach leads in some cases to a very significant improvement of the results (Fig. 6), but its implementation requires some external control, slows down the execution time and so conflicts with the assigned objectives, which explains why it has not yet been introduced to the commercial implementation.

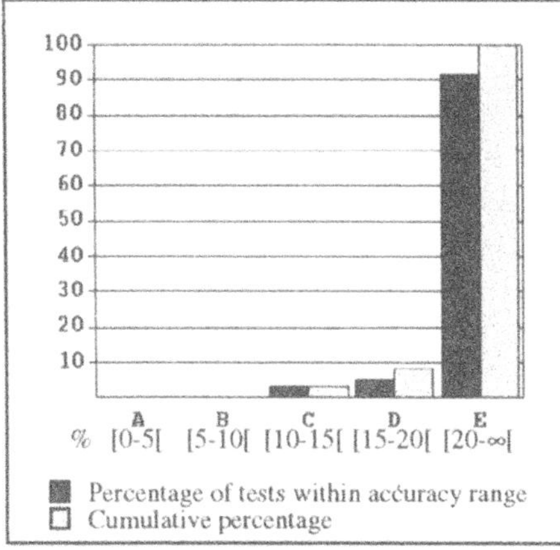

Figure 5. Forced convection

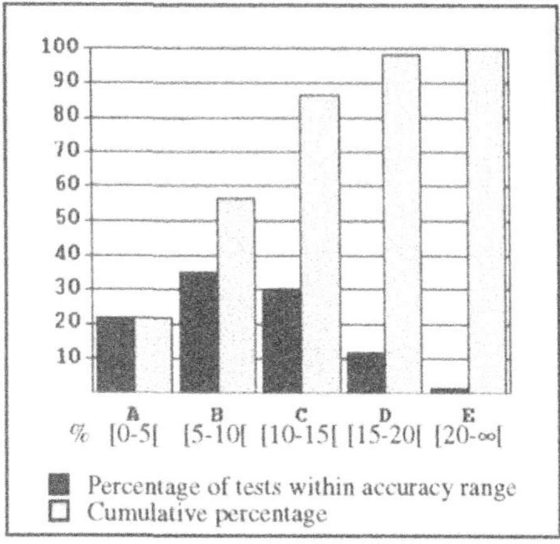

Figure 6. Forced convection. Source rate correction

4.3 COMPONENT MODEL INFLUENCE

R_{jc} values provided by manufacturers are widely available and give useful indications, but they cannot claim to be perfectly accurate [7]. This is illustrated in Fig. 7 where models based on local resistive networks are employed in a FD-based simulator [8]. These models are significantly more accurate, but their creation is far from being a trivial task.

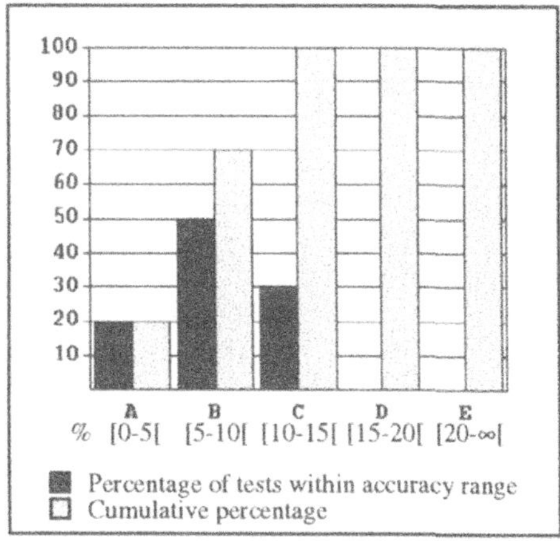

Figure 7. Model influence (FD simulation)

5. Conclusion

Ease of use, interactivity and speed are the goals of the approach proposed for the thermal analysis of electronic boards. Ease of use is illustrated by a heat transfer coefficient assisting tool allowing an estimation of the heat transfer coefficients by a quick graphical characterization of the board environmental conditions. Interactivity is emphasized by the instantaneous update of the board thermal map as the component placement is modified. Speed, underlying to interactivity, is achieved by a numerical method based on Fourier series expansion. These combined features give a non-thermal specialist user the ability to appreciate easily and quickly the influence of data change on the board thermal behavior during the pre-design phase of a project.

6. References

[1] Ottavy, N., Bourhrara, M., Le Jannou, J.P., Paris, P., *Thermal study of a laser diode using a finite element method associated with a meshing superimposition method*, Proceedings of Eurotherm Seminar No. 29, pp. 129–138, Kluwer Academic Publishers (June 1993).

[2] Kos, A., *New method of thermal placement in hybrid circuits*, SPIE vol. 1783 International conference of Microelectronics, pp. 398–409 (1992).

[3] Krum, A.L., *Thermal spreadsheets for hybrid analysis*, Hybrid Circuit Technology, pp. 45–48 (March 1988).

[4] Rottiers, L, De Mey, G., *Hot spot effects in hybrid circuits*, IEEE Trans. on Components, Hybrids and Manufacturing Technology, vol. CHMT-11, pp. 274–278, 1988.

[5] Arpaci, V.S., *Conduction heat transfer*, Addison-Wesley publishing company, 1966.

[6] Lasance, C.J.M., *Thermal management of air-cooled electronic systems: new challenges for research*, Proceedings of Eurotherm Seminar No. 29, pp. 3–24, Kluwer Academic Publishers (June 1993).

[7] Le Jannou, J.P., Huon, Y., *Representation of thermal behavior of electronic components for a databank*, IEEE transactions on hybrids and manufacturing technology, vol. 14, no. 2, pp. 366–373 (June 1991)

[8] Blanchard, J.L., Morelle, J.M., *Overall thermal simulation of electronic equipment*, EURO DAC'94 proceedings, pp. 420–425 (Sept. 1994).

[9] Le Jannou, J.P., *Etude thermique expérimentale de cartes montées côte à côte*, Journées d'étude sur les problèmes thermiques dans les systèmes électroniques, S.F.T., Paris, (March 1984).

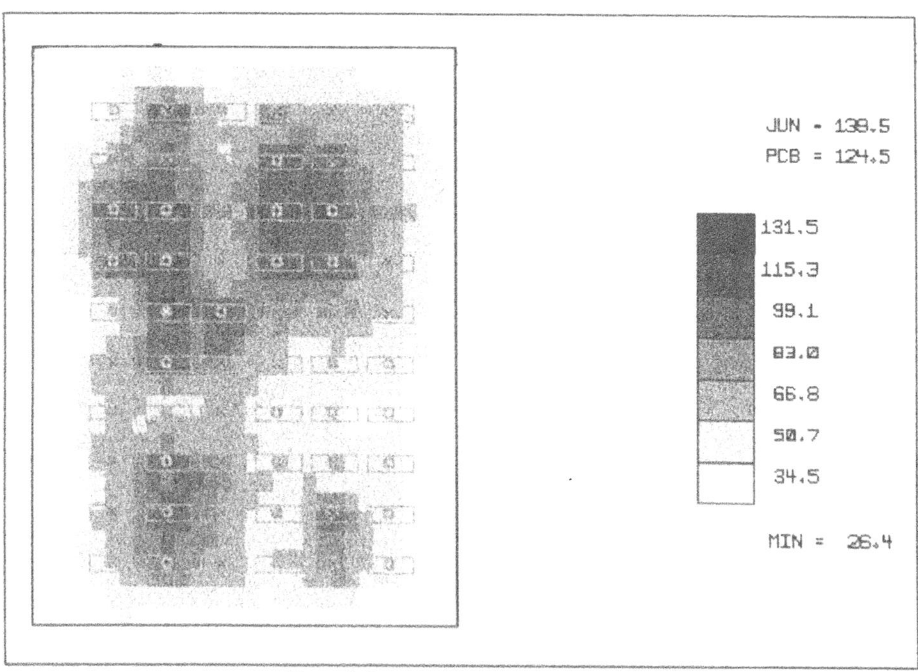

Figure 8. Isotherm map. Current approach

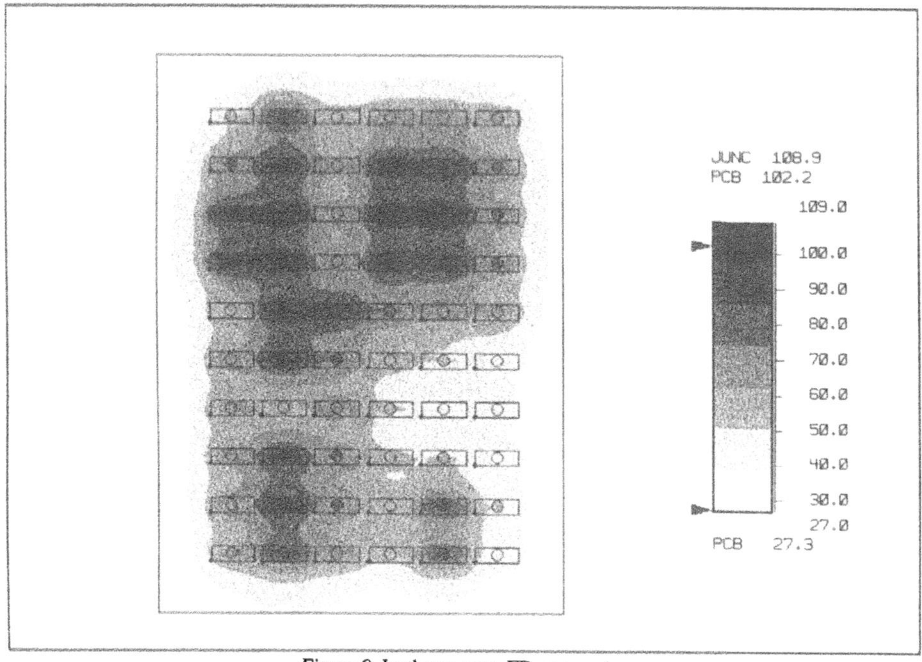

Figure 9. Isotherm map. FD approach

3. THERMAL CHARACTERISATION

THERMAL CHARACTERIZATION OF ELECTRONIC DEVICES BY MEANS OF IMPROVED BOUNDARY CONDITION INDEPENDENT COMPACT MODELS

H. VINKE and C. LASANCE
Philips Centre for Manufacturing Technology
P.O. Box 218
5600 MD Eindhoven, The Netherlands

Abstract
In this paper improvements on the thermal characterization of electronic devices by means of so-called 'compact models' are described. A compact model is a simple network comprising a limited number of thermal resistances (typically 7), connecting the critical part of the device (usually the junction) to the outer parts of the device and is independent of the boundary conditions applied. This method of thermal characterization is suitable for embedding in the design environments that are employed by the electronics industry, and the compact models can be incorporated in the component libraries linked to PCB thermal analysis software packages. It is demonstrated that the thermal behaviour of electronic devices such as a 208-PQFP and a CPGA can be approached within typically 6% of the full model values.

1. Introduction

The accurate prediction of operating temperatures of temperature-sensitive electronic devices at the component, board and system level is hampered by the poor thermal characterization of these critical electronic devices. This thermal characterization usually consists of experimental data which describes the thermal behaviour of electronic devices under a set of standardized and idealized conditions, e.g. the junction-to-case thermal resistance or the junction-to-ambient resistance according to MIL or SEMI standards.

Although the thermal characterization by means of a single thermal resistance can be accurate for certain types of electronic devices and some boundary conditions, it is obvious that a single thermal resistance cannot accurately describe the essentially three-dimensional heat transfer in a package.

A growing number of designers and researchers are also aware of the need for a more accurate thermal characterization of electronic devices. Too often [3,4,5] the thermal characterization, which is obtained under standardized and idealized conditions, deviates from the thermal behaviour encountered in practice.

E. Beyne et al. (eds.), Thermal Management of Electronic Systems II, 115-124.

In a previous paper [7] a new methodology was introduced to describe the thermal behaviour of electronic devices by means of 'compact models'. A compact model is a simple network comprising a limited number of thermal resistances (typically 7), connecting the critical part of the device (usually the junction) to the outer parts of the device and is independent of the boundary conditions applied. It was shown that with this kind of thermal characterization the *junction temperatures* of a so-called Validation Chip and a 208-PQFP can be accurately described. In this paper this methodology is extended and attention is paid to several other aspects.

Topics that will be discussed in this paper are:
- outline of the method,
- the derivation of a compact model,
- boundary conditions,
- definition of the cost function.

Results will be presented for the Validation Chip, a Plastic Quad Flat Pack (208-PQFP), and a Ceramic Pin Grid Array (CPGA).

2. Compact models

2.1. OUTLINE OF THE METHOD

The method consists of the following steps.

1. Creation of a 'full' model of an electronic device. This 'full' or complete model can be made with any type of software package. We used Pstar (a Philips proprietary E-SPICE type code) and Flotherm (from Flomerics, Inc.).
2. A set of 38 combinations of boundary conditions, representing conditions which a device may encounter in practice, is imposed on all faces of interest. The junction temperature and heat flow rate through each side is then calculated using the full model for each combination of these boundary conditions.
3. The junction temperatures and heat flow rates thus obtained are implemented in Optimize, a proprietary software code that essentially minimizes a user-defined cost function using optimization routines from the NAG library. The implementation of heat flow rates in the optimization procedure is an improvement with respect to previously published optimization runs [7].
4. Definition of a user-defined cost function that is to be minimized in the optimization procedure.
5. Incorporation of a thermal network, i.e. the 'compact model', which should be able to characterize the thermal behaviour of the package in Pstar.
6. The actual optimization. Pstar and Optimize work together as a single package: Pstar is the network solver and Optimize the optimization routine. By varying the resistances in the network the user-specified cost function is minimized. If the results are within the required accuracy range, a compact model which is independent of the imposed boundary conditions is obtained.

It can be envisaged that compact models are suitable for embedding in the design environments that are employed by the electronics industry. Compact Models in the form of a simple resistance network can be incorporated in the component libraries linked to PCB thermal analysis software such as Cadence, Mentor Graphics, Pacific Numerix. The use of compact models in CFD codes is a bit more difficult but the solution of this problem is part of the ESPRIT-project DELPHI.

2.2. CREATION OF A COMPACT MODEL

Although the derivation of a compact model comprises some trial and error we have developed a sort of methodology which usually produces good results. In the future, it is expected that this procedure will have a self-learning effect, so that the creation of compact models is based on physical judgement.

The compact model consists of a network of thermal resistances that is partially based on the Bar-Cohen network [1, 2, 3] which connects the junction to the top, bottom, side and leads of the package. As a result, each surface is represented by a single node. The extensions consist of extra resistances connecting each side of the packages to the other sides, e.g. resistors connecting the top and bottom to the leads. If the top or the bottom of the package is made of a poor heat-conducting material, i.e. a hot spot could occur on these surfaces, then these sides have to be subdivided into two parts.

Starting with this network, which actually consists of a relatively large number of resistances (say at most 30), the optimization procedure is performed. In this optimization procedure, the values of the thermal resistances are changed so that the user-defined cost function is minimized. At the end of this procedure, resistances with a high value (relatively to the other values) are often found. The optimization run is then repeated without these thermal resistances using the previously found values as starting values in the optimization procedure. This procedure is repeated until no resistances with high values are found. By following this procedure, the compact model is reduced to, typically, six or seven resistances.

2.3. BOUNDARY CONDITIONS

In theory, an electronic device can be subjected to an infinite number of boundary conditions. In practice however, a device will only be exposed to a limited range. A set of 38 boundary conditions was selected which is thought to cover these practical applications (see TABLE 1). It is however recognized that this chosen set of boundary conditions might not cover all conditions encountered in practice and that the selected combinations are debatable. More research will be devoted to this subject in the future.

For the sake of simplicity, the conductive heat transfer through the leads, the cold plate and the heatsink is re-calculated in terms of a convective boundary condition, i.e. to a certain heat transfer coefficient which is applied to the cross-sectional area of that side. Following this approach for heatsinks, it is implicitely assumed that the temperature averaging effect of the groundplate can be incorporated in an 'enhanced' heat-transfer coefficient.

TABLE 1. List of 38 combinations of boundary conditions which an electronic device may encounter in practice

cooling method	α_{TOP} (W/m²K)	α_{BOT} (W/m²K)	α_{SIDE} (W/m²K)	α_{LEADS} (W/m²K)
free convection	10	10	10	10;100;1000;10000
	10;100	10;100	10	100;1000
	30	30	30	30
	50	50	50	50
	100	100	100	100;500
forced convection	100	100	100	1000;10000
	1	100	100	1000
	100	1	100	1000
	200	200	200	1000;10000
	50	50	50	1000;10000
fluid bath	10^9	10^9	10^9	10^9 (*)
	10000	10000	10000	10000
	1000	1000	1000	1000
	500	500	500	500
single cold plate	10000	10	10	100;1000
	10	10000	10	100;1000
	1	1000	1	10000
	10000	1	1	10000
heat sink	500	10	10	100;1000
	1000	10	10	100;1000
	10	500;1000	10	100;1000

(*) This boundary condition cannot be achieved in practice, but it gives the lowest junction temperature possible

The number and especially the range of boundary conditions affect the number of resistances in the compact model and the the accuracy of the model. Therefore it is conceivable that a compact model with less resistances than the ones proposed in this paper could be created for a device that is cooled by natural convection only. Of course, in the ongoing documentation of a compact model the application area should be mentioned and checked.

At this moment, it is not clear which approach has to be followed. Should the compact model be generally applicable to all practical applications, or should the compact model be applicable to a limited number of practical applications only? In the first case the accuracy will be reduced and the compact model will be more complex. In the second case the compact model can be quite simple and a high degree of accuracy to the full model values can be achieved.

2.3.2. *Definition of the cost function*

A considerable amount of effort was devoted to the definition of the cost function, since this function considerably affects the results. It was found that a cost function as defined in Eq. (1) produces the best results:

$$F = \left(\frac{T_{j,c} - T_{j,f}}{T_{j,f} - T_{amb}} \right)^2 + W \sum_{i=1}^{n_{sides}} \left(\frac{\phi_{i,c} - \phi_{i,f}}{\phi_{tot}} \right)^2 \qquad (1)$$

in which W is a weighting factor. The junction temperatures are scaled by the temperature difference between the junction and ambient temperature, the heat flow rates by the total heat flow rate. In this way, emphasis is placed on the lower values of the junction temperatures, and the lower heat flow rates are considered less important than higher heat flow rates. This definition is motivated in Section 5.

By specifying the value of the weighting factor W, emphasis can be placed on the relative importance between junction temperatures and heat flow rates. If W=0, only junction temperatures are optimized. Higher values of W account for more significance on the heat flow rates. In this paper W is varied between 0 and 10.

The best value of the weighting factor W may depend on the type of application. It is conceivable that for fluid bath applications accurate junction temperatures are more important than accurate heat flow rates because all the generated heat is locally transferred to the liquid. However, for free convection applications, it is conceivable that the accurate description of both junction temperatures and heat flow rates is necessary, because the heat flow rates thermally affect other parts of a system.

3. Description of the Electronic Devices

Results will be presented for a Validation Chip, a Plastic Quad Flat Pack (208 PQFP), and a Ceramic Pin Grid Array (CPGA) each of which are described briefly below.

3.1. THE VALIDATION CHIP

The Validation Chip is an idealized package that includes the principal features likely to be found in most packages yet simple enough to have an analytical solution (Fourier series). The Validation Chip features a centrally located die with layers of low and high thermal conductivity to emulate the following typical structural components of mono-chip packages: mould compound, lead frame and heat spreader.

3.2. THE PLASTIC QUAD FLAT PACK (208-PQFP)

A full model of this package was made using FLOTHERM (Flomerics, Inc). The calculated results were validated against experimental data as reported in [6]. The junction temperature and the heat flow rates through all the sides are calculated with the full model as a function of a selected set of boundary conditions.

3.3. THE CERAMIC PIN GRID ARRAY (CPGA)

A full model of this package was also made using Flotherm. This device is an Intel Pentium type processor and features a ceramic housing (the ceramic having a thermal conductivity of 23 W m^{-2} K^{-1}), a cavity filled with air, and a number of pins which are located at the bottom of the package. The same calculation procedure is followed as for the 208-PQFP.

4. Results

In this paragraph, the following results will be presented:
- the compact models obtained,
- the effect of the weighting factor W on both the accuracy of the compact model and on the values of the thermal resistances,
- intermediary boundary conditions.

4.1. COMPACT MODELS

According to the procedure described in Section 2.2 the following compact models were obtained for the Validation Chip, the PQFP and the CPGA.

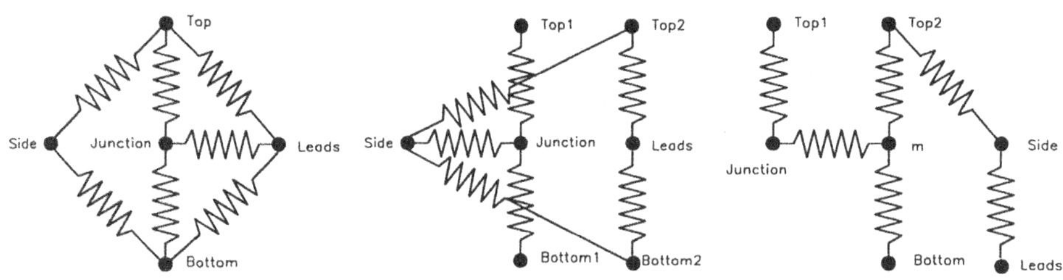

Figure 1. Compact models for the Validation Chip, the 208-PQFP and for the CPGA.

Many other compact models with equal performance and accuracy are possible and have been tested. It is believed, however, that the compact models presented combine high accuracy with a relatively low number of thermal resistances.

4.2. COST FUNCTION

TABLE 2 gives an overview of the maximum error of the three compact models as a function of the weighting factor W. Since the errors are not Gaussian distributed, the use of average errors with their standard deviation gives unrealistic values. Instead only the maximum errors will be supplied. The 'mean' error is typically a factor of 3 to 5 times lower than the maximum error. Usually, the maximum error occurs for the impossible boundary conditions with $\alpha_{top}=\alpha_{bot}=\alpha_{side}=\alpha_{leads}=10^9$ $Wm^{-2}K^{-1}$. Omitting this boundary condition set gives results of which the maximum error for the heat flow rates is typically about a factor of two lower.

TABLE 2. The effect of the weighting factor W on the error in junction temperature and heat flow rates

Weighting factor W	error on(*):	Device		
		Validation Chip	208-PQFP	CPGA
		max(%)	max(%)	max(%)
0	Junction temp:	2.1	2.5	3.9
	heat flow rates:	8.5	3.9	6.9
0.1	Junction temp:	2.0	2.3	4.0
	heat flow rates:	4.9	3.8	7.0
1	Junction temp:	2.1	3.3	4.0
	heat flow rates:	4.4	3.4	6.0
10	Junction temp:	2.2	9.6	6.3
	heat flow rates:	4.3	3.1	4.5

(*) errors are defined as:
for junction temp: $100*(T_{j,c}-T_{j,f})/T_{j,f}$
for heat flow rates: $100*(\phi_c-\phi_f)/\phi_{tot}$

Since the error on the junction temperature is considered to be more important, the relative error between the compact model and the full model junction temperature values is used. For the heat flow rates, the error between the compact model and the full model is related to the total heat flow rate. This definition, instead of the relative definition, is chosen because relatively low values of the heat flow rate compared to the total heat flow rate (e.g. 10 mW to a total of 1000 mW) are considered to be of little importance.

TABLE 3. Overview of the thermal resistances and fractional surface areas found for W=0 and W=10.

Validation Chip		208-PQFP		CPGA	
parameter	value range	parameter	value range	parameter	value range
R_{JT}	70.0-88.6	R_{JT1}	11.5-10.8	R_{JT1}	0.31-0.28
R_{JB}	42.5-45.1	R_{JB1}	8.5-8.1	R_{JM}	2.9-2.9
R_{JL}	47.8-39.7	R_{JS}	21.5-28.0	R_{T2M}	0.2-0.04
R_{TL}	19.8-18.2	R_{T2L}	7.7-1.8	R_{BM}	0.1-0.08
R_{TS}	47.0-34.0	R_{B2L}	7.7-1.8	R_{SL}	0-0
R_{BL}	9.3-8.4	R_{T2S}	0-0	R_{T2S}	0.9-0.8
R_{BS}	42.2-23.6	R_{B2S}	0-0	A_{TOP1}/A_{TOP}	0.1-0.09
		$A_{TOP1}/A_{TOP}=A_{BOT1}/A_{BOT}$	0.16-0.23		

It is found that optimizing both junction temperatures and heat flow rates, i.e. increasing the value of W, does change the values of some of the thermal resistances considerably (see TABLE 3). This is due to the fact that the thermal network is overdetermined (there are more solutions possible) if only junction temperatures are treid out. By incorporating the heat flow rates into the optimization this overdeterminition of the thermal network diminishes. Apparently there is an optimum value for the weighting factor W, which depends on the type of device. Future research will partly be devoted to the sensistivity of the network.

The zero thermal resistances are a consequence of the fact that for all boundary conditions tested, the temperature differences between some of the nodes are negligible. By combining nodes T2, B2, and S into a single node and connecting the three external resistances in parallel, the 208-PQFP could be represented by only five resistances with the same accuracy.

4.3. INTERMEDIARY BOUNDARY CONDITIONS

For the Validation Chip, two extra series of boundary conditions were studied. The first series contained 81 boundary conditions comprising all permutations of α=1, 100, 10000 $Wm^{-2}K^{-1}$. The second set contained all permutations of α=10, 100, 1000 $Wm^{-2}K^{-1}$. In these studies, the thermal resistances of the compact model were fixed to check if the compact model was able to predict the thermal behaviour of other intermediary boundary conditions equally well. Comparison of full model and compact model results showed that this was indeed the case.

5. Discussion

In the previous section it was shown that the proposed compact models are able to describe the thermal behaviour of the three electronic devices accurately to the full model values. The relative error in the junction temperature is typically within 3%, the error in the heat flow rates is typically less than 6%. As is to be expected the maximum error in the junction temperature increases and the error in the heat flow rates decreases with increasing value of the weighting factor W. It should be noted that the compact models presented in this paper are just one of the many possible types. Future research will be devoted to the improvement on the physical basis of these models.

During the research other types of cost functions were also evaluated. A cost function in which the relative error between the heat flow rates should also be minimized, i.e:

$$F = \left(\frac{T_{j,c} - T_{j,f}}{T_{j,f} - T_{amb}} \right)^2 + W \sum_{i=1}^{n_{sides}} \left(\frac{\phi_{i,c} - \phi_{i,f}}{\phi_{i,f}} \right)^2 \tag{2}$$

is found to give unsatisfactory results. The compact model and this cost function were unable to describe the lower values of the junction temperature *and* the heat flow rates in particular.

A cost function in which the absolute error between the junction temperature and the temperature of all sides, i.e.

$$F = \left(T_{j,c} - T_{j,f}\right)^2 + W \sum_{i=1}^{n_{sides}} \left(T_{i,c} - T_{i,f}\right)^2 \tag{3}$$

was also found to give good results. With this type of cost function the errors in both the junction and face temperatures are considered to be equally important for all boundary conditions. Another advantage is that all terms in the cost function have the same dimension. A drawback of this cost function is the foilowing. Suppose that calculations show that, for specific operating conditions, the junction temperature is well below a critical value. In this case it may be tempting to raise the power (e.g. increase of the clock speed) of this device, in order to operate it near the critical value of the junction temperature. If the compact model is made using this definition of the cost function, it is inaccurate when the power of a device is increased since the power and junction temperatures are proportional. This is the reason why relative junction temperatures are used (see Eq. 1). The relative definition ensures that the emphasis is placed on the lower values of the junction temperatures, i.e. in that range of boundary conditions where power increase is possible, and that the error is significantly smaller compared to the definition as in Eq. (3).

Finally, it should be noted that optimization with heat flow rates only resulted in maximum errors in the junction temperature of more than 25 %.

6. Conclusions

Electronic devices such as a 208-PQFP and a CPGA can be characterized by so-called compact models, typically consisting of seven thermal resistances, which describe the thermal behaviour, characterized by junction temperatures and heat flow rates leaving the faces of a device, within 6% for a specified set of boundary conditions.

At this moment, it is not clear which approach has to be followed. Should the compact model to be generally applicable to all practical applications, or should the compact model be applicable to a limited number of practical applications only. In the first case the accuracy will be reduced and the compact model will be more complex. In the second case the compact model can be quite simple and a high degree of accuracy to the full model values can be achieved. It is conceivable that, for instance, two relatively simple compact models are derived, one of which is suitable for natural and forced convection only, and the other one suitable for liquid cooling and cold plates.

The definition of the cost function largely affects the accuracy and the values of the thermal resistances of the compact model. It seems that a cost function, as defined in Eq. (1), produces the most satisfactory results.

7. Acknowledgment

The partial support of the Commission of European Communities under the ESPRIT contract DELPHI, and the useful discussions with H. Rosten, J. Parry and B. Ali from Flomerics are acknowledged.

8. Notation

A	[m^2]	surface area
F	[-]	cost function
R	[K/W]	thermal resistance
T	[K]	temperature
W	[-]	weighting factor (see Eq. 1)
n	[-]	number

greek letters

α	[$Wm^{-2}K^{-1}$]	fluid side heat-transfer coefficient
ϕ	[W]	heat flow rate

Subscripts

amb	ambient temperature
i	referring to side i
i,f, i,c	referring to side i, calculated with full model or compact model
j,f, j,c	junction temperature, calculated with full or compact model
sides	referring to (the number of) sides
BOT, BOT1, TOP, TOP1, LEADS, SIDE	(part of) surface area of the bottom, top, leads or side
BL, B2L, BS, B2S, BM	from (part of) bottom to the leads, to the side or to point m
JB, JB1, JT, JT1, JS, JM	from junction to (part of) bottom, top, sides or to point m
TL, T2L, TS, T2S,T2M	from (part of) top to the leads, to the side or to point m
TOT	total

9. References

[1] A. Bar-Cohen, T. Elperin and R. Eliasi: θ_{jc} Characterisation of Chip Packages-Justification, Limitations, and Future, *IEEE Trans. CHMT*, vol.12, pp. 724-731, 1989.

[2] A. Bar-Cohen and W. Krueger: Thermal Characterization of a PLCC - Expanded R_{jc} Methodology, *Trans. CHMT*, vol. 15, no. 5, pp 691-698, Oct 1992.

[3] A. Bar-Cohen and W. Krueger: Cooling of Electronic systems, *Kluwer Academic Publishers*, 1994, pp. 789-810.

[4] C.J.M. Lasance: About the validation of CFD analyses of electronic systems, presented at the Third *International Flotherm user conference*, Londen, England, sept 22-23, 1994.

[5] C.J.M. Lasance: Cooling of Electronic systems, *Kluwer Academic Publishers*, 1994, pp. 859-898.

[6] H.I. Rosten and C.J.M. Lasance: The Development of Libraries of Thermal Models of Electronic Components for an Integrated Design Environment, in *Proceedings of the conference of the Int. Elec. Pack. Soc.*, sept. 1994, Atlanta GA.

[7] C. Lasance, H. Vinke, H. Rosten, K.L. Weiner: A Novel Approach for the Thermal Characterization of Electronic parts, *Proc. Semi-Therm*, 1995, pp. 1-9, San José, U.S.A.

EXPERIMENTAL VALIDATION METHODS FOR THERMAL MODELS

W. TEMMERMAN, W. NELEMANS, T. GOOSSENS, E. LAUWERS
Alcatel Bell Telephone
F. Wellesplein 1, 2018 Antwerpen, Belgium

C. LACAZE
Alcatel Espace
26 Av.J.F. Champollion,31037 Toulouse, France

Abstract

The present paper evaluates the most important experimental thermal characterisation techniques for electronic components. Weak points are indicated in the standardised methods for the still air, cold plate and fluid bath environment. Improvements have been developed: Suitable measures determine well-controlled surface temperature profiles over the whole component surface in the double cold plate. Two impinging jets impose reproducible heat transfer coefficients in the fluid bath method. Hence, the reproducibility, the scope and the accuracy of the measurements are enhanced.

1 Introduction

1.1 Thermal Modelling

Due to the impact of ever progressing integration and speed in electronic systems, which is not completely offset by the reduction in power dissipation per function, the heat load per unit area or volume continues to grow, reaching critical values in many application areas. This has brought thermal problems from an afterthought to an early system design issue.

Consequently, recent years have seen a sharp rise in the use of enclosure-level and PCB-level thermal analysis software. Enhanced computer power allows calculation of local air temperatures and heat transfer coefficients. If the analysis packages are supplied with good thermal models of the components then it becomes possible to calculate the junction temperature with sufficient accuracy to serve as an input for later reliability analyses. The PCB-level thermal analysis packages all contain thermal models of components, of varying degrees of sophistication, ranging from a single thermal resistor using the manufacturers value of R_{jc} through to quite complex thermal

E. Beyne et al. (eds.), Thermal Management of Electronic Systems II, 125-136.

resistor networks. However, the accurate prediction of the temperature of critical electronic parts is still seriously hampered by the lack of standardized, reliable, input data: two different networks for the same part cannot be expected to produce the same results. The conclusion is that both enclosure- and PCB-level analysis require correct thermal models from a component database, according to an internationally-agreed format.

This paper discusses model validation as part of a structured approach to develop libraries of models for thermal simulation. It is based on work done in DELPHI, a CEC-sponsored project under the ESPRIT III Programme. For further details, we refer to [1]. Empirical results, specific simulations and some general project results are used to compare and enhance existing measurement methods. An actual systematic comparison between the simulations and the experiments is presented in [2].

1.2 DELPHI

The project title DELPHI stands for DEvelopment of Libraries of PHysical models for an Integrated design environment. It was proposed to create and validate generic thermal models of electronic parts. Thermal analysis is considered at all packaging levels by providing at least 2 different levels of accuracy:

- detailed models which represent sufficient physical features to permit accurate calculations of the temperature distribution throughout the package, either finite-element or finite-difference conduction models; and
- compact models for use as an approximate simplified representation (typically resistor networks modelling the heat flow paths inside).

The package itself should be modelled, nothing else. In general, one cannot assume that the component manufacturer, who eventually should supply the models [1], knows the environment to which his part will be exposed. Therefore, his responsibility should be restricted to supplying a model of the package valid for all practical environments, whereas the equipment around it is system-level engineering responsibility.

Standard R_{ja} methods, as well as useful enhancements, where R_{ja} is complemented by supplementary info (such as the sensitivity of T_j to T_{brd}, Andrews, Mahalingam, [3], [4], [5]) do not comply with this requirement. To respect this principle, a methodology has been defined, at least suitable for mono-chip packages. The 208-pin plastic quad flat pack (PQFP208) served as pilot device. As reported in [1] and [6], its compact model achieves an accuracy of 1% over 38 "practical" boundary conditions.

Clearly, one of the important objectives of the DELPHI project is to define a thermal characterization philosophy which is essentially independent of all practical boundary conditions, in other words, one which really provides a universal thermal model of the package. Accordingly, this paper describes experiments which focus on the validation of thermal conduction models for the package alone. The experiments must provide boundary conditions to be applied in simulations. This imposes more requirements then a performance comparison test, such as described in [7].

1.3 Thermal resistance concept

Thermal characterization is typically based on some R_{th} -concept, defined as :

$$R_{th} = (T - T_{ref})/P \qquad (1)$$

where P is usually the steady-state dissipation of the component, T is some critical temperature (generally the junction temperature) and T_{ref} some reference temperature, either of the case or the ambient air temperature, for which the corresponding thermal resistances are customarily denoted by R_{jc} and R_{ja}, respectively.

The definition of thermal resistance requires a greater degree of generality and care than its electrical analogue, because physical surfaces, such as the component case, are rarely isothermal and, even if they are by design, it is difficult to create a controlled region between them without heat losses to the outside world.

1.4 Device under test (DUT)

This study is based on a thermal test die (SGS-Thomson P655) packaged in a 208-pin plastic quad flat pack (PQFP208). Other test dice are also possible and are being used within the project. They all allow to spread the dissipation equally over the die and feature at least 1 centrally located temperature measurement diode.

Because of the high thermal conductivity of silicon compared to that of plastic and to a lesser extent ceramic, the variation of temperature over the silicon is often (but not always) small compared to the variation from the silicon to the case. The assumption of an isothermal silicon surface is then a good approximation. So, ideally, we have to devise an isothermal reference surface and some control of the heat fluxes.

Theoretically, the use of test boards for mechanical and electrical connection of the packages is undesirable: Acting as heat sinks they introduce additional uncertainties and distort the thermal characterization. However, as all fine pitch SMT components (0.5mm pitch), our DUT has very fragile leads. In this case, test boards are inevitable, and the parasitic effects must be minimized by special designs.

1.5 Calibration

During the calibration, the electrical values of the temperature sensitive parameter (TSP) are determined as a function of the temperature. In our case, the TSP is a forward diode voltage at a constant diode current. The acceptable calibration current range was established to be between 10μA and 1mA for our DUT. Below this range, the temperature sensitivity of the forward diode voltage tends to increase. Both 4- and 2-point measurements result in a linear calibration curve. When using 2-point calibrations however, one must always do the measurement and the calibration exactly the same way, with the same DUT, to keep the series resistance identical. Even small processing variations can generate on-chip resistance differences.

In principle the calibration is not much more than a measurement without power applied. But, since any error generated in it is introduced in all following results, we must pay special attention. An accurate temperature measurement is needed, other than the TSP. Typically, the same sensor as used to monitor the environment provides this value. The advantage is that a constant systematic error in the reference temperature value is cancelled out when considering temperature differences.

2 Model validation requirements

In general, test methods for validation of a thermal model should satisfy the following 6 criteria (listed more or less in order of importance):

1. They must provide well-defined boundary conditions, easy to simulate numerically: convective effects must be eliminated, because the heat transfer coefficient (HTC) is not known accurately enough.
2. They must provide all possible boundary conditions, necessary for a real validation: this means that all the boundary conditions applicable to a component at any one time must be simulated. In other words, they must be able to address all practically important heat flow paths individually or in combinations allowing mathematical extraction of those paths' separate characteristics.

The relevance of this requirement is illustrated by the "Expanded R_{jc} Methodology" (Bar-Cohen et al. [8], [9], [10]), where the analysis is also restricted to the package itself. It is characterized by a limited number of well-chosen thermal resistances in a star topology, where the centre corresponds to the junction, e.g. by placing nodes on all surfaces that take part in the heat transfer, and additionally at some lead groups.

Unfortunately, despite the fact that the Bar-Cohen method generates remarkably accurate results for a range of boundary conditions uniformly applied to all faces, such as occur in practice when testing the package in a fluid bath or (without a PCB) in still air, it failed to predict the junction temperature for other boundary conditions of interest, such as for forced convection and heat sinks. This is due to the fact that the model validation and parameter estimation are based on (numerical) experiments which don't represent all the boundary conditions adequately.

The boundary conditions can be characterized by the HTC applied to different exposed parts of the component: top, bottom, sides and leads. In [6] 38 HTC combinations are used in as many numerical experiments to develop the compact model. They represent practical component environments, in which the compact model must fit the reality. For an empirical approach, it is impractical to impose 38 different experiments for each device. Therefore we need to reduce the number of tests to a representative set, which includes the extreme (i.e. difficult for the model) conditions.

3. The tests must be reproducible: results should not be too sensitive to the details of the test implementation, and must certainly remain the same when replicated after removal of the device under test (DUT). If this is not the case it is impossible to repeat tests in different laboratories.

The following quote from SEMI G43-87 (Fluid bath measurements of moulded plastic packages) shows this is not trivial : "... Due to the thermophysical properties of the heat transfer fluids used and the effects of the variable nature of the fluid-stirring and package-mounting procedures, this test should only be used for comparing the thermal characteristics of plastic packages in the **same** fluid bath system. ...".

The next 3 requirements, although not of the same scientific nature, are equally relevant if we are to obtain a widely used specification, suitable for standardization.

4. The test set-ups must be cost-effective to build and run: they must be easy to use, automation must be possible and measurements should not take too long.
5. The methods must be flexible : adaptation to a new device (type) should be easy.
6. They must be readily explicable and understandable, allowing efficient numerical simulation (without irrelevant features).
7. They must provide accurate calibration possibilities.

3 Measurement methodologies

The following approaches, based on existing and emerging standards are considered.

1. Well stirred fluid bath (FB, Semiconductor Equipment and Materials International - SEMI Standard G30-88, G43-87) to mimic the isothermal case.
2. Temperature controlled cold plate (CP, SEMI G30-88): unidirectional heat flow.
3. Still air (SA, SEMI G38-87, JC15.1)

Clearly, different methods generate different values for R_{ja} for the same package. For one specific PQFP208, FB measurements in different laboratories showed values between 7.4 and 18.5K/W. CP results on the same component ranged between 8.3 and 12.1K/W and SA results between 34.5 and 38.5K/W. For R_{jc}, a totally different picture was obtained: 4.4K/W for SA, 7.2K/W for FB and 11.3K/W for CP. Of course, the latter results only have a very limited, relative, value since the case is not isothermal, as explained earlier, and T_c measurements are unreliable in general.

3.1 Fluid bath (FB)

The simplest possible surface temperature distribution is the isothermal one but it is difficult to contrive experimentally. It requires very high HTCs : the temperature drop across the region between isothermal ambient (bulk fluid) and component case must be negligible. The necessary HTC depends on the DUT. For an isothermal case, the HTC needed to reduce the R_{ca} to a certain fraction of R_{jc} is calculated in table 1. An outer area of $2.10^3\,m^2$ is assumed, typical for a PQFP208, or a PGA.

This is the intention of the fluid bath, in which the part is suspended in an agitated fluid: the high specific heat of the fluid coupled with the agitation will lead to a very high heat transfer coefficient at all points of the case. If the target from table 1, last column is not met, the HTC must be known and reproducible. In this case, the set-up has to be specified in full detail (bath dimensions, test board and clamps, stir-

Table 1 : HTC required to reduce R_{ca} to a certain fraction of R_{jc}

R_{jc} (K/W)	HTC (W/m²K) required to make	
	$R_{ca} = R_{jc}$	$R_{ca} = 0.01\ R_{jc}$
10	50	5000
5 (PQFP)	100	10000
1 (PPQFP)	500	50000

rer/pump type, rate and pipes), which is highly undesirable (not flexible, not robust), or a set-up has to be devised which forces the fluid flow, independent of the DUT.

The two fluids mostly used in a fluid bath, are fluorinert and de-ionized water. Fluorinert is often preferred because it is electrically insulating by nature. However, a thermally better choice is de-ionized (DI) water, which has a 4 times bigger specific heat. Furthermore, the viscosity of fluorinert depends on temperature, resulting in different thermal resistances at different fluid temperatures (powers). Unfortunately, DI water is more difficult to handle (to keep it DI to avoid leakage currents and corrosion). The SEMI standard does not define the kind of fluid to be used nor the degree of agitation.

Measurements were performed in a fluid bath for a range of dissipations with those two fluids, in both natural convection and forced convection, the latter using stirring and recirculation pumps. All these tests turned out to be hard to interpret (HTC unknown) at best, and mostly not even reproducible between different laboratories. More details can be found in [11]. The main problems were due to the influence of the DUT on the flow, with usually a very unsteady flow, and a HTC that wasn't high enough, so there was also significant dependence on the fluid stirring rate.

In [11], the enhancements from the standard fluid bath to the submerged double jet impingement test (SDJI) are described. The use of 2 impinging fluid jets offers the highest HTCs and solves the other mentioned FB problems. When the bath is suffi-

Table 2 : Heat transfer coefficients for fluidum-based methods

Application		P (W)	T_{wall} (°C)	HTC (W/m²K)	Remark
SA	Nat. conv.	1-5	40-100	12-15	
FB, natural convection	FC-70	5-10-20	34-42-54	96-119-150	
	DI water	5-10-20	23-25-28	418-495-609	
SDJI	DI water	10-20	20.37-20.51	13800-14800	V = 10l/min
	optimized	-	-	40000	optimized jets
Tested with aluminum plates glued to heater foil, size 50mm x 50mm.					

ciently large, the outlet nozzles determine the flow and the HTC, not the bath and DUT geometry. With a flow rate of 10l/s HTCs around 40000W/m²K can be obtained. Table 2 illustrates the potential of some fluidum-based measurement methods.

For calibration, the situation is totally different: as there is no significant heat flow, all fluid bath systems provide a stable and uniform environment. No differences could be observed between fluorinert or DI water baths, [11].

3.2 Natural convection

Natural convection is one of the predominant modes of cooling. Provided the ambient is isothermal, the R_{ja}-definition can be applied straightforward, for all the heat dissipated at the junction ends up in the environment. However, R_{ja} depends strongly on the application method: very different values apply to, for example, an isolated component in still air compared with a board-mounted component with a heat sink. Therefore a very specific standard is needed to obtain reproducible results. This is achieved by the JEDEC-standard (no heat sink) and the corresponding SEMI standard (vertical test board only). The test set-up is very cheap and flexible, since the main contributors to hardware and cost are also needed in every other method.

The main disadvantage of the method is its insensitivity to the actual device conduction model: When splitting R_{ja} into two components, the conduction resistance R_{jc} and the resistance from case to ambient R_{ca}, the latter will typically predominate (since the package is not isothermal, the meaning of both resistances is not precisely defined). This term is mainly determined by the environment: convection, conduction in the test board and radiation. R_{ja} varies between 35 and 38K/W for the PQFP. Clearly, conduction is only responsible for less than 30% of this. In [13], the still-air measurement is simulated : 75% of the generated heat entered the test board trough the leads (and is then convected), about 10% was radiated and about 10% was directly convected in the air. Still-air measurements only characterize the complete assembly, component and board. In [13] it is only applied to the detailed model, and is complemented by infra-red thermography, to map the case temperature contours.

3.3 Cold plate

The purpose of the cold plate method is to extract all the heat through the copper or aluminum cold plate. The cold plate should force the component case to a known constant temperature, which defines R_{jcp}. The 2 main issues associated with the cold plate are heat losses and interface resistances.

Interface resistances can add a major term in R_{jcp}, as shown by Kozarek in [12] and move the isothermal surface away from the component. The effect of a layer of air (k=0.0261W/mK),0.050mm thick, is detrimental : assuming 1-dimensional heat flow it adds 2.5K/W. To reduce their effect, it is imperative to use some kind of adhesive and apply pressure on the device. The same layer, now with k=2W/mK has a more than 80 times lower thermal resistance, which falls well within the manufacturing tolerances, [2]. We have used uncured thermally conductive adhesive, thermal paste and Indium foils. No significant difference was found. Also when comparing enhanced cold plate tests (DCP-1, see further) with SDJI tests, only minor differences were found. At least for the PQFP, interface resistances can be neglected when using good quality interface material and a well-built cold plate clamp, see [2]. We also tried to measure the case temperature directly, in stead of the temperature of the copper block below it, but found this not accurate and not reproducible. As demonstrated in [12] for a ceramic PGA, the thermocouple size is critical : unless a surface thermocouple is used one may read some temperature in the adhesive. For plastic packages, the situation is worse because the thermocouple itself will evacuate some heat, locally reducing the plastic temperature.

Heat losses could be minimized in 2 ways : insulating the component thermally and creating an ambient at the junction temperature. Certainly the latter cannot be done perfectly, as the test is based on a temperature difference. As there is no perfect insulator (lowest practical conductivities: 0.2-0.02W/mK), we must accept some heat losses. This is not covered in G30-88, the SEMI standard for ceramic packages, which specifies neither the insulation, nor boundaries for the cold plate environment.

The cold plate temperature varies typically between 60ºC (a hot PCB), and 20ºC. For materials with significantly variable conductivity, the first environment is the most relevant one, as it represents the worst case. Low cold plate temperatures could be chosen to lower the junction temperature to the ambient, so as to maximize the heat flow in the cold plate. A T_{cp} around, or below 0ºC will be needed, resulting in condensation, and even freezing of the coolant, which makes this setting very impractical.

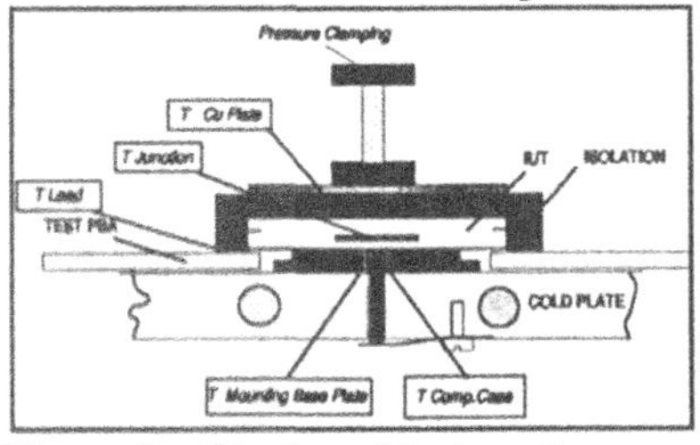

Figure 1 : Single cold plate fixture

Where power losses away from the (isothermal) cold plate cannot be avoided (i.e. the boundary is not adiabatic), we must provide for another temperature-controlled boundary completing the DUT enclosure. So the cold plate environment must be adapted to provide a well-known temperature distribution on the full surface. Although, in theory, an IR scan of this area should suffice [12], comparison with simulation becomes cumbersome. A copper or aluminum box, dimensioned to be isothermal, around the insulator is characterized by just 1 temperature. Our PQFP208 only needs a top plate, as the sides have a negligible influence. Simulations have shown a sensitivity of T_j to the top boundary temperature T_t of 16% (for P = 2, 6W and T_{cp} = 40, 60ºC) to 20% (for P = 1W and T_{cp} = 20, 40, 60ºC) : a lowering of 10ºC in T_t results in a lowering of 1.6 to 2ºC in T_j. This is quite acceptable for performance comparisons. For validation purposes, i.e. in combination with a simulation model, the added accuracy from a completely characterized outer boundary is essential. In the following figures such a top cover is always assumed.

Figure 2 shows the power leaving the cold plate in function of the cold plate temperature, and figure 3 the corresponding temperature difference (which would be the thermal resistance if the ambient were isothermal). Both graphs were obtained by simulation. The PQFP (3.6mm thick) sits in an insulator block of 5mm height, with a thermal conductivity of 0.15W/mK, and dissipates 1W. The insulator material is covered by the copper plate. Normally, this plate is freely cooled in the air, with natural convection in a 20ºC ambient (HTC= 15W/m²K). The fraction of power entering the cold plate decreases with increasing T_{cp}, and so does ΔT_{cp}.

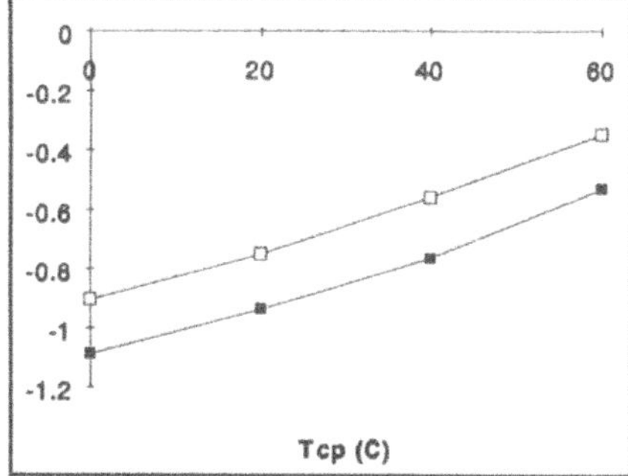

Figure 2 : Cold plate power

■ : top cold plate free, HTC=15W/m²K - □ : top fixed at app. 5ºC below its simulated value

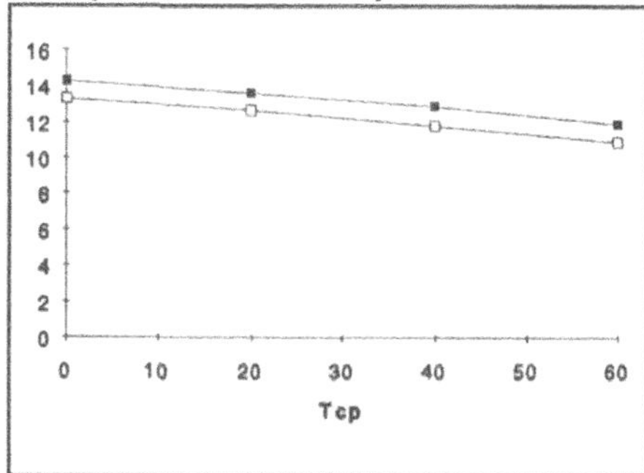

Figure 3 : ΔTjcp in function of Tcp

■ : top cold plate free, HTC=15W/m²K - □ : top fixed at app. 5ºC below its simulated value

For calibration, the interface resistances are not relevant, but heat losses are. If the (local) ambient is not fully raised to the cold plate temperature, a temperature gradient develops across the component, resulting in differences between T_j and T_{cp}.

3.4 Double cold plate (DCP)

An obvious extension from the isothermal insulator casing is to equip it with a temperature control device. The device under test is sandwiched between two cold plates with a spacer between the package top and the upper cold plate, as illustrated in figure 4. Analysis with FLOTHERM reveals that a low conductivity spacer results in a temperature distribution on the top of the package akin to what is found there in natural convection conditions. Alterations of spacer conductivity alter the path of heat escape from the package with more or less leaving via top, bottom or leads. Nothing changes with respect to the interface resistances. There are 3 alternatives for the top.

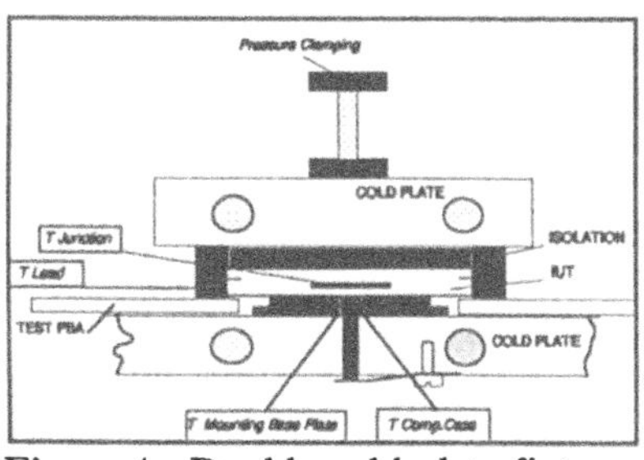

Figure 4 : Double cold plate fixture

1. The top cold plate is floating and acts as a more or less efficient heat sink in air. It can be modelled as a fixed iso-thermal boundary because it provides a uniform surface temperature, although it is not enforced actively.

Figure 5 shows simulated temperature differences between junction and cold plate for our DUT, dissipating 1W, and with a 3mm spacer (k = 0.15W/mK), as compared to the 1.4mm in previous graphs. The HTC at the top area is assumed $15W/m^2K$. Figure 6 shows the bottom cold plate power.

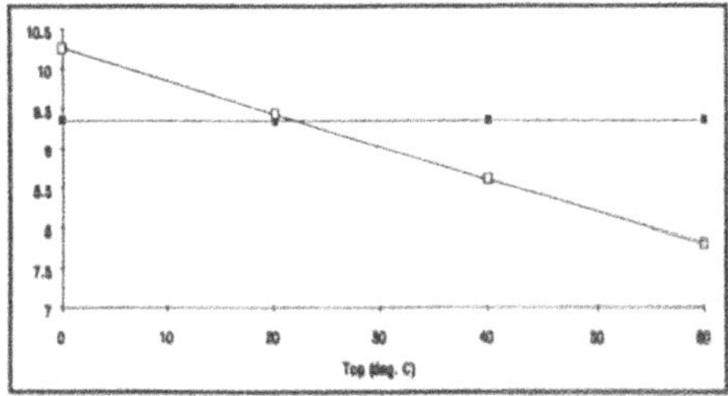

Figure 5: Double cold plate: ΔT_{jcp} (K)
□: top cold plate free, $HTC = 15W/m^2K$
■: top fixed at bottom temperature

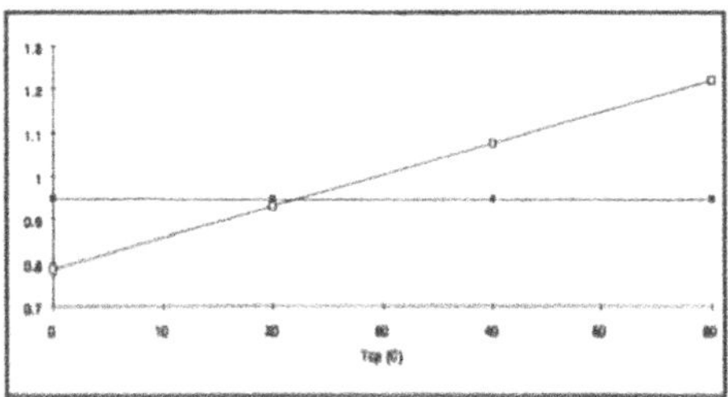

Figure 6 : DCP - plate power, P_{cp}
□: top cold plate free, HTC = 15W/sqmK
■: top temperature same as bottom.

2. Both cold plates are at the same reference temperature. Then it is possible to use only 1 cooling system with the same coolant. Of course, both plates are monitored independently. The difference in heat flow is caused by inserting an insulator between the component and the top cold plate. Practical spacer materials have thermal conductivities of 0.2-0.02 W/mK for an insulator and 200-400 W/mK for metals (Cu or Al).

Those tests are illustrated in figures 5 and 6, curves "fixed top". They show that this mode of operation ensures independence from the reference temperature for the most important physical quantities involved in the test. Furthermore, it guaranties that there is really only 1 reference temperature, as required in equation (1). Validation consists then in comparing always 1 thermal resistance value (junction to isothermal ambient) between measurement and simulation. For the low HTC, with insulating spacer, test results are fairly insensitive to its thickness and conductivity [2]. Also the cold plate calibration accuracy improves, as the environment is quasi iso-thermal.

3. The last mode of operation of the dual cold plate system is with two cold plates independently controlled. The investment is higher in this case: typically the most expensive part of the set-up is the temperature control equipment. If the validations call not only for a difference in heat flow but also for a highly non-isothermal top surface, we still have to insert a thermally insulating spacer.

Clearly, the last approach yields no additional advantage for a linear system (thermal conductivity constant). Hence we prefer alternative 2. A test sequence was defined that seems to validate all important features of the PQFP, [2]:

DCP-1: No spacers, thermal interface resistances minimized. This test resembles SDJI.

DCP-2: Insulating spacer on top, 5mm thick. This test mimics use of a heat sink.

DCP-3: Insulating spacer on bottom (same as 2, but with component flipped upside down, i.e. top heat sink).

DCP-4: Insulating spacer on top, thin insulator on bottom, between leads (air). Good thermal contact between leads and cold plate. This test represents component

Table 3 : Heat transfer coefficients in the tests

Application		HTC_{TOP} (W/m^2K)	HTC_{BOT} (W/m^2K)	HTC_{SIDE} (W/m^2K)	HTC_{LEAD} (W/m^2K)
SDJI		>10000	>10000	500	500
Double Cold Plate	DCP-1	10000	10000	40	40
	DCP-2	40	10000	40	40
	DCP-3	10000	40	40	40
	DCP-4	40	40	40	10000
Still air		30 - 10 (22.5)	30 - 10	30 - 10	10 - 30

Under the assumption of a 5mm thick isolator with 0.2W/mK thermal conductivity (canvas)

mounting without heat sink, but with an important thermal path to the board. Table 3 lists the HTC combinations that are realized experimentally.

4 Conclusions

The single cold plate is not an accurate calibration environment. All others are, but the oven typically requires more than 6 times longer settling times than the rest of them. The settling times for FB (SDJI) and DCP are determined by the cooling aggregate : high capacity systems need more time.

The DCP with both plates at the same temperature promises the best overall performance. A range of well-defined boundary conditions for verification of the corresponding conduction-only simulations is available. SDJI avoids the interface

Table 4 : Test method performance summary

Test method		BC well-def.	all BC	repr.	low-cost	flex.	underst.	cal
Fluid bath	FC-70	x	-	-	-	x	-	+
	DIW	x	-	-	+	x	-	+
	SDJI	+	x	+	+	+	+	+
Cold plate	open	-	x	-	+	+	x	-
	closed	+	+	+	+	+	+	-
	DCP	+	+	+	+	+	+	+
Free air		-	-	x	+	+	-	x

resistances which may trouble cold plate measurements, but it is not trivial to realize the low HTCs in this test. The following table summarizes the performance of the investigated measurement methods with respect to the aforementioned requirements.

5 Acknowledgements

The authors acknowledge the support for this work by the Commission of the European Communities under the ESPRIT contract DELPHI (9197). They also wish to thank the other members of the project: P. Zemlianoy from Alcatel Espace, H.I. Rosten and J. Parry from Flomerics, C. Cahill from NMRC, C. Lasance from Philips CFT and T. Gautier and Y. Assouad from Thomson CSF. At IMEC a lot of work was done on fluid bath tests, especially by F. Christiaens and E. Beyne.
FLOTHERM is a registered trademark of Flomerics Limited.

6 References

[1] H. I. Rosten, C. J. M. Lasance, " The Development of Libraries of Thermal Models of Electronic Components for an Integrated Design Environment", IEPS 1994.

[2] W. Temmerman, W. Nelemans, T. Goossens, E. Lauwers, C. Lacaze, P. Zemlianoy, "Validation of Thermal Models for Electronic Components", to be presented in the 1995 IEPS conference

[3] J. A. Andrews, "Package Thermal Resistance Model: Dependency on Equipment Design", IEEE Trans. on CHMT, Dec. 1988, pp. 528-537, Vol. 11, N°4.

[4] J. Andrews, M. Mahalingam and H. Berg , "Thermal Characteristics of 16- and 40-Pin Plastic DIP's", IEEE Trans. CHMT, 1981, vol.4, pp.455-461.

[5] M. Mahalingam, "Surface-Mount Plastic Packages - An Assessment of their Thermal Performance", IEEE Trans. CHMT, 1989, vol.12, pp. 745-752.

[6] C.J.M.Lasance, H. I. Rosten, H. Vinke and K-L Weiner, "A Novel Approach for the Thermal Characterization of Electronic Parts", Proc. 11th IEEE SEMI-THERM Symp., San Jose, Feb 1995.

[7] B. Joiner, "Evaluation of Thermal Characterisation Techniques", 1994 IEPS conference, pp. 413-420

[8] A Bar-Cohen and W Krueger, "Thermal Characterization of a PLCC - Expanded R_{jc} Methodology", Trans. CHMT, Oct 1992, vol. 15, no. 5, pp. 691-698.

[9] A. Bar-Cohen, T. Elperin and R. Eliasi, "R_{jc} Characterisation of Chip Packages - Justification, Limitations, and Future", IEEE Trans. CHMT, 1989, vol.12, pp. 724-731.

[10] A. Bar-Cohen and W. Krueger, "Determination of the Weighted-Average Case Temperature for a Single Chip Package", in Cooling of Electronic Systems, ed. S. Kakac et al., Kluwer Academic Publishers, 1994, pp. 789-810.

[11] F. Christiaens, E. Beyne, W. Temmerman, K. Allaert, W. Nelemans, "Experimental Thermal Characterization of Electronic Packages in a Fluid Bath Environment", Eurotherm Seminar Nº 45 "Thermal Management of Electronic Systems", September 20-22, 1995, Leuven, Belgium

[12] R. Kozarek, "Effect of Case Temperature Measurement Errors on the Junction-to-Case Thermal Resistance of a Ceramic PGA', Proc. 7th IEEE SEMI-THERM Symp., 1991, pp.44-51

[13] H. I. Rosten, J. Parry, S. Addison, R. Viswanath, M. Davis, E. Fitzgerald, "Development, Validation and Application of a Thermal Model of a Plastic Quad Flat Pack', Proc. 45th IEEE ECTC, 1995

EXPERIMENTAL THERMAL CHARACTERISATION OF ELECTRONIC PACKAGES IN A FLUID BATH ENVIRONMENT

F. CHRISTIAENS, E. BEYNE
IMEC
Kapeldreef 75, B-3001 Heverlee, Belgium

W. TEMMERMAN, K. ALLAERT, W. NELEMANS
Alcatel Bell
Francis Wellesplein 1, B-2018 Antwerpen, Belgium

Abstract
This paper focuses on the application of a fluid bath system for experimental thermal characterisation of electronic packages. Problems related to the standard method for R_{jc} and R_{ja} measurements in a fluid bath environment are addressed. In order to overcome these problems and to increase the accuracy and repeatability of R_{jc} fluid bath measurements, a submerged liquid jet impingement cooling scheme is proposed. Using impinging liquid jets, a quasi isothermal boundary condition over the package surfaces can be obtained, minimising the external convective temperature difference. As a result, R_{jc} approximates R_{ja}, which can be measured without system disturbance effects. The advantages of this new experimental boundary condition for thermal characterisation of electronic packages will be illustrated with practical thermal resistance measurements on a Plastic Quad Flat Package (PQFP) and a Ceramic Pin Grid Array (CPGA).

1. Introduction

Over the years several analytical, numerical, and experimental techniques have been developed to determine thermal resistance values of electronic packages for well-defined mounting configurations. Analytical solution of the energy equation yields very useful thermal models for packages in which the heat flow is mainly one-dimensional. Finite element methods, finite difference methods, and boundary element methods are popular techniques and may provide accurate temperature distributions. Though, as the thermal properties of the materials used in electronics packaging are not always available for numerical analyses, thermal characterisation experiments still form an essential part of the thermal design phase. The experimental approach allows fine tuning of the material properties and validation of the model accuracy [1, 2]. In addition, unknown thermal contact resistances at interface layers may be identified.

The starting point of this work was to validate thermal models under different experimental boundary conditions [3]. Furthermore, the attention was mainly focused on

E. Beyne et al. (eds.), Thermal Management of Electronic Systems II, 137-148.

the application of a thermostatic fluid bath for experimental thermal characterisation of electronic packages. A thermostatic bath allows complete immersion of the test package within a dielectric fluid and is ideally suited for calibration of the on-chip temperature sensitive electrical parameters. Uncertainty analysis has been applied to the fluid bath calibration algorithm in order to estimate the accuracy of subsequent thermal resistance measurements.

Existing standard methods for measuring thermal resistance do include a fluid bath test environment. Unfortunately, none of these standards do specify any flow and mounting condition. This lack of flow definition may lead to serious problems in measuring R_{jc} and R_{ja}. As the convective heat transfer coefficient between component and fluid heavily depends on the flow field around the component, the junction-to-fluid temperature difference varies according to the bath equipment (stirring speed, circulating pump, bath dimensions, position of pump inlet and test package, size of test board, ...). Due to an unsteady flow field around the package, large case temperature variations were observed during junction-to-case measurements. A few experiments in a still fluid environment revealed that the case temperature was stable, but not reproducible for repeated thermocouple attachments. Installation of a thermocouple introduces large system disturbance and system-sensor interaction errors that have to be corrected [4, 5].

In order to overcome these problems, a submerged multiple jet impingement cooling configuration is proposed. Optimisation of the number of nozzles and nozzle diameter leads to convective heat transfer coefficients as high as 50000 W/m^{2}K for de-ionised water, even with a modest flow rate of a few litre per minute. The external temperature drop between case and fluid can thus be neglected and the package outer surfaces are assumed to be at the same temperature as the fluid (ambient). Experiments were carried out on a Plastic Quad Flat Package (PQFP) and a Ceramic Pin Grid Array (CPGA), using fluorinert liquid FC-70 and de-ionised water as the cooling medium.

2. Calibration of the TSEP in a Fluid Bath

Prior to measurement of the junction temperature, the on-chip temperature sensitive electrical parameters (TSEPs) have to be calibrated in a well-defined temperature chamber. An important characteristic of this temperature controlled environment is the agreement between the environmental temperature which can be measured, and the actual junction temperature which has to be estimated. At least four different thermal environment control systems can be considered [6]: dielectric fluid bath, temperature controlled heat-sink, natural and forced convection air oven.

When the component is immersed in an appropriate heating and refrigerating fluid bath with circulation pump, the calibration can be done very fast in comparison with the natural and forced air oven. Furthermore, a fluid bath unit provides a stable, uniform, and accurate temperature chamber. In this work, all calibrations were done using a HAAKE D8-GH thermostatic bath unit, filled with the fluorinert liquid FC-70. The calibration procedure is performed by a computer controlled thermal measurement set-up. A block diagram of the measurement set-up, for simultaneous calibration of 3 diodes, is depicted in figure 1. Temperature of the liquid is continuously monitored by a Pt-100 thermometer while the package temperature is measured by a thermocouple of type T (AWG size 36). If the test chip contains multiple diodes, the value of each diode is monitored by switching the current and voltage terminals.

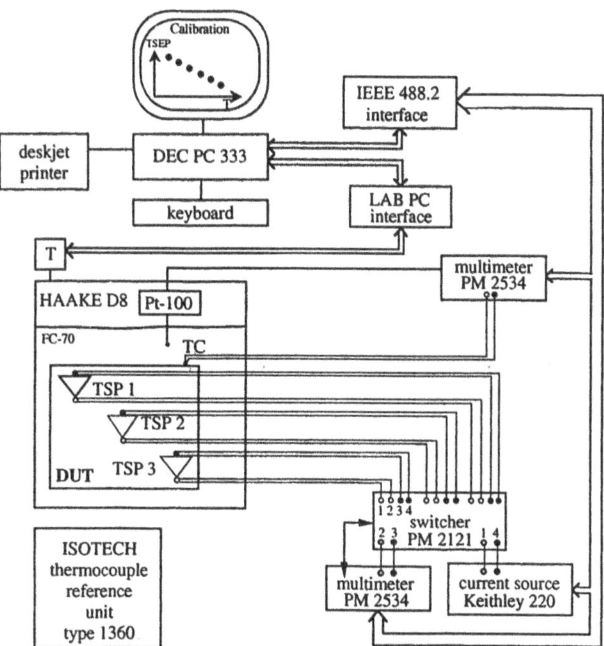

Figure 1. Schematic representation of the measurement set-up for calibration in a fluid bath.

After thermal stabilisation at each setpoint temperature, a multiple sample measurement methodology [10] is used to estimate the variable error of the measurands, caused by random effects and process instabilities. 31 samples of the case temperature (thermocouple type T), fluid temperature (Pt-100 thermometer), and diode voltage are recorded. Statistical analysis of the sample data provides the mean value which is further interpreted as the measured value, and the precision index. The variable error for a 95 % confidence interval is then calculated from the precision index and the Student parameter t = 2 (30 degrees of freedom and 95 % confidence). This statistical post-processing is performed for the fluid temperature, the case temperature and the different diode voltages.

The calibration procedure yields a data set of n temperatures T_{ic} and voltages V_{ic}, that can be related by a linear curve.

$$T_j = a\,V_j + b \tag{1}$$

T_j and V_j denote the junction temperature and junction voltage respectively. Relationship (1) will be used to calculate the junction temperature during a subsequent thermal resistance measurement. The slope a and the intercept with the T-axis b is governed by:

$$a = \frac{\sum_{i=1}^{n} V_{ic} T_{ic}}{\sum_{i=1}^{n} V_{ic}^2} \qquad b = \frac{\sum_{i=1}^{n} T_{ic}}{n} \tag{2}$$

The formulas (2) for a and b are based on the assumption that $\sum_{i=1}^{n} V_{ic} = 0$. This condition can be easily obtained by a transformation of the measured voltages V_{ic}. The curve fit is calculated by minimising the temperature deviations. Note that the calculated T_j depends on 2n+1 variables:

T_{ic} : n independent temperatures (case or fluid)
V_{ic} : n independent diode voltages
V_j : 1 diode voltage during power dissipation

The error function ΔT_j can be expressed as the total differential of the function T_j.

$$\Delta T_j = \sum_{i=1}^{n} \frac{\partial T_j}{\partial V_{ic}} \Delta V_{ic} + \sum_{i=1}^{n} \frac{\partial T_j}{\partial T_{ic}} \Delta T_{ic} + \frac{\partial T_j}{\partial V_{dj}} \Delta V_j \tag{3}$$

Each voltage error ΔV_{ic}, ΔV_j and temperature error ΔT_{ic} can be divided into a fixed (systematic) part and a variable (random) part. The fixed errors of variables measured by the same instrument are dependent of each other and are assumed to be equal in this study. All variable errors are independent. Applying the variance rule on the function ΔT_j (3), the variance of the junction temperature error $\sigma^2(\Delta T_j)$ can be written as the sum of the independent variances of the error components, multiplied by the square of their sensitivity coefficients [10].

TABLE 1. Representative values for variable and fixed errors of temperature and voltage measurands.

	Variable error (95 %)	Fixed error (95 %)
Fluid temperature during calibration ΔT_{ic}	0.05 °C	0.33 °C
Case temperature during calibration ΔT_{ic}	0.08 °C	0.67 °C
Diode voltage during calibration ΔV_{ic}	0.040 mV	0.089 mV
Diode voltage during resistance measurement ΔV_j	~ R_{th} method	0.089 mV

Table 1 lists some representative values for variable and fixed errors which are valid within a calibration temperature range from 20 °C to 90 °C. The resulting junction temperature error obtained with the linear curve fit is depicted in figure 2 as function of diode voltage. The resulting error was calculated for a variable voltage error of 0.020 mV (95 % confidence) during the thermal resistance measurement. Note that the temperature uncertainty is minimal for junction temperatures in the middle of the calibration range.

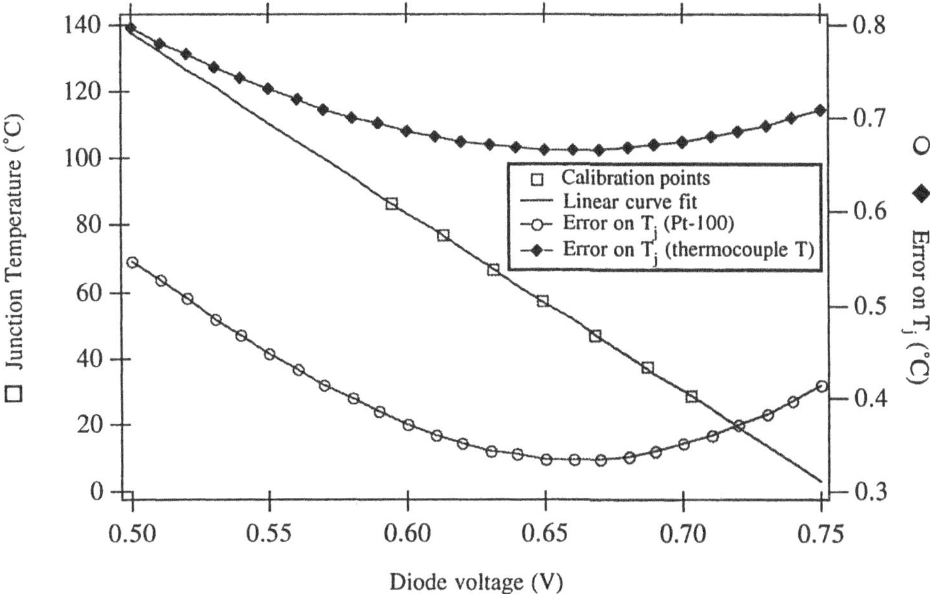

Figure 2. Junction temperature and its error versus diode voltage.

3. Thermal Resistance Measurement in a Fluid Bath

3.1. JUNCTION-TO-CASE THERMAL RESISTANCE R_{jc}

In a fluid bath thermal resistance evaluation experiment, an electronic package is completely immersed in a dielectric liquid which creates a quite uniform cooling environment. The SEMI and JEDEC standards both include a fluid bath method for measuring R_{jc}. These methods recommend to attach a thermocouple to the package side at a specified reference point. According to the SEMI G30-88 thermal test method, the case-to-fluid temperature difference at the case reference point of interest should be less than or equal to 20 °C. Otherwise, accuracy and repeatability difficulties may occur due to a large variable temperature gradient in the fluid boundary layer at the package-fluid interface. The thermocouple hot junction has to be welded in order to form a bead. In addition the fluid should be stirred or agitated.

All measurements described in this work were done using a computer controlled thermal resistance evaluation system. The instrumentation system is equipped with two thermostatic baths, both with integrated heating and refrigerating system and circulation pump. A HAAKE D8-GH bath is filled with the fluorinert liquid FC-70 and also serves for calibration purposes, while a second bath (NESLAB RTE 221) is used for thermal characterisation experiments in (de-ionised) water. The fluid temperature is monitored by a platinum resistance thermometer. Prior to starting the actual thermal resistance evaluation procedure, the calibration data file is loaded and will be used to convert the measured diode voltages into the corresponding temperature.

Repetitive observation of the junction-to-case temperature difference revealed that the flow in the thermostatic baths is very turbulent and unsteady when the fluid is circulated by the internal pump. The boundary layer at the fluid-solid interface is continuously interrupted and causes significant variations in the local convective heat transfer coefficient and case temperature. Although the junction temperature reaches a stable value, the case temperature is oscillating with a magnitude of a few degrees during the whole experiment, resulting in a standard deviation of 1.3 K/W for R_{jc}.

In order to prevent oscillations in the case temperature, the circulation pump of the thermostatic bath unit was switched off (free convection liquid cooling). The mean and standard deviation of the measured R_{jc}-values for 6 indentical 208 lead PQFPs are summarised in table 2. It can be concluded that, as long as the thermocouple was not removed (i.e. repetitive measurements on the same sample), the thermal resistance measurements were quite reproducible and linear. Though, a significant discrepancy between the mean values of the 6 samples was observed.

TABLE 2. Mean R_{jc} values and standard deviation for 6 identical 208-PQFPs measured in still FC-70 (20°C) for P=1.5 W.

Sample	R_{jc} (K/W)	σ_R (K/W)
A	7.91	0.22
B	8.89	0.05
C	9.79	0.40
D	8.00	0.05
E	7.57	-
F	5.54	0.26

TABLE 3. Mean R_{ja} values and standard deviation for 6 identical 208-PQFPs measured in still FC-70 (20°C) for P=2.5 W.

Sample	R_{ja} (K/W)	σ_R (K/W)
A	18.53	0.16
B	18.15	0.12
C	17.87	0.11
D	17.71	0.29
E	17.94	-
F	17.11	0.50

The spreading in the mean data cannot be explained by encapsulation process variations, but has to be attributed to the thermocouple attachment. Other investigators [4, 5, 7] also experienced problems in measuring case temperatures for fluid bath and cold plate assemblies. It is not easy to install the thermocouple in perfect mechanical contact with the case after it has been epoxied. If there is a poor thermal contact, the case temperature is measured somewhere in the adhesive and yields a lower value, resulting in a larger R_{jc}. A second reason for measuring a lower temperature at the case is the influence of the surrounding fluid on the thermocouple bead. The bead has a point contact with the package surface while the thermocouple wire is exposed to the measuring ambient, at a much lower temperature, and may act as a heat-sink. Contactless measurement of the case temperature is thus required.

3.2. JUNCTION-TO-AMBIENT THERMAL RESISTANCE R_{ja}

Since it is very difficult to measure the junction-to-case thermal resistance accurately, one may take the junction-to-ambient thermal resistance as a validation number for numerical models. R_{ja} consists of an internal (mainly) conductive part and a convective part which is determined by the fluid type and flow field around the package. Due to the bad thermal properties of most dielectric liquids, a substantial external temperature difference between case and fluid will always exist. Table 3 summarises R_{ja}-values for the 208 lead PQFP family measured in still FC-70 at 20°C. Repetitive measurements on the same component mounted in different fluid bath systems resulted in R_{ja} values that largely differ. These discrepancies can be explained as follows:

1) The flow field heavily depends on the position of the component in the bath. This makes the actual R_{ja} measurement method prone to repeatability problems.
2) The flow field is determined by the particular equipment (stirring speed, circulating pump, bath dimensions, position of pump inlet and test package, size of test board, ...). Comparison of the results from several experimenters is difficult.

In addition, R_{ja} also depends on the amount of power dissipation, since the fluid properties of fluorinert liquids vary with temperature. From the results of these experiments, we may conclude that the complex and unpredictable flow field around an immersed component makes the R_{ja} an irrelevant parameter for validation of numerical models. Fortunately, the problems encountered in the present R_{ja} method could be solved by increasing the convective heat transfer coefficient and by defining a proper flow field around the package, i.e. a flow field that can be easily created and reproduced.

4. Liquid Jet Impingement Cooling: the Way to Standardisation ?

In the last few years, a lot of attention has been directed to liquid jet impingement cooling of electronic systems. An attractive option for obtaining large heat transfer coefficients involves the use of one or more impinging jets. In an extensive review on liquid jet impingement cooling, Incropera et al. [8] revealed that convective heat transfer coefficients of in the order of 10000 W/m^2K may be achieved using liquid impinging jets with fluorinert liquids. These high convective heat transfer coefficients are well-suited to improve the thermal characterisation of electronic packages in a fluid bath environment. A simple calculation, assuming a heat transfer coefficient of 40000 W/m^2K (water), shows that the external convective thermal resistance for a 2 cm by 2 cm surface area is as low as 0.0625 K/W, which is much smaller than the experimental measurement accuracy obtained by the present R_{jc} standard methods. In order to find the optimal nozzle outlet configuration, i.e. number of parallel nozzles, nozzle diameter, and nozzle-to-package separation distance, a fundamental understanding of the hydrodynamic and thermal conditions of an impinging jet flow is required [8, 9].

Liquid jet impingement cooling schemes can be divided into two categories: free surface jet impingement and submerged jet impingement. A free surface jet is discharged into an ambient gas, while a submerged jet is discharged into a stagnant fluid of the same type. These two types of impinging jets significantly differ in their hydrodynamic and thermal performance. A free surface jet flow is characterised by a very small shear stress at the liquid/gas interface. In the absence of gravitational acceleration, the jet diameter and the nozzle exit velocity is preserved until impingement on the target surface. This hydrodynamic behaviour causes the convective heat transfer between fluid and surface to be independent of the nozzle-to-surface separation distance. A submerged liquid jet is characterised by a shear layer between jet and the surrounding fluid, as shown in figure 3. Experiments with both water and FC-77 [8, 9] have shown that the convective heat transfer coefficient is higher for a submerged jet with S/d=3 (S, d see figure 3) than for a free surface jet, for $Re_d > 4000$. The higher Nusselt numbers obtained with the submerged jet can be attributed to heat transfer enhancement by turbulence generated in the free shear layer of the jet. For submerged jets, the Nusselt number (and heat transfer coefficient) is quasi independent of S/d for 1 < S/d < 4. This can be explained by the fact that, for 1 < S/d < 4, the heat source is still located within the potential core of the jet. Furthermore the Nusselt number decreases with increasing S/d ratio for large separation distances. The highest heat transfer coefficients are thus obtained for submerged jets and small S/d values, i.e. S/d ≤ 4.

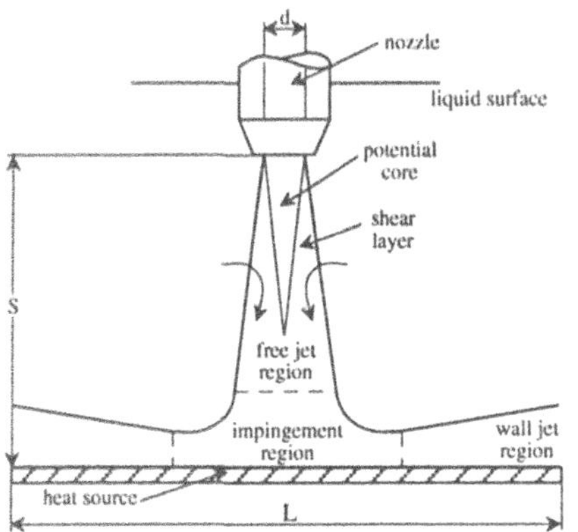

Figure 3. Unconfined submerged liquid jet.

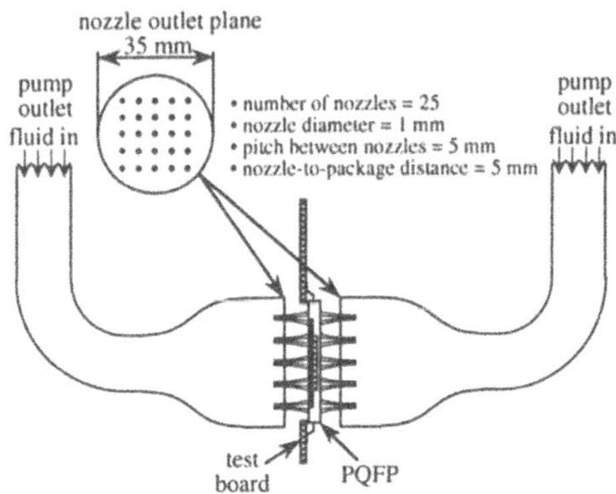

Figure 4. Multiple submerged jet configuration.

For an unconfined, submerged circular jet, the local and average convective heat transfer coefficients are affected by the nozzle exit velocity profile, jet turbulence, jet Reynolds number, nozzle-to-surface distance, and the size of the target surface. At this point, the ratio S/d is set at a fixed value S/d = 4...5. The size of the target surface depends on the chip and package size. A target surface with length L = 25 mm is presumed. Several correlations for the average Nusselt number are available in literature [8, 9]. Womac et al. proposed an area-weighted combination of correlations for the two surface regions: the impingement region was assumed to extend to a radius of 1.9d and to be followed by transition to a turbulent wall jet region.

$$\frac{Nu_L}{Pr^{0.4}} = C_1 Re_d{}^m \frac{L}{d} A_r + C_2 Re_{L*}{}^n \frac{L}{L*} (1-A_r) \tag{4}$$

where

$$A_r = \frac{\pi\,(1.9d)^2}{L^2} \tag{5}$$

$$m = 0.5$$
$$n = 0.8$$
$$C_1 = 0.785$$
$$C_2 = 0.0257$$

and L* denotes the average length of the wall jet region:

$$L^* = \frac{0.5\,(1+\sqrt{2})L - 3.8d}{2} \tag{6}$$

For multiple, unconfined submerged jet configurations, the same correlation can be used to calculate the average convective heat transfer coefficient. In this case, the length L in equations (4 - 6) has to be replaced by the pitch P between the nozzles. The use of multiple jets has the potential to maintain a greater degree of temperature uniformity across the surface. However, jet interactions may influence the local heat transfer conditions.

The Womac correlation (4) for single and multiple unconfined circular jets was then used to calculate the optimal number of parallel jets and the optimal nozzle outlet diameter. The actual flow rate was determined as the intersect of the circulation pump characteristic (thermostatic bath) and the load curve of the tubes and nozzle contractions. The outlet flow rate of the circulation pump is split into two parallel flows which are impinged perpendicular to the top and bottom surfaces of the package. A schematic view of the proposed mounting arrangement is given in figure 4. The optimal nozzle diameter for both water and FC-70 can be obtained from the graphs in figure 5.

It is very clear that water has much better thermal properties than fluorinert liquids. When using water as the cooling fluid, heat transfer coefficients in the order of 40000 W/m^2K can be obtained for an array of 5 x 5 nozzles with diameter = 0.5 mm. Also for fluorinert FC-70, an array of multiple small nozzles yields the highest heat transfer coefficients. The optimal nozzle outlet diameter is somewhat larger than for water; and the accompanying average heat transfer coefficient is about ten times smaller. Though, comparing to an initial value of about 100 to 200 W/m^2K for free FC-70, the heat transfer coefficient has increased significantly and more importantly, the flow field is well-defined, resulting in reproducible values for R_{ja} which will approximate R_{jc}. The graphs below were calculated for a circulation pump with a maximum flow rate of 15 litre/min and maximum pressure drop of 480 mbar.

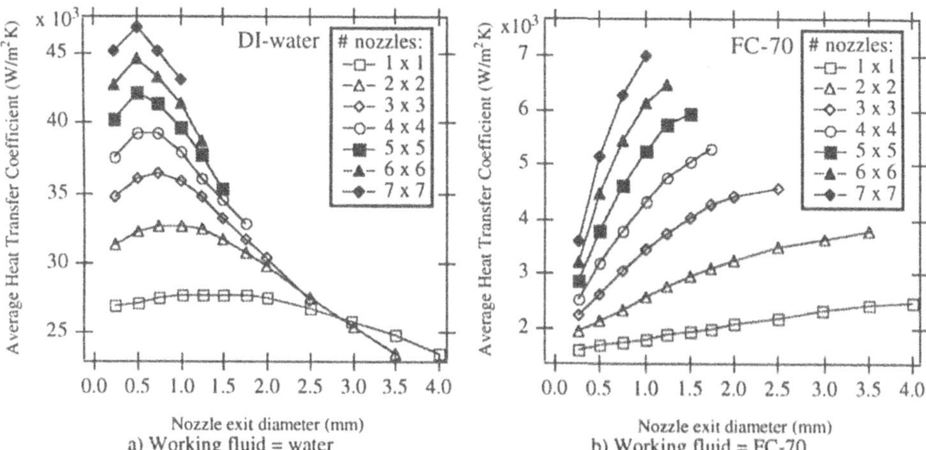

Figure 5. Average convective heat transfer coefficient as function of the number of outlet nozzles and nozzle diameter.

5. Practical Measurement Results

In order to investigate the practical cooling capabilities of liquid jet impingement schemes for thermal characterisation of packages, the mounting configuration depicted in figure 4 was experimentally tested for a 208 lead Plastic Quad Flat Pack and a 224 pin Ceramic Pin Grid Array. It was intended to create an isothermal case by impinging an array of 5 x 5 liquid jets to both sides of the package, which is completely submerged in the fluid. All relevant geometrical and flow parameters are summarised in figure 4. A minimal diameter of 1 mm was chosen, to prevent blockage of the nozzles. Thermal resistance experiments were performed in fluorinert FC-70 and de-ionised water. Care should be taken when using de-ionised water because it very rapidly loses its insulating property due to contamination and absorption of ions.

5.1. PLASTIC QUAD FLAT PACK (PQFP)

The first family of investigated test vehicles consists of a 9.1 x 9.1 mm^2 thermal test die which is packaged in a 208 lead PQFP. The power is dissipated by a metal resistor; a diode in the centre of the chip senses the temperature. The packages were soldered on a test board to provide an electrical interface to the measurement set-up.
Table 4 summarises the measured R_{ja}-values and standard deviations for liquid jet impingement experiments in a water bath. Two types of errors are reported. The experimental error, called σ_{exp}, is a measure for the overall reproducibility of the cooling scheme and mounting arrangement. It is obtained by repeating the whole thermal resistance experiment, including installation of the component. The measurement uncertainty on the R_{ja}-values, called σ_{meas} is a measure for the error caused by instrument inaccuracy and measurement instabilities. σ_{meas} was determined using an error propagation model that relates the error on the resistance value to all independent fixed and variable errors of the individual measurands. This error model assumes a perfect repeatability of the environmental cooling conditions. The variable error was obtained from multiple sample analysis of each measurand (31 samples).

TABLE 4. R_{ja} for the 208-PQFP family measured in water, submerged liquid jet impingement cooling.

component	Power (W)	R_{ja} (K/W)	σ_{exp} (K/W)	σ_{meas} (K/W)	# experiments
B	5	7.23	0.05	0.015	4
C	5	7.31	0.07	0.015	4
E	5	8.04	0.007	0.015	2
H	5	7.28	0.04	0.015	5

The measured thermal resistances were independent of power dissipation. Parasitic electrical effects (leakage currents) in water were not observed for the PQFPs. In contrary to the general R_{jc} and R_{ja} methods for fluid bath environments, the repeatability of this new characterisation approach is much better, due to a unique mounting arrangement and cooling condition which can be easily reconstructed.

Table 5 reveals the bad thermal properties of the fluorinert liquids. An external thermal resistance of about 1 K/W still exists under impingement cooling conditions. One may conclude that the measured R_{ja}-values are much smaller than the R_{ja}-values reported in table 3 which were obtained in still FC-70. Impingement cooling experiments in FC-70 were not repeated, but spreading between components is very small.

TABLE 5. R_{ja} for the 208-PQFP family measured in FC-70 (40 °C), submerged liquid jet impingement cooling.

component	Power (W)	R_{ja} (K/W)	σ_{meas} (K/W)
B	5	8.43	0.025
C	5	8.43	0.025
E	5	9.37	0.025
H	5	8.53	0.025

5.2. CERAMIC PIN GRID ARRAY (CPGA)

A second family of packages was the 224 pin cavity-up Ceramic Pin Grid Array (CPGA), which has a much better thermal performance, i.e. smaller thermal resistance, than the PQFP. The thermal test vehicle consists of a 10 x 10 mm^2 test die, bonded into the cavity of the 40 x 40 mm^2 ceramic package. Due to huge leakage currents in water, the CPGA could not be tested in a two side submerged liquid jet impingement configuration. To cope with this electrical problem in water, a free jet configuration was constructed at the top side of the package, in such a way that the pins were not in contact with the liquid. Additional measurements were performed in fluorinert liquid, as tabulated in table 6. Since R_{ja} approximates R_{jc} for water jets, the thermal resistance between junction and package top side is 4.82 K/W.

TABLE 6. R_{ja} for a 224-CPGA measured under three different cooling conditions.

Cooling scheme	R_{ja} (K/W)	Power (W)	σ_{meas} (K/W)
water jet at top side	4.82	5	0.01
fluorinert jet at top side	5.86	5	0.015
fluorinert jet at both sides	3.28	5	0.01

6. Conclusions

This paper was devoted to experimental thermal characterisation techniques for electronic packages in a fluid bath environment. A thermostatic bath allows complete immersion of the test package within a dielectric fluid and is ideally suited for calibration of the on-chip temperature sensitive electrical parameters. Uncertainty analysis has been applied to the fluid bath calibration algorithm in order to estimate the accuracy of subsequent thermal resistance measurements.

Problems related to the standard method for measuring R_{jc} and R_{ja} in a fluid bath were addressed and illustrated with practical measurement examples. A new cooling environment, consisting of a submerged multiple jet impingement configuration has been proposed. The optimal number of nozzles and nozzle geometry was calculated using an existing correlation for the average convective heat transfer of impinging jets. This optimisation study resulted in heat transfer coefficients in the order of 40000 W/m^2K for water and 5000 W/m^2K for fluorinert FC-70.

Next, this new thermal characterisation environment was experimentally investigated for a 208 lead PQFP and a 224 pin CPGA. Repeatability of the measured thermal resistance values was much better than the existing R_{jc} standards. Theoretical calculations, as well as experiments in de-ionised water revealed that R_{ja} approximates R_{jc}. Though, for the CPGA electrical problems were observed in water. The next challenge is to find a dielectric fluid with thermal properties comparable to those of water.

References

[1] Temmerman, W., Nelemans, W., Goossens, T., Lauwers, E.., and Lazace, C.: "Validation of Thermal Models for Electronic Components", to be published in Proceedings IEPS, 1995.

[2] Rosten, H.I. and Lasance, C.J.M.: "The Development of Libraries of Thermal Models of Electronic Components for an integrated Design Environment", Proceedings IEPS, 1994.

[3] Temmerman, W., Nelemans, W., Goossens, T., Lauwers, E.., and Lazace, C.: "Experimental Validation Methods for Thermal Models", Proceedings EUROTHERM Seminar No 45, 1995..

[4] Sweet, J.N. and Cooley, W.T.: "Thermal Resistance Measurements and Finite Element Calculations for Ceramic Hermetic Packages", Proceedings of the Sixth IEEE SEMI-THERM Symposium 1990, pp. 10-16.

[5] Dutta, V.B.: "Junction-to-case Thermal Resistance - Still a Myth ?", Proceedings of the Fourth IEEE SEMI-THERM Symposium 1988, pp. 8-11.

[6] Shope, D.A., Fahey, W.J., Prince, J.L., and Staszak, Z.J.: "Experimental Thermal Characterization of VLSI Packages", Proceedings of the Fourth IEEE SEMI-THERM Symposium 1990, pp. 19-24.

[7] Kozarek, R.L.: "Effect of Case Temperature Measurement Errors on the Junction-to-Case Thermal Resistance of a Ceramic PGA", Proceedings of the Seventh IEEE SEMI-THERM Symposium 1991, pp. 44-51.

[8] Incropera, F.P. and Ramadhyani, S.: "Single-phase, Liquid Jet Impingement Cooling of High-performance Chips", Proceedings NATO-ASI Cooling of Electronic Systems 1994, pp. 457-506.

[9] Womac, D.J., Ramadhyani, S., and Incropera, F.P.: "Correlating Equations for Impingement Cooling of Small Heat Sources with Single Circular Liquid Jets", ASME Journal of Heat Transfer Vol. 115, February 1993, pp. 106-115.

[10] Moffat, R.J.: "Describing the Uncertainties in Experimental Results", Experimental Thermal and Fluid Science, 1988, No 1:pp. 3-17.

MODELLING OF IC-PACKAGES BASED ON THERMAL CHARACTERISTICS

I. EWES
Philips Kommunikations Industrie AG
Technology and Product Engineering WTK21
Thurn-und-Taxis-Str. 10
D-90411 Nürnberg, Germany

Abstract

When modelling the thermal behaviour of electronic systems with Computational-Fluid-Dynamics (CFD), the problem usually occurs of how to treat IC-packages. Especially with finite-difference-type codes it is found that a detailed modelling of all interesting components, e.g. on a printed circuit board, is not feasible with regard to the number of grid cells, whereas crude block-models are unable to yield the desired results.
This paper describes a method of building compact models of IC-packages which meet the demands of an economical grid and at the same time show a thermal behaviour almost identical to the detailed model. It is further shown that such a compact model can be deduced directly from the data-sheet values of the thermal resistances $R_{th,JA}$ and $R_{th,JC}$, provided the available data are well specified and based on meaningful test methods. From this point of view the often-criticized thermal resistance values appear quite useful, which adds an aspect to the current debate about thermal characterization of IC-packages.

1. Introduction

The use of CFD-programs is characterized by the need to find a compromise between a detailed rendering of the considered system and a feasible number of grid cells. Especially when looking at board or system-level calculations, it is important to include the relevant influences attributed to the air flow and conduction phenomena and avoid unnecessary refinement.

Figure 1 depicts the numerical model of a bouyancy-cooled circuit board as it may be found in a typical telecom equipment. The board is modelled with all boundaries present if installed in a shelf, i.e. enclosing walls, EMC-screens, profiles and a simplified representation of neighboring boards. The calculation domain consists thus of a slice cut out of a fully equipped shelf and may therefore be classified somewhere between board and system level. Figure 2 shows a zoomed detail of the calculation grid in two vertical planes as chosen for the numerical simulation with a finite-volume program. Being very closely related to the finite-difference method, the gridlines extend over the whole domain, in contrast to finite-element methods. They are in this example 3-dimensional and cartesian. The reasoning on which this grid was based is discussed in the following.

E. Beyne et al. (eds.), Thermal Management of Electronic Systems II, 149-158.

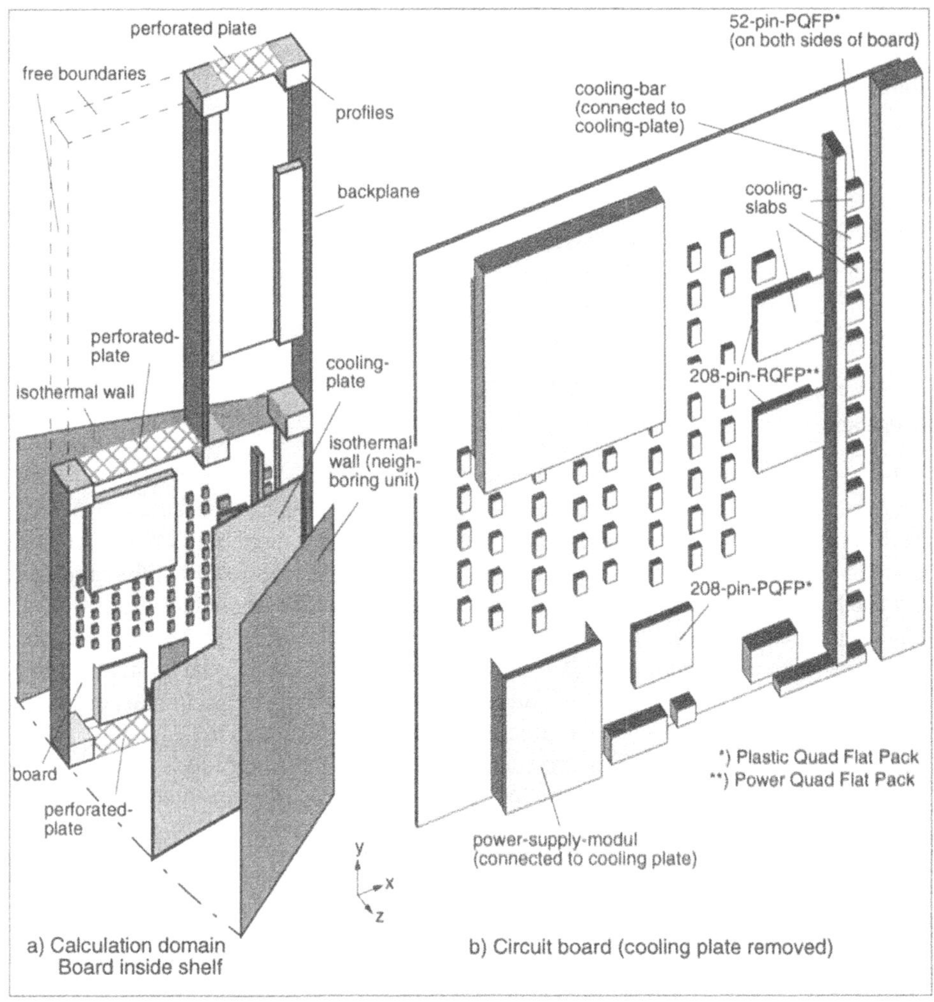

Figure 1. Numerical model of a typical telecom unit

Since the air flow is bouyancy driven and the scale of the shelf renders Rayleigh numbers below the critical value, laminar flow conditions can be assumed. Thus, the distance of the first grid-line adjacent to a surface is not as critical with regard to the heat transfer rate as in turbulent flows. More importantly, the calculated flow field should exhibit the overall shape of the velocity and temperature profiles in any of the three coordinate directions. It is the interaction of the 3-dimensional air stream and the heat conduction inside the board and other solids (especially the cooling plate), which is considered crucial for the thermal behaviour, whereas e.g. local eddies around the edges of solids are regarded as less important. In the direction in which the steepest gradients of the field variables occur, in this case the z-direction, the grid lines have to be denser than in the other directions. Figure 3 shows the flow pattern and the temperature distribution as obtained with the described grid in the two cross sections corresponding to figure 2.

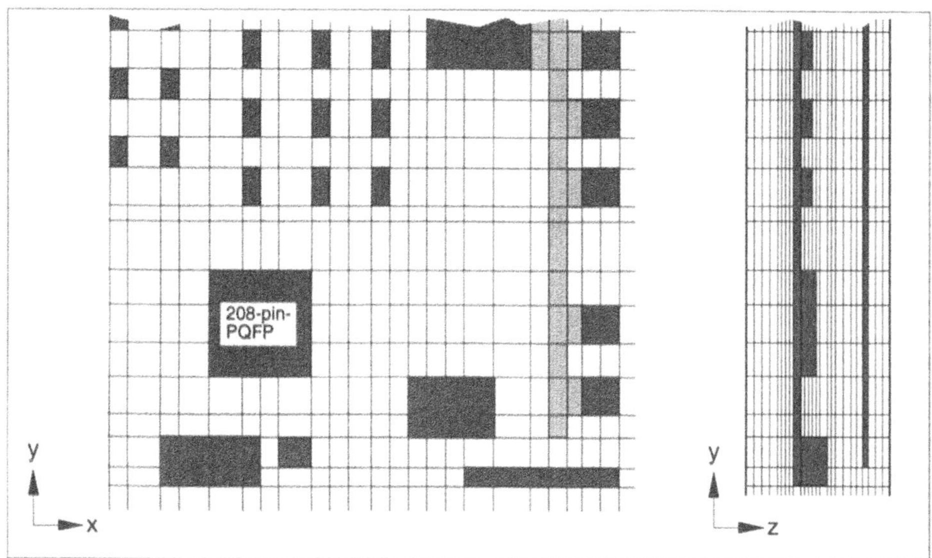

Figure 2. Detail of the calculation grid in two planes

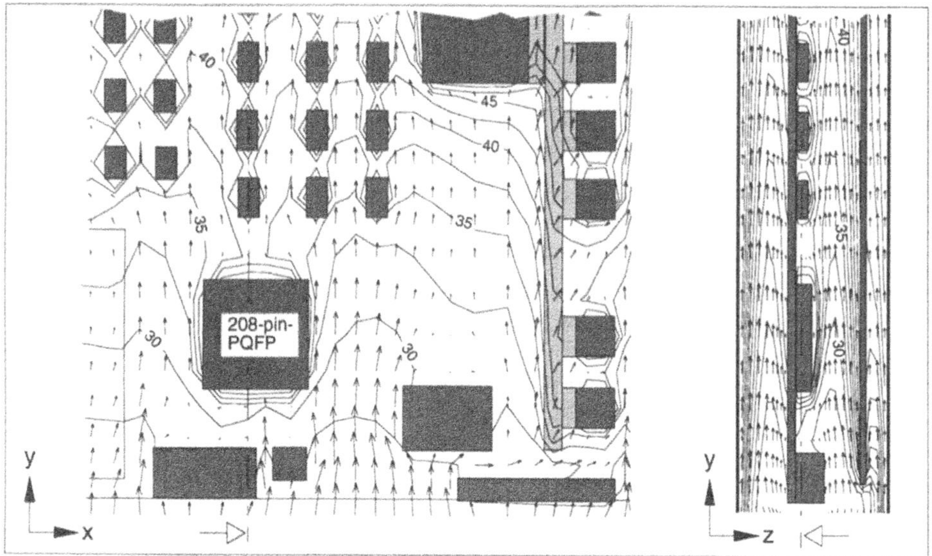

Figure 3. Flow vectors and isotherms (°C) in sections corresponding to figure 2

The flow field, together with the temperature distribution in the board can thus be obtained as desired. The main question, however, arises when looking at the results which are primarily expected from the calculation, namely the junction temperatures of the components, especially the ICs. With the knowledge of these temperatures a worst-case analysis of the system as well as lifetime calculations can be performed. So the question remains how to treat the ICs in the numerical model.

2. The Detailed Model

With the knowledge of the internal structure of the IC-package a detailed model of the package can be built up. The figures 4a,b show the detailed model of a 208-pin-PQFP on a test-board as it is used to evaluate the JA-resistance. It is reasonable to introduce some simplifications, e.g. combining the pins to homogeneous blocks with an effective thermal conductivity. As it has been demonstrated before [1], a numerical simulation of the IC-package on a test board is able to yield a JA-resistance in good accordance to the measured value, in this case 36.6 K/W.

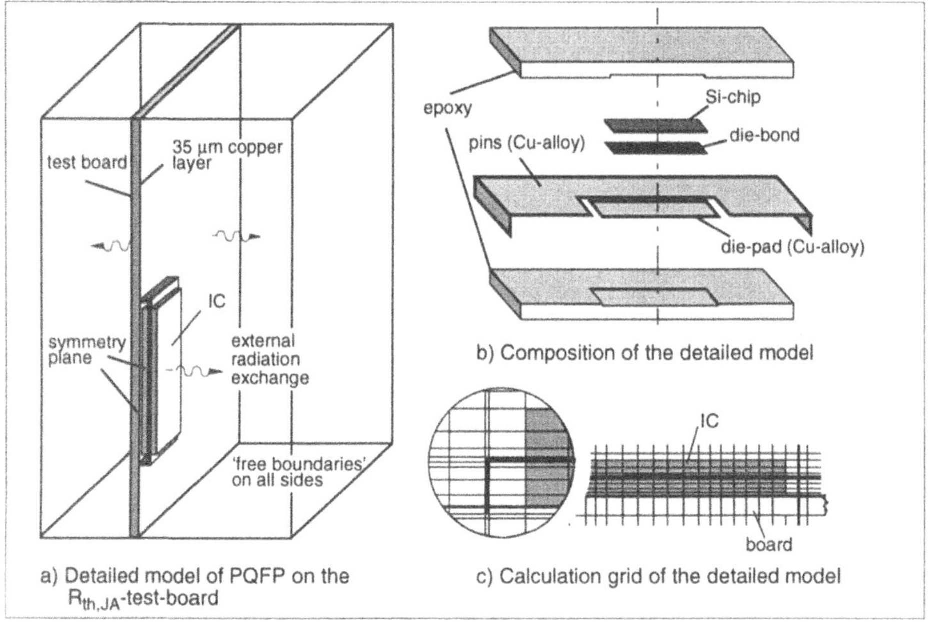

Figure 4. Detailed model of the 208-pin-PQFP

However, the calculation grid of this package alone, as shown in figure 4c, already creates more than 5000 additional grid cells. So it becomes obvious that the implementation of a detailed model for each IC in a typical unit, as e.g. in the example case of section 1, would result in far to many grid cells to be handled by a usual workstation. This problem is made worse by a typical feature of finite-difference-type codes, that the gridlines extend over the whole domain, resulting in distorted grid cells and a very uneven distribution of grid lines. A much simpler representation of IC-packages has therefore to be found.

3. The One-Block Model

A method, widely used to avoid the mentioned difficulties, is the modelling of IC-packages as single uniform blocks with a heat source either homogeneously distributed across the block or concentrated in the center. Between block and board a plane with an effective thermal resistance is inserted, roughly equivalent to the thermal effect of the leads. Besides being based largely on assumptions, the major problem is, that this model does not yield a definite result, as will be discussed in the following.

There are two possible ways of choosing the thermal conductivity of the block. If it is set to a value according to the main body of the package (e.g. epoxy resin), steep temperature gradients within the block will result from the calculation. Except for the case of a very fine grid - which, however, is contradictory to choosing a simplified model - it is impossible to assess any true surface temperature of the block since the temperatures are given at the cell centers.

If in the second case the thermal conductivity is set to a very high value, a uniform temperature results throughout the whole block and thus on its entire surface. The temperature of the block might then be interpreted as 'case temperature' and the JC-resistance, as given in the data sheet, used to estimate the junction temperature. However, a uniform temperature on the entire surface is not consistent with the real physics. The impact on the air flow will be incorrect and, moreover, the temperature will be lower than the real temperature measured e.g. at the top of the case. Further, it is undesirable to depend on the JC-resistance, knowing that this value should not be transferred from the test situation to any real environment, not to mention the vagueness of its definition (see also section 6).

4. The Compact Model

To avoid the disadvantages of both the detailed model and the one-block model, a concept for a compact model is proposed in this paper. The major requirements for this model may be summarized as follows.

- The model must work with a minimal number of gridlines in each coordinate direction, yet being insensitive to additional gridlines within its domain
- Various IC-packages have to fit into the same grid without creating too small or distorted grid cells
- The outer shape of the package and the impact on the air flow should be similar to the real package
- A clearly defined point within the model must unambiguously represent the junction temperature
- As an additional feature, the temperature at the center top of the case shall be given, corresponding to the temperature as measured with a sensor connected to the real package
- In any situation in which the IC may be found, both the junction temperature and the temperatures of a sensor connected to the top surface have to assume the same values as the detailed model would produce

The compact model, designed to meet these requirements is shown in figure 5a, again taking the 208-pin-PQFP as example. The two blocks denoted 'chip' and 'temperature sensor' have a very high thermal conductivity and thus assume uniform temperatures, regardless of any splitting up by additional gridlines. The block denoted 'lead-frame' was given an effective thermal conductivity to account for the heat-spreading effect of the lead-frame material. The shape and the dimensions of the package and its constituting blocks, however, are not intended to coincide exactly with the real structure. Rather, they were fit into a predetermined grid, figure 5b. As far as possible, the gridlines are shared by all other IC-packages on the modelled board. The underlying idea of the compact model is not to try simulating the real package, but to adjust the simplified model to the detailed model. This is done by setting two remaining parameters, the thermal conductivity of the package body $k_{package}$ and the thermal resistance of the contact plane between package and board $(\delta/k)_{contact}$.

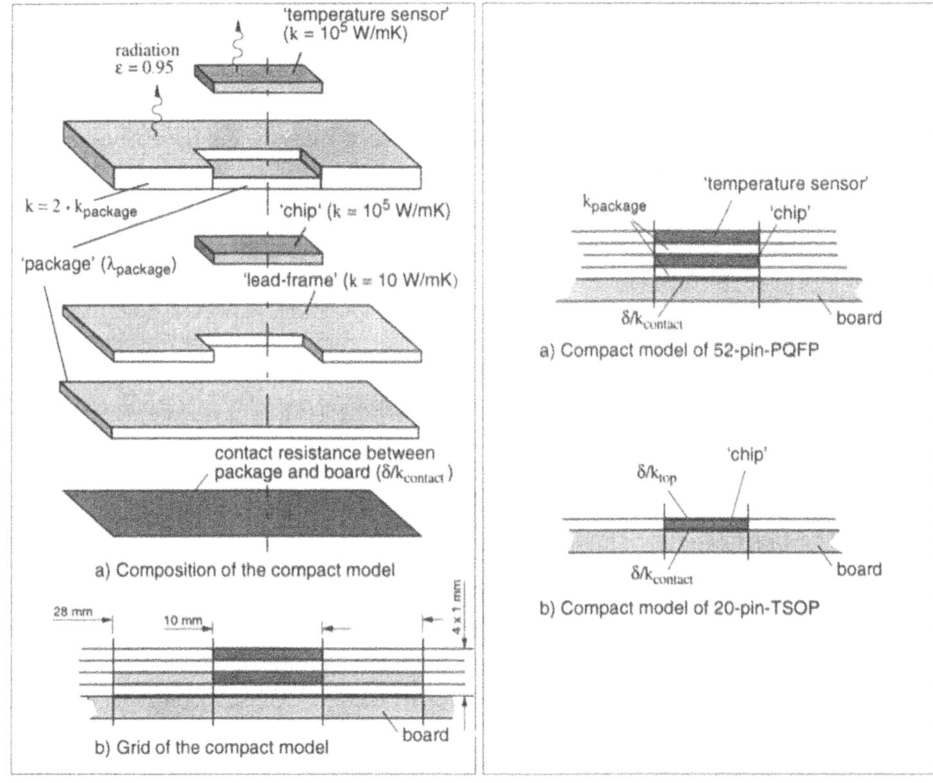

Figure 5. Compact model of the 208-pin-PQFP *Figure 6.* Compact models of small packages

The adjustment is done by performing two computations with the detailed model under two distinct conditions and then doing the same calculations with the compact model. It suggests itself to use a boundary condition similar to the $R_{th,JA}$-test-board as shown in figure 4 for the first calculation. The second calculation can readily be set up by adding a block with a prescribed and uniform temperature connected to the top of the package, acting as a heat sink. This accords with one of the methods to measure the JC-resistance. The two parameters of the compact model are then varied until the junction temperature of both models are identical in both situations. In the considered case this procedure results in $k_{package} = 0.32$ W/mK and $(\delta/k)_{contact} = 0.022$ Km^2/W. (The upper part of the 'package' body was shaped in a way to maintain a plane surface of the case. To reduce unwanted thermal effects from the different thickness, the outer region of this block was given twice the conductivity of the 'package'.)

Since the bounds of the two calculations represent very different thermal situations, the assumption is that similarity between the compact and the detailed model exists in all situations. This means that the compact model can be used in any numerical simulation as a substitution for the detailed model.

For other package-types appropriate compact models can be built in a similar fashion. Small sized packages like a 52-pin-PQFP may be represented as shown in figure 6a, comprising only one grid cell in the lateral direction. For very thin packages it may be adequate to use an even simpler model omitting the 'temperature sensor' and the package body, as shown in figure 6b. In this case the two adjustable parameters consist of the 'contact resistance' and a resistance at the top of the block. It should be noted that this representation of a compact model is still essentially different from the one-block-model as described in section 3.

5. Verification of the Compact Model

The compact model was based so far on two distinct test board situations in which conformity with the detailed model was established. To prove that it is justified to assume conformity also in any other situation or environment, a parameter study was performed for the 208-pin-PQFP. In this parameter study the calculation domain was kept to the test board environment as in figure 4a, the major influencing parameters, however, were varied in a broad range.

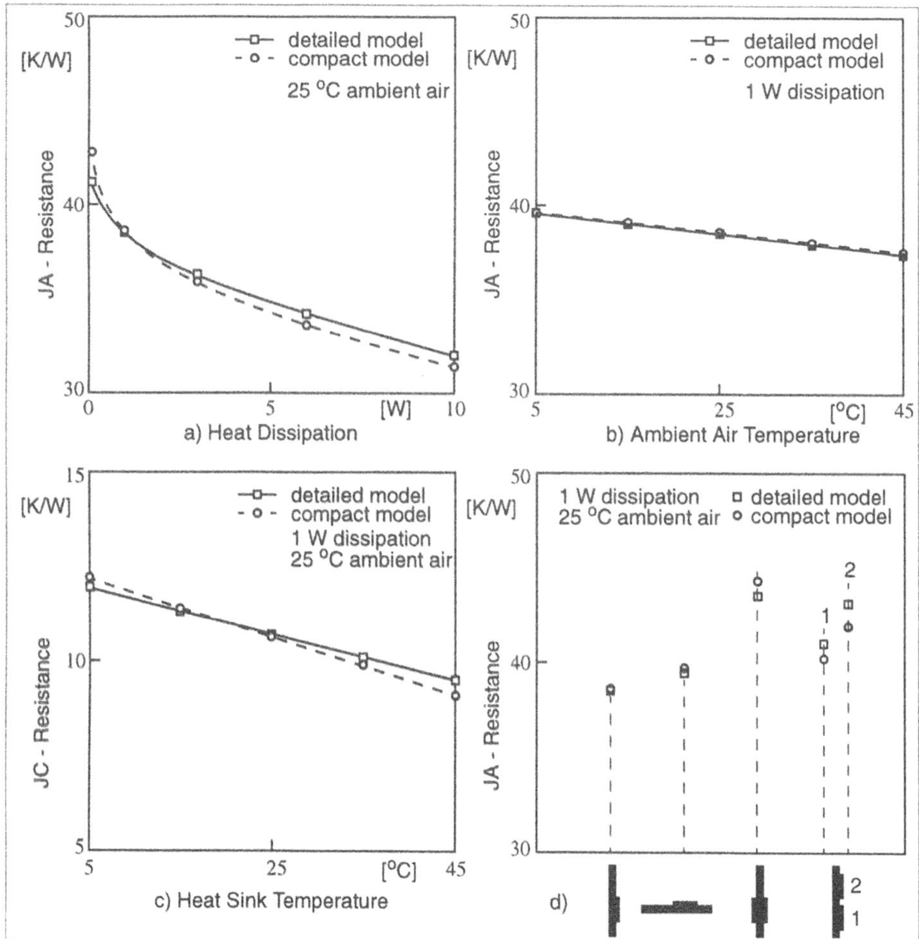

Figure 7. Comparison of the detailed and the compact model

The Figures 7a,b depict the influence of heat dissipation and ambient air temperature on the JA-resistance, comparing the results of the detailed and the compact model. Figure 7c shows the influence of the heat sink temperature in the $R_{th,JC}$ test configuration. Figure 7d compares the results of both models for the vertical and horizontal board orientation and for the vertical board with two ICs on the same side of the board as well as on opposite sides of the board. The deviation between the two models is in all these cases less than 3 %.

6. Deduction of the Compact Model from the Thermal Characteristics of the IC-Package

The choice of the two thermal resistances for adjusting the compact model to the detailed model already anticipated the basic idea of this paper. Instead of first building up a detailed model, the compact model can be deduced directly from the resistance values $R_{th,JA}$ and $R_{th,JC}$ given in the data sheets of the IC in question. The procedure remains essentially the same as before. The test board and the testing environment in which the compact model of the IC is modelled have to coincide with those of the experimental set-up. Unfortunately, there are some problems to be mentioned which impede the application of this method.

First, it is usually very difficult to obtain the complete specification of the tests from the IC-manufacturer. A parameter which has a considerable influence on the JA-resistance is for instance the dissipated power, as could be seen in figure 7b. This value, however, is mostly unknown, which makes it necessary to use an assumption.

The second, more severe problem is that there are various different methods to measure, and thus define, the JC-resistance. Some methods, applicable to PQFPs, are listed below.

- Test-IC mounted on $R_{th,JA}$ test board. Case temperature is defined as temperature at bottom of package under $R_{th,JA}$ test conditions.
- Test-IC (without board) is immersed in a fluid bath with the intention of imposing a uniform temperature on the entire package surface.
- Test-IC (without board) is connected with one face of the package to a temperature-controlled heat sink. On the other face there is usually a spacer with unknown temperature and some mechanism to press the package against the heat sink.
- Test-IC mounted on $R_{th,JA}$ test board and is connected with top of package to a temperature-controlled heat sink.

Of these test methods the first is the least suitable for the purpose of adjusting the compact model because instead of subjecting the IC to different situations in the $R_{th,JA}$ and $R_{th,JC}$ measurement, two temperatures at two locations are measured in the same situation. The next two methods contain some uncertainties as to heat transfer between fluid and package or thermal behavior of the spacer. The last method is the one which is proposed in this paper as the standard method. Besides the advantage that it can be modelled easily, it is clearly defined and resembles an IC in a possible real situation, i.e. mounted in a normal fashion on a board and connected at the package top to some kind of cooling device.

7. Thermal Evaluation

With two subsequent simulations, the junction temperatures of all interesting components under worst-case conditions and under typical conditions can be calculated, which provides the relevant information for a thermal evaluation of the system, which should comprise two parts, a lifetime calculation and a worst-case analysis. The differences between the two simulations are mainly made up by the component heat dissipation, the ambient temperature and the temperature of neighbouring boards. In both parts of the thermal evaluation process, however, there is still much need of improvement.

The lifetime (MTBF) calculation, based on rather simplified physical models, expressed in form of Arrhenius formulas, is often criticized and has very little in common with the variety of failure mechanisms observed in reality [2]. Moreover, the input data have to be either estimated for whole families of components or are based on often inconsistent information.

The worst-case analysis suffers from an even worse deficiency. While e.g. for most customized ICs and power devices a rough but clearly defined guideline is given, that a certain maximum junction temperature must not be exceeded, for most commercial ICs an 'operating' or 'ambient' temperature is specified, usually classified into a 70 °C and a 85 °C category. Since this specification is obviously based on a selection process, it cannot be ignored. On the other hand the test procedure which leads to the selection is characterized by so many uncertainties, as to transient behavior or thermal environment, that an accurate evaluation of maximum junction temperatures, corresponding to the ambient temperature classes, seems impossible.

These difficulties associated with the thermal evaluation could only be discussed very briefly in this paper. The intention was to point at the fact that alongside with increasingly sophisticated tools to calculate the component temperatures, improved methods and criteria for the evaluation have to be made available to ensure the progress in thermal analysis as a whole.

8. Summary and Conclusions

It could be shown that it is possible to create a compact model of an IC-package which is simple enough to allow the CFD simulation of systems with a multitude of components. Though simplified, it yields the crucial information expected from a thermal analysis, i.e. the junction temperatures of the components.

The underlying idea is to desist from modelling all the heat conduction phenomena inside the IC in detail, but instead to build up a composite of blocks and plates that, in the simplest case, gives the junction temperature as the only result. A methodology was outlined of how the compact model can be adjusted to the detailed model of the IC by performing two distinct CFD simulations in a confined test environment.

The compact model can be based on the values of two thermal resistances $R_{th,JA}$ and $R_{th,JC}$ alone, i.e. without the need to set up a detailed model first, if all parameters of the testing procedure for obtaining these values are known and an appropriate test method for the JC-resistance was applied. In this sense, the thermal characteristics are seen as a mere representation of the results of two distinct measurements, rather than as resistances in its original meaning.

A worldwide obligatory standardization of the thermal characteristics would improve the accuracy of building compact models after the method outlined in this paper, as well as the applicability of these figures for the purpose of comparing different package-types. As a matter of fact, today most of the manufacturers of semiconductors use customized test boards to measure the JA-resistances and various methods to measure the JC-resistance.

A more elaborate thermal characterization of IC-packages, e.g. in form of a resistance network, might conflict with the outlined procedure of deducing compact models if the original data of the measurments cannot be regained. It has to be bourne in mind that a resistance network can not be directly plugged into a finite-difference CFD code as a thermal representation of an IC. Future work on improved thermal characterisation of IC-packages should thus regard two functions the thermal characteristics have to fulfill:

- Serve as a figure-of-merit for comparison of thermal performance of different packages.
- Give sufficient data to build up compact models to be plugged into CFD codes.

9. Abbreviations and Symbols

CFD	computational fluid dynamics	k	thermal conductivity [W/mK]
EMC	electromagnetic compatibility	δ	thickness [m]
IC	integrated circuit	R_{th}	thermal resistance [K/W]
JA	junction-ambient		
JC	junction-case		
MTBF	mean time between failures		
PQFP	plastic quad flat pack		
TSOP	thin small outline package		

10. References

[1] Rosten, H. and Viswanath, R.: *Thermal Modelling of the Pentium Processor Package* 3rd Int. Flotherm User Conf. (1994)

[2] Bar-Cohen, A. and Witzman, S.: *Analysis and prevention of thermally induced failures in electronic equipment*, Proc. Eurotherm Sem. 29 (1993), pp. 25-42

CHARACTERISATION OF THERMALLY ENHANCED PLASTIC PACKAGES

ARE BJORNEKLETT and GUNNAR GUSTAVSSON
Ericsson Components AB
S-164 81 Kista-Stockholm
Sweden

1. Abstract

Thermally enhanced plastic packages with integrated metal heat spreaders are an interesting type of package for telecommunication systems. This paper presents results from a thermal characterisation of 15 different thermally enhanced plastic packages, 2 standard plastic packages and 1 metal package. The characterised package types are Quad Flat Packs (PQFP) and Leaded Chip Carriers (PLCC) with pin counts from 44 to 208. The packages were obtained from five different vendors and they have been characterised according to the 3-parameter method developed and used at Ericsson.

2. Introduction

The telecommunication network is evolving towards a more decentralised structure. The large central office switching equipment will gradually be replaced by smaller switching units closer to the customers. Important trends addressing building practice and packaging of decentralised telecommunication equipment are high volume manufacturing, low cost packaging, compact packaging, outdoor environmental conditions, closed cabinets without air exchange and high reliability. Stringent requirements are put on thermal design of such equipment because of large environmental temperature range and limited possibilities for designing with active cooling such as fans and cooling machines. Reliability which is depending on component temperature, must be high because maintenance and repair cost will be high for distributed equipment.

It will be important to find cheap ways of handling the power dissipation. Ceramic packages have been the traditional way of handling high power dissipation, but the cost is generally too high. Metal packages are another possibility, but the cost may still be 10 times higher than for plastic packages. Many package suppliers offer so-called thermally enhanced plastic packages with integrated metal heat spreaders. The cost of these packages are higher than that of standard plastic packages, but they are much cheaper than the alternatives. They are therefore obviously interesting from a thermal management point of view in cost sensitive equipment, where as the power dissipation is one of the dimensioning parameters.

E. Beyne et al. (eds.), Thermal Management of Electronic Systems II, 159-166.

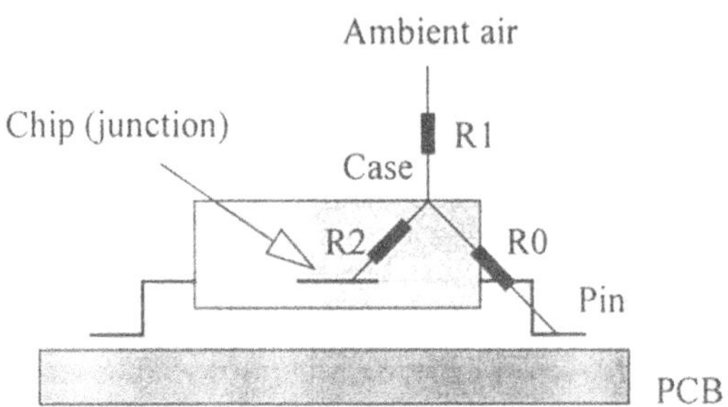

Figure 1. The three-parameter model

3. Thermal Characterisation of Components

The thermal characteristics of an electrical component is often given as one or more thermal resistances, with the unit [K/W]. There are a large number of different models, mostly based on different thermal resistance networks. The resistance values given by the component vendors are often completely different when using the component in real electronic applications. It is therefore often necessary for the component user to predict, by measurements or finite element simulations, the thermal characteristics of the components in order to be able to calculate junction and board temperatures in the application.

3.1 THE 3-PARAMETER MODEL

The model used in this study is the 3-parameter method [1,2,3] pioneered and used in thermal analyses of printed circuit boards at Ericsson. The network of the model is shown in figure 1 and the basic equation is:

$$T_j = R_s \cdot P + S \cdot (T_p - T_a) + T_a \tag{1}$$

Where Tj is the junction temperature, Tp the pin temperature defined as the average temperature where the leads of the component meet the PCB and Ta the ambient temperature of the inlet air. Rs and S are defined as:

$$R_s = \frac{R_0 \cdot R_1}{R_0 + R_1} + R_2 \tag{2}$$

and

$$S = \frac{R_1}{R_0 + R_1} \tag{3}$$

One of the resistances (R1) is external and independent of the internal properties of the package, while the other two (R0 and R2) turn out to be constants within a reasonable range of air velocities and board conductivities. These two resistances are in other words specific to the component. R1 can be calculated by board level simulation program and together with the board conductivity and the thermal characteristics of the other components on the board, makes it possible to calculate accurate temperatures.

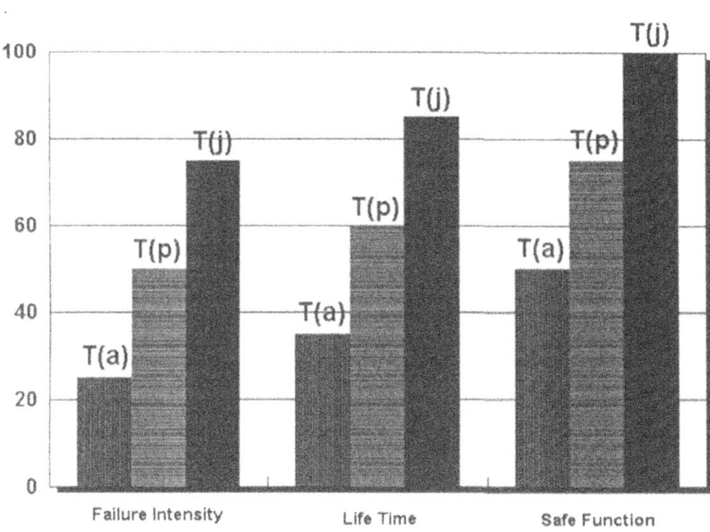

Figure 2. Temperature limits

4. Thermally Enhanced Plastic Packages

Thermally enhanced plastic packages facilitate the use of plastic packages in electronics where power dissipation is a main concern, but these kind of packages also provide enhanced electrical performance. In most of the cases the chip is directly attached to a metal block integrated in the plastic mold compound (see figure 3). Another version is to use a substrate, encapsulated in the mold compound, in thermal contact with the lead-frame.

Thermally enhanced packages are especially well suited to be used together with component heat sinks, but the heat flow path through the leads down to the board is also improved. The latter fact is very important when natural cooling of the components mounted on a PCB is considered, due to the fact that the power in this case mainly is dissipated through the leads into the board, spread out in power and ground layers and convected into the air.

The temperature limits used today at Ericsson Telecom are illustrated in figure 2, including a pin temperature limit to protect temperature sensitive discrete components and to reduce solder joint fatigue. However, a 10°C increase of the pin temperature limit (Tp) while keeping the present junction limit (Tj) is considered. Such an action might lead to 60% increased allowed power dissipation on the board level in a natural convection cooled case. To gain benefits of this possibility, components with low junction to pin thermal resistance values have to be used, due to the fact that the internal temperature budget is decreased from 25°C to 15°C. Thermally enhanced plastic packages fit in well into this scenario also.

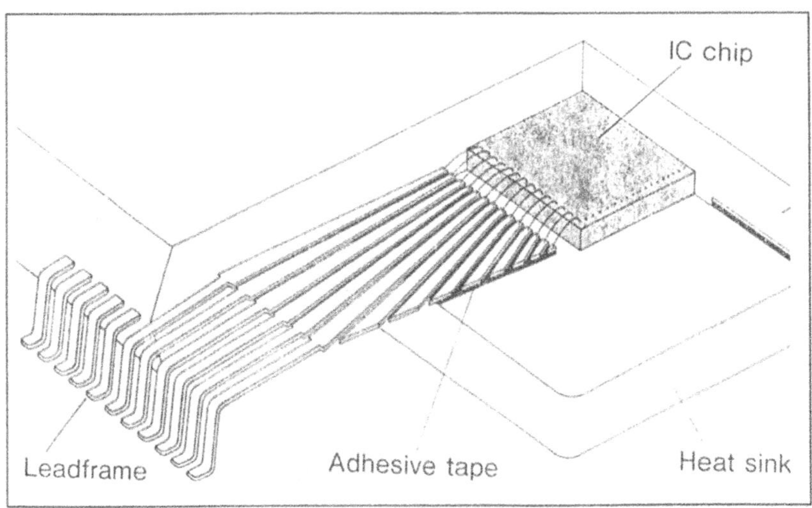

Figure 3. Thermally Enhanced Plastic Package

5. Measurements

A total number of 15 thermally enhanced plastic packages, 2 standard plastic packages and 1 metal package, were characterised according to the measurement procedure of the 3-parameter method. The temperatures of interest (Tj, Tp and Ta) were measured for 4 different air velocities, 0.2, 0.5, 1.0, 1.5 m/s. The air velocity of 0.2 m/s is equivalent with natural convection in a typical telecom building practice.

5.1 EXPERIMENTAL SETUP

The experimental setup consists of a wind-tunnel supporting the range of air velocities used in this study, 0.2, 0.5, 1.0 and 1.5 m/s. The air velocity was measured by means of the pressure drop in a laminar flow-element. This method of air velocity measurement is generally considered to be more accurate than anemometer sensors. The air velocity was continuously monitored and adjusted during the measurement by a closed loop regulation system implemented in the computer controlling the measurement.

Printed circuit boards with dimensions 265 x 285 mm, were manufactured with a standard Ericsson process. The number of layers were 4, including two inner layers of 35 μm thick copper with no electrical function. There is also a layer for mounting and electrical connection of the packages. Finally there is a layer with heating resistors, implemented by means of meandering tracks, for the board heating procedure used in the three-parameter characterisation, see ref. 3.

Packages with thermal test chips (ref. 4,5) were bought and mounted on the boards. The junction temperature measurement was made by means of the forward voltage drop in a diode on the chip and power was dissipated in ohmic resistors. A chip power level of approximately 2 W was used throughout the measurements.

The pin-temperature was measured with T-type thermocouple soldered to unused pins on the package. Care was taken to place the thermocouple as close as possible to the board so that the pin temperature where the pins are in contact with the board, was measured. This thermocouple was also used for calibration of the forward diode voltage used for junction temperature measurement, on the test chip. Calibration took place in a Heraeus VMT04/16HT thermal chamber. The accuracy of the calibration process was estimated to be ±1°C.

The ambient temperature was measured at the outlet of the wind-tunnel with a T-type thermocouple. This ambient temperature actually never departed more than 1 K from temperature measurements in the laboratory.

The measurement setup was completely automated by means of a LabView program running on a PC and IEEE-488 communication with the various instruments.

5.2 RESULTS

The measured ambient-, pin- and junction-temperatures were used to calculate the thermal resistances in the three-parameter model, ref. 3. The figures are given in table 1.

5.3 COMPARISION WITH VENDOR DATA

The component thermal data from the vendors for these packages is given as a junction to ambient thermal resistance, R(ja), as a function of different parameters. In some cases also a R(jc)-value is given, however not comparable with R2 in this report due to different definitions. A comparison between calculated R(ja)-values based on the measurements and data given by Vendor D is shown in figure 4. The data in the figure exemplifies the fact that the R(ja)-values given by the vendors are very dependent of parameters like board conductivity, board size and cooling environment.

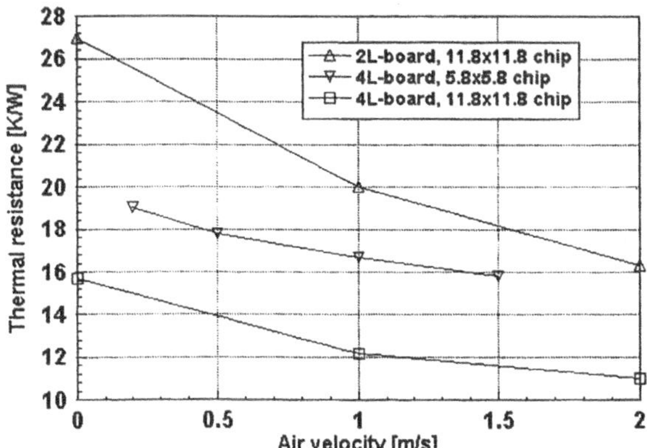

Figure 4. Plot showing the thermal resistance from junction to ambient for a thermally enhanced QFP208 (Vendor D). The effect of board construction and chip size is shown. The data for the 11.8x11.8 mm chip was obtained from the vendor. The data for the 5.8 x 5.8 mm chip was gathered in this work.

Table 1. Measured thermal resistance values. R1 is given for different air velocities, 0.2, 0.5, 1.0 and 1.5 m/s.

Package	R0	R2	R1 (0.2)	R1 (0.5)	R1 (1.0)	R1 (1.5)
Vendor A						
PLCC44A	6.8	3.4	755	399	222	148
QFP100A	7.6	2.3	431	224	138	110
QFP208A1	4.4	1.2	116	76	65	59
QFP208A2	4.9	3.3	251	185	128	86
Vendor B						
QFP100B	4.9	2.3	461	368	171	78
QFP160B	3.2	2.0	125	51	26	20
QFP208B	3.1	2.6	120	71	50	28
Vendor C						
PLCC44C1	3.0	2.1	653	274	145	128
PLCC44C2 (std.)	16.6	8.7	823	473	279	169
QFP44C	6.8	1.9	811	436	279	198
QFP100C	6.4	4.8	620	319	189	133
QFP208C1	8.4	3.1	91	39	28	19
QFP208C2 (std.)	11.6	5.8	166	130	76	60
Vendor D						
QFP160D	5.9	5.2	164	88	53	33
QFP208D1	8.1	4.7	158	106	51	43
QFP208D2	5.9	3.8	122	51	37	22
Vendor E						
QFP160E	3.8	2.2	126	46	27	21
MQFP160E	2.2	1.5	71	59	28	22

6. Discussion

A dominant feature of the results is that R1 is much larger than R0 and R2 for all air velocities. This indicates an efficient thermal conduction between the chip and the leads in line with the expectations from thermally enhanced packages. The measured values for R0 and R2 are considerably lower than typical values for standard packages.

A first rough comparison between different packages of the same size can be accomplished by using the sum of R0 and R2 as a performance parameter. These thermal resistances are the smallest ones in the model and they will dominate the heat flow in most cases. However, if the pin temperature is much higher than the ambient temperature, for instance on boards that are densely populated with heat dissipating components, then this comparison is not valid because the heat is forced to flow through R1.

The measured values for R0 and R2 are reasonable consistent considering the differences in package design. The values for R1 show relatively large variation, even for packages of the same physical size. This variation is probably caused by the difficulty of measuring the temperature difference between junction and pin when the heat is generated outside the package. This difference was in the 0.5 to 2 K range for most of the measurements reported here. However, this variation will not impact on the usefulness of the model due to the fact that R1 is much larger than the other measured resistances. A reasonable accuracy for the junction and pin temperature measurement is +/- 0.2 K and the accuracy in the measurement of the difference is therefore not very good. This is probably the main reason for the spread in the R1 values.

No attempt has been made in this work to analyse the effect of the chip size on the thermal parameters. A larger chip size should in particular decrease the R2 thermal resistance. A 5.8 x 5.8 mm test chip was used in all packages except for those obtained from Vendor C. The chip size in the packages from Vendor C is 5.08 x 5.08 mm for the 44 and 100 lead packages. The QFP 208's have a 10.2 x 10.2 mm chip. Care should be taken when comparing the QFP 208's from Vendor C with the other ones because they have a much larger chip. The effect of the chip size is also evident in figure 4 where a comparison is given in the thermal resistance from junction to ambient for different chip size and board construction for a QFP208 from Vendor D.

The standard plastic packages, Vendor C; PLCC44 and QFP208, show considerable higher R0 and R2 values than the thermally enhanced counterparts.

The metal package, Vendor E; MQFP160E, shows the best thermal performance, but this type of package is considerably more expensive than the thermally enhanced type of packages and the relative improvement from thermally enhanced packages is not very large.

7. Conclusion

A total of 18 packages from 5 different vendors have been characterised thermally according to the 3-parameter model. The packages were 15 thermally enhanced plastic packages, two standard plastic packages and one metal package. The metal package shows the best thermal performance, but some of the thermally enhanced plastic packages are not far behind and at a much lower cost. Thermally enhanced plastic packages show a considerable improvement over standard plastic packages.

8. References

1. Malhammar A. (1991), Thermal Models for Circuit Boards and Microcircuits Today and Tomorrow, CERT 91 Tutorial, Electronic Components Institute International Ltd, Crowborough, England.
2. Flodman G., A Systematic Approach to Thermal Management and Reliability Prediction in a Telecom System. Seventh Annual International Electronic Packaging Conference, Boston, Mass.
3. Malhammar A., A Three Parameter Model for Thermal Characterization of Microcircuit Packages. Eurotherm seminar 29, 1993.
4. Sweet J.N., Peterson D.W., Chu D., Bainbridge B.L., Gassman R.A., Reber C.A., Analysis and Measurements of Thermal Resistances in a Three-Dimensional Silicon Multichip Module Populated with Assembly Test Chips. SEMI-THERM symposium 9, 1993.
5. Rodkey D.L., Manual for Using Delco Electronics Thermally Sensitive Die. Delco Electronics.

THERMAL CHARACTERIZATION OF MULTISTRIPE HETEROSTRUCTURE LASERS.

V. LEPALUDIER Y. SCUDELLER
Groupe de Thermique des Interfaces et Microthermique
Laboratoire de Thermocinétique URA CNRS 869
I.S.I.T.E.M.
La Chantrerie C.P. 3023
44087 NANTES Cedex 03

1. Introduction

Thermal behavior is one of important fields in semiconductor laser diode development, because it is related to threshold current, output optical power, wavelength of oscillating modes, as well as device life-time. Optical efficiencies are very weak, so heat generation, localized in very small volumes of few μm^3, can reach several hundreds of MW/m^3. It leads to important temperature rises very closed to P-N junctions which reduce device reliability. So, experimental methods for thermal analysis become fundamental for high power devices design.

This paper concerns microscale heat transfer study in GaAlAs/GaAs laser diodes with non-uniform heat generation. It deals with the development of a theoretical model and a specific thermal characterization method.

Theoretical temperature response to a step or a pulse current drive is evaluated by means of Laplace and Fourier transforms. A steady and transient states analysis is presented. Junction temperature is measured using the thermodependence of the terminal voltage across the laser supplied with a small current, after switching off. So, it concerns transient temperature measurements, which lead to determine some thermal parameters, such as N-GaAs substrate diffusivity and thermal contact resistance between the laser and its submount.

2. Description of the laser diode

The laser studied is a conventional double heterostructure, oxide isolated, grown by liquid-phase epitaxy *(figure 1)*. Its dimensions are presented in *table 1*. It owns ten stripes centered on the laser width, and separated each other by proton-isolated areas.

TABLE 1. Laser diode and submount dimensions (μm)

Laser diode width	2b	350	Submount width	2b'	1240
Cavity length	L	620	Submount length	L'	740
Laser diode thickness	a	88	Submount thickness	H	300
Heat sources width	b_0	4	Protoned areas width	b_s	6

The active region of P-GaAs is bounded by $Ga_{0,6}Al_{0,4}As$ clad layers. Adding Aluminum decreases refractive indexes, so these layers have a role of waveguide for the laser beam.

E. Beyne et al. (eds.), Thermal Management of Electronic Systems II, 167-177.

The capping layer P-GaAs is metallized with PtAu. Thermal properties of each medium are given in *table 2*. Heat sources are assumed to be mainly located at each P-N junctions of the laser, uniform along the cavity axis. Radiative effects, due to photon absorption outside the active area, may be taken into account as a volumic heat generation rate [1, 2, 3]. Thermal properties are assumed to be uniform along the laser cavity, independent on temperature, and two-dimensional heat flow will be treated. Radiative and convective heat transfer is negligible. The device is mounted P-side up on a copper submount, in which heat transfer may be treated in a three-dimensional way. The die attach is an electroconducting glue of a few μm, modeled by an uniform thermal contact resistance, independent on temperature. Submount is connected to a heat sink, cooled of a constant and uniform temperature, by a thermostated fluid flow.

TABLE 2. Thermophysical properties of laser layers and submount

	a_i (μm)	doping (cm^{-3})	λ_i (W/m.K)	α_i (m^2/s)
PtAu	2	—	300	1,25 10^{-4}
P+-GaAs	1	10^{19}	45	2,5 10^{-5}
P-$Ga_{0.6}Al_{0.4}As$	2	5.10^{17}	11	0,6 10^{-5}
N-$Ga_{0.6}Al_{0.4}A$	1,5	10^{18}	11	0,6 10^{-5}
N-$Ga_{1-x}Al_xAs$	1,5	10^{18}	15	0,8 10^{-5}
N-GaAs	80	2.10^{18}	45	2,5 10^{-5}
Cu	300	—	380	1,16 10^{-4}

3. Theoretical analysis

Thermal characterization requires the calculation of the laser diode cooling down θ(M,t). If heat sources are independent on temperature, θ(M,t) is obtained by difference between the steady-state $\theta_p(M)$ reached when the device is submitted to a heat generation rate P, and the overheating T(M,t) following a step or a pulse of heat flux variation P. Calculation of T(M,t) is presented, considering a heat generation q_i at each interface. The very heterogeneous feature of the structure requires a specific methodology, based on Integral Transforms technique [4].

With x and y respectively perpendicular and parallel axis to layers, conduction system in each laser diode layer i is writen :

$$\lambda_x^i \frac{\partial^2 T_i}{\partial x_i^2} + \lambda_y^i \frac{\partial^2 T_i}{\partial y^2} = \rho_i C_i \frac{\partial T_i}{\partial t} - P_i\left(x_i, y\right)$$

$$\begin{array}{ll}
y = 0,b & \dfrac{\partial T_i}{\partial y} = 0 \\
x_1 = 0 & \dfrac{\partial T_1}{\partial x_1} = 0 \\
x_i = 0 & T_i(0) = T_{i-1}(a_{i-1}) \\
 & -\lambda_x^i \dfrac{\partial T_i}{\partial x_i}(0) = -\lambda_x^{i-1} \dfrac{\partial T_{i-1}}{\partial x_{i-1}}(a_{i-1}) + q_{i-1}(y) \\
x_i = a_i & T_i(a_i) = T_{i+1}(0) \\
 & -\lambda_x^i \dfrac{\partial T_i}{\partial x_i}(a_i) = -\lambda_x^{i+1} \dfrac{\partial T_{i+1}}{\partial x_{i+1}}(0) - q_i(y) \\
x_n = a_n & T_n(a_n) = T_c(y,t) + R_c \varphi_c(y,t) \\
t = 0 & T_i = 0
\end{array} \quad (1)$$

And, in the submount :

$$\nabla T - \frac{1}{\alpha}\frac{\partial T}{\partial t} = 0$$

y=0,b' $\quad \frac{\partial T}{\partial y} = 0$

z=0,L' $\quad \frac{\partial T}{\partial z} = 0$

x = 0 $\quad -\lambda\frac{\partial T}{\partial x} = \varphi_c(y,z,t)$ if $0 < y < b$ and $0 < z < L$ (2)

$= 0$ elsewhere

x=H $\quad T(H)=0$

t=0 $\quad T=0$

A Laplace transform on temperature is applied in each medium (1) and (2) :

$$\bar{T}_i(x_i,y,p) = \int_0^\infty T_i(x_i,y,t)\exp(-pt)dt \tag{3}$$

$$\bar{T}(x,y,z,p) = \int_0^\infty T(x,y,z,t)\exp(-pt)dt \tag{4}$$

where p is the Laplace variable.

And, a Finite Fourier Transform is also applied to get a reduced system with one space variable x, direction of homogeneous boundary conditions.

$$\bar{\bar{T}}_i(x_i,\alpha_m,p) = \int_0^b \bar{T}_i(x_i,y,p)\cos(\alpha_m y)dy \tag{5}$$

$$\bar{\bar{T}}(x,\beta_n,\gamma_q,p) = \int_0^{L'}\int_0^{b'} \bar{T}(x,y,z,p)\cos(\beta_n y)\cos(\gamma_q z)dydz \tag{6}$$

with $\alpha_m = \frac{m\pi}{b}$, $\beta_n = \frac{n\pi}{b'}$ and $\gamma_q = \frac{q\pi}{L'}$.

The second transformed system leads to introduce the mean temperature T_c at the contact laser/submount as a function of interfacial heat flux φ_c, in Laplace space :

$$\bar{T}_c(p) = \bar{F}(p)\bar{\varphi}_c(p) \tag{7}$$

F(p) is a transfer function which depends on geometry and thermal properties of submount.

Thus, a third kind boundary condition appears in the equation relative to $x_n=a_n$ of the system (1). So, it may be solved in Laplace-Fourier space. Temperatures and fluxes are determined at each interface, using a quadripole formulation [4]. The real expressions are obtained, first by a numerical Laplace inversion [5], and inverse Fourier Transform :

In the laser :

$$\bar{T}_i = \sum_{m=0}^{\infty} \bar{\bar{T}}_i \frac{\cos(\alpha_m y)}{\int_0^b \cos^2(\alpha_m y)dy} \tag{8}$$

In the submount :

$$\bar{T} = \sum_{n=0}^{\infty}\sum_{q=0}^{\infty} \bar{\bar{T}} \frac{\cos(\beta_n y)}{\int_0^{b'} \cos^2(\beta_n y)dy}\frac{\cos(\gamma_q z)}{\int_0^{L'} \cos^2(\gamma_q z)dz} \tag{9}$$

This hybrid method presents a lot of advantages for heat transfer analysis. It leads to obtain junction temperature profiles in a very short time, and permits to study easily sensitive geometrical and thermal parameters. When temperature fields have to be plotted, a finite element model leads to win CPU time. In this case, the obtention of temperature fields in the laser diode and submount requires only a few minutes.

Figure 2 presents the evolution of the temperature field, obtained by a finite element code, at different times of the cooling down. In steady state, the overheating appears very closed

to P-N junctions. All the temperature falling occurs inside the N-GaAs substrate and particularly in the clad layer N-GaAlAs which undergoes a quarter of the falling and represents less than 4% of the substrate thickness. In steady state, the increasing of the metallization thickness reduces the mean temperature and the temperature variations along the y-axis *(Figure 3)*. Since the first microsecond of the cooling down, multijunction laser diode behavior is equivalent to an unijunction one having a stripe of B_0 width *(figure 4)*. This duration corresponds to establishment of microscale constriction where heat flux spreads out from b_0 to B_0/N, N denotes the number of junctions. At a few hundreds of microseconds, temperature distributions become monodimensional. It corresponds to establishment time of macroscale constriction where heat flux spreads out from B_0 to b *(figure 5)*. Thus, there are two scales of constriction. The macroscale constriction determines the mean temperature along B_0. And, the microscale constriction determines the correction to bring to this mean temperature to take into account the non-uniformity of flux along B_0.
Sensitivity coefficient on a parameter w is defined by :

$$S_w = \frac{w}{\theta_p}\frac{\partial T(w,t)}{\partial w} \tag{10}$$

Their calculation on N-GaAs and N-GaAlAs diffusivity *(figure 6)* shows that, during the first microseconds, temperature evolution depends on the clad layer diffusivity, and progressively, with heat diffusion, they become more sensitive to substrate diffusivity. It may be considered that since 80 µs, only substrate diffusivity defines cooling behavior. The sensitivity coefficients on thermal contact resistance, calculated on temperature increasing, are presented in *figure 7*. It shows that overheating behavior is very sensitive to R_c beyond several milliseconds. It means that R_c determines greatly the temperature level, but does not influence cooling down before one millisecond.
In conclusion, transient state on microsecond scale is very sensitive to clad layer N-GaAlAs properties. At about 80 µs, the structure is equivalent to an homogeneous laser with a single junction of B_0 width. From a few milliseconds, heat transfer is monodimensional, the laser is equivalent to an homogeneous one, in imperfect contact with submount.

4. Experimental results

The temperature measurement is based on the thermodependence of the terminal voltage across the laser. The experimental arrangement is presented in *figure 8*. The laser diode[3] is supplied with a constant current I above threshold, over a duration getting steady-state. This primary circuit is driven by a pulse generator[1] at 0.1 Hz, with a pulse duration of 9,9 seconds. A standard resistor R[2] of a few ohms tests the current intensity I. Then, this circuit is switched off on the measurement circuit. This one is driven by a battery[5] which supplies a small current I_0, at about 1 mA. A resistor R'[4] of a few hundreds of ohms allows to control this intensity level. This current must be sufficiently low to avoid overheating and sufficiently high to reduce noises and capacity effects due to the high impedance of the semiconductor device. A silicon diode[6], reverse biased after switching off, insulates the primary circuit, during the measurement sequence. The laser diode [3] is set in an isothermal cell, controlled by a thermostated water flow. Temperature is tested by several K-thermocouples of 80 µm diameter. A microthermocouple of 25 µm diameter is fixed at the back of the submount.

Junction temperature cooling down is obtained, measuring terminal voltage evolution across the device $\Delta V(\theta)$, with a transient recorder (bandwidth 25 MHz). A offset circuit[7] permits, removing the terminal voltage corresponding to thermal steady state, to operate on the highest resolution (full scale 100 mV, resolution 100 μV). The terminal voltage is composed of the potential barrier at the junctions $\Delta V_j(t)$, the diode series resistance $V_r(t)$, and the thermoelectric voltage due to non-uniform temperature distribution $V_{th}(t)$. Diode series resistance is negligible in comparison with the two other contributions. It follows that:

$$\Delta V(t) = \Delta V_j(t) + V_{th}(t) \tag{11}$$

It may be shown that :

$$\Delta V_j(t) = a(I_0).\theta(t) + b(I_0) \tag{12}$$

$$V_{th}(t) = \varepsilon_{GaAs}.G(t).[\theta(t) - \theta_c(t)] \tag{13}$$

a, b are two coefficients depending on I_0, with a logarithmic law. They are determined previously in isothermal conditions. ε_{GaAs} is the thermoelectric power of the N-GaAs substrate. G(t) is a function depending on geometry and thermal properties, as also on time because temperature distributions evolve in transient state. $\theta_c(t)$ denotes the temperature decreasing at laser/submount interface. Then, the mean junction temperature is given by:

$$\theta(t) = \frac{\Delta V(t) - b(I_0) + \varepsilon_{GaAs}.G(t).\theta_c(t)}{a(I_0) + \varepsilon_{GaAs}.G(t)} \tag{14}$$

Figure 9 gives an example of terminal voltage measurement, current evolution and junction temperature decreasing during the 500 first microseconds.
Conforming to sensitivity analysis, substrate diffusivity governs the transient phase during about 100 μs. It is the reason why this parameter may be identified over a sequence of $[t_0,t_1]$ near 100 μs. Adimensional experimental and theoretical temperatures $\theta^*(t) = \frac{\theta(t_0) - \theta(t)}{\theta(t_0) - \theta(t_1)}$ and $T^*(\alpha,t) = \frac{T(\alpha,t_0) - T(\alpha,t)}{T(\alpha,t_0) - T(\alpha,t_1)}$ are adjusted by minimization of a criterion defined by $J(\alpha,t) = \left\|\theta^*(t) - T^*(\alpha,t)\right\|^2$. Identification for various current levels *(table 3)* leads to α=2,59.$10^{-5}m^2/s$. Thermal contact resistance between laser and submount may be also identified over a sequence between t_0=3 ms and t_1 = 4 ms, conforming to the sensitivity analysis. Identified values are presented in *table 4*, and lead to R_c=1,55.10^{-5} $K.m^2/W$. This result shows the major importance of the contact quality between the laser diode and its submount [6].

TABLE 3. N-GaAs diffusivity for various current levels

I (mA)	α_i (m^2/s)
189	2.58 10^{-5}
178	2.46 10^{-5}
172	2.72 10^{-5}

TABLE 4. R_c for various current levels

I (mA)	R_c ($K.m^2/W$)
179	1.61 10^{-5}
204	1.52 10^{-5}

The steady-state temperature determination requires N-GaAlAs thermal properties knowledge. The five first microseconds do not get any information on temperature because of perturbations induced by switching and capacity effects. It the raison why, to estimate the steady-state temperature, thermal properties given by Adachi [7] are used. In order to free ourselves of the heat generation rate knowledge P, experimental cooling $\theta(t)$ is written under the form :

$$\theta(t) = \theta_p - P.T_u(t) \quad (15)$$

Where $T_u(t)$ the overheating due to a unit flux step. Writing this expression at two moments t and t_0 in the very first times of the cooling down leads to an expression of θ_p depending only on T_u and θ:

$$\theta_p = \frac{\theta(t) \cdot T_u(t_0) - \theta(t_0) \cdot T_u(t)}{T_u(t_0) - T_u(t)} \quad (16)$$

The model used to calculate $T_u(t)$ is a bidimensionai one, with two layers N-GaAlAs (8 μm) and N-GaAs (80 μm). The initial values of temperatures, presented in *table 5*, appears proportional to input power. It shows that optical efficiency has a low dependence on temperature at these supply levels. A good agreement is observed between experiments and model *(table 6)*. It proves that radiative effect is not really significant on heat sources distributions at these intensity levels, it is due to spontaneous saturation [8] which is usually observed above threshold. Moreover, these results show that temperature dependence of heat sources can be negliged.

TABLE 5. Steady state temperatures for various current levels

V.I (mW)	θ_p (°C)
231	40.2
217	39.2
205	37.9
198	36.7
187	35.4

TABLE 6. Experimental and theoretical cooling down

t (μs)	θ (°C)	θ_{theo} (°C)
5	38.2	38.7
10	37.8	38.3
20	37.1	37.7
30	36.6	37
40	36.1	36.7
50	35.7	36.3
70	35.2	35.7
80	35	35.5
90	34.8	35.3
100	34.6	35.1
2000	27.3	27.2
2500	27.1	26.9
3000	27	26.7

5. Conclusion

A comprehensive study of heat transfer in multistripe GaAlAs/GaAs heterostructure laser on submount, based on simulations and experiments, have been presented. An hybrid model based on Integral transforms has been developed. This method appears very suitable for heterogeneous systems composed of several microsystems of various sizes, with many heat sources. A sensitivity analysis has brought to the fore the governing geometrical and thermal parameters in steady and transient states.

Experimental temperature measurements have been presented. A method, based on the thermodependence of the terminal voltage has been set up. Analysis of various sequences of transient phase located over a few microseconds to a few milliseconds duration, supported by a model, has allowed to identify simultaneously the thermal diffusivity of the substrate, the thermal contact resistance laser/submount, and the steady-state temperature of junctions.

NOMENCLATURE

a	Layer thickness (μm)
b	Laser half-width (μm)
b_0	Junction width (μm)
b_s	Proton-isolated area width (μm)
b'	Submount width (μm)
B_0	Equivalent junction width for an unijunction laser diode (μm)
H	Submount thickness (μm)
L	Cavity length (μm)
L'	Submount length (μm)
N	Number of P-N junctions
P	Heat generation rate (W)
T	Temperature rise for a heat flux step P (°C)
T_u	Temperature rise for an unit heat flux step (°C)
T^*	Adimensional theoretical temperature
θ	Cooling down temperature (°C)
θ_p	Steady state temperature (°C)
θ_c	Contact temperature laser/submount (°C)
θ^*	Adimensional experimental temperature
R_c	Thermal resistance of die attach ($K.m^2/W$)
S	Sensitivity coefficient
ΔV	Terminal voltage (V)
ΔV_j	Junction potential barrier (V)
V_{th}	Thermoelectric voltage (V)
α	Thermal diffusivity (m^2/s)
ε	N-GaAs substrate thermoelectric power (μV/K)
λ	Thermal conductivity (W/m.K)

REFERENCES

[1] Joyce, W B and Dixon, R W (1975) Thermal resistance of heterostructure lasers, *J. Appl. Phys.* **48**, 855-862

[2] Kobayashi ,T and Furukawa, Y (1975) Temperature distributions in the GaAs-AlGaAs double heterostructure laser below and above the threshold current. *Japenese J. Appl. Phys.* **14**, 1981-1986

[3] Steventon, A.G and al. (1977) Low threshold current proton-isolated GaAlAs double heterostructure lasers, *Opt. Quant. Electr.* **9**, 519-525

[4] Lepaludier, V. (1995) Contribution à l'étude des champs thermiques et électriques dans les lasers semiconducteurs : Application à la caractérisation thermique et à l'optimisation des dispositifs à double hétérostructure de type GaAlAs/GaAs. *Ph. D. Thesis, University of Nantes*, France

[5] Stehfest Gaver, H (1970) Comm. A.C.M. **13**,47-49

[6] Lepaludier, V. and Scudeller Y. (1996) A transient method of thermal characterization of double heterostructure laser diodes, *Microelectronics Journal* **27**, accepted to be published

[7] Adachi, S. (1985) GaAs, AlAs,and $Ga_{1-x}Al_xAs$: Material parameters for use in research and device applications, *J. Applied Physics* **58**, R1-R29

[8] Ito, M. and Kimura, T. (1981) Stationary and transient thermal properties of semiconductor laser diodes, *IEEE J. of Quantum Electronics* **QE-17-5**, 787-795

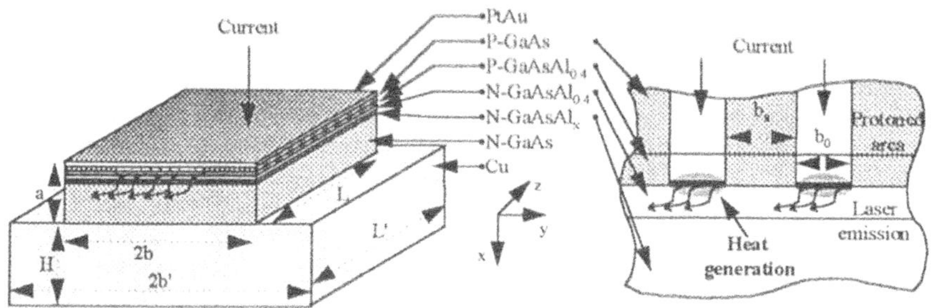

Figure 1 : Ten-stripes laser diode

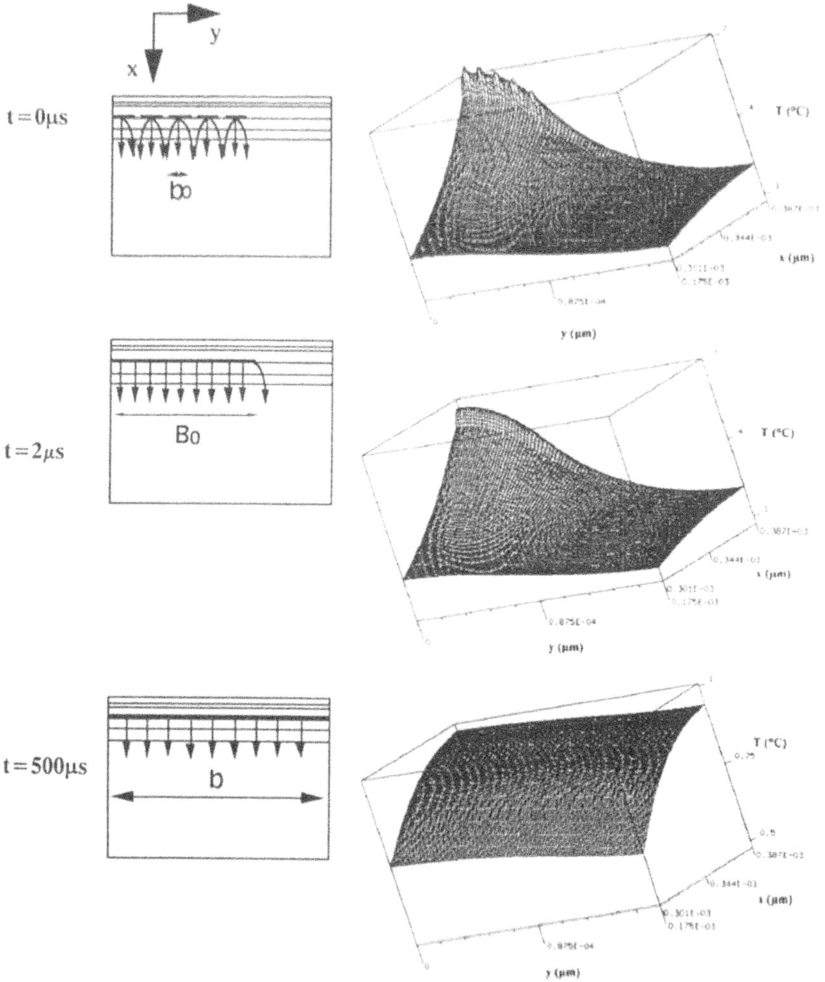

Figure 2 : Temperature fields

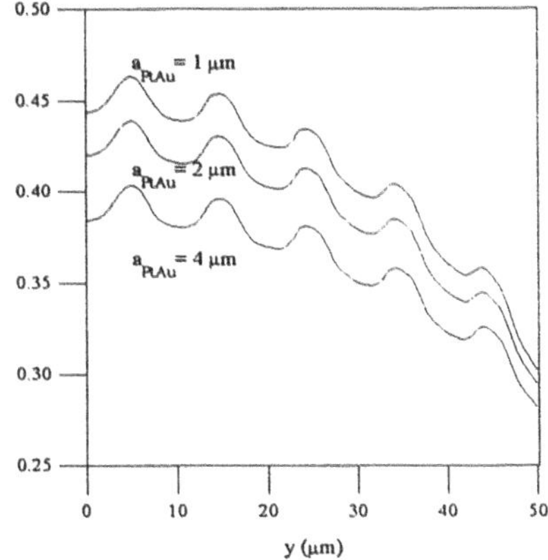

Figure 3 : Temperature profiles as a function of metallization thickness

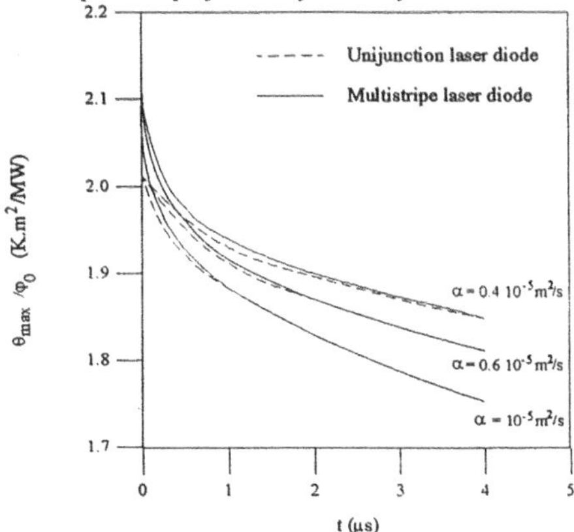

Figure 4 : Cooling down behaviors

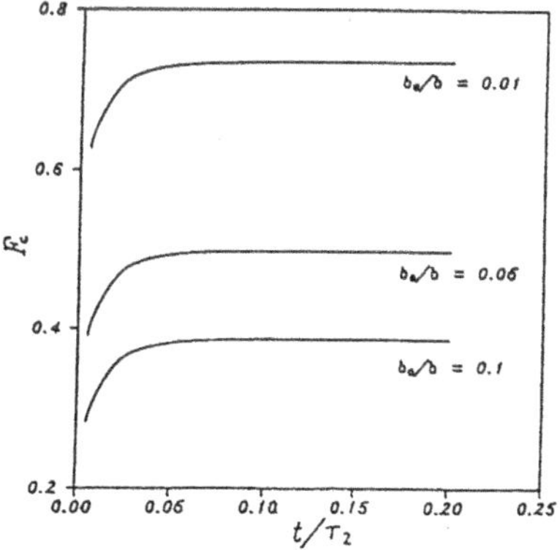

Figure 5 : Evolution of the constriction function ($\tau_2 = \dfrac{b^2}{\alpha_m \pi^2}$)

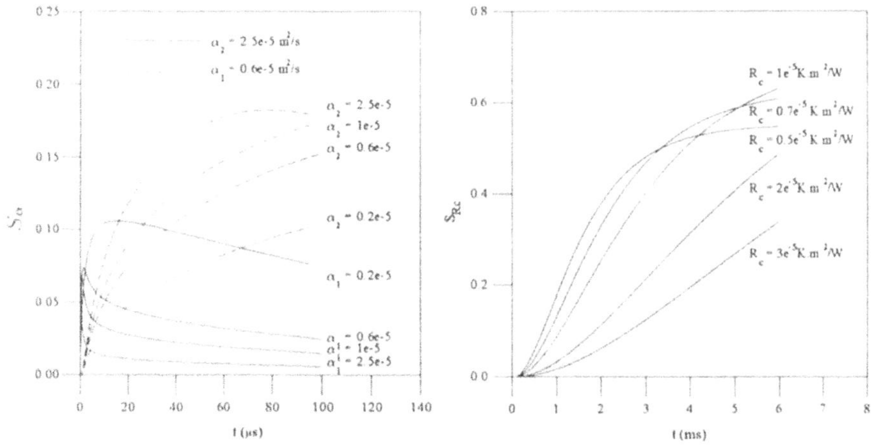

Figure 6 : Sensitivity coefficients on diffusivities

Indexes 1 : Clad layer - 2 : Substrate

Figure 7 : Sensitivity coefficients on thermal contact resistance

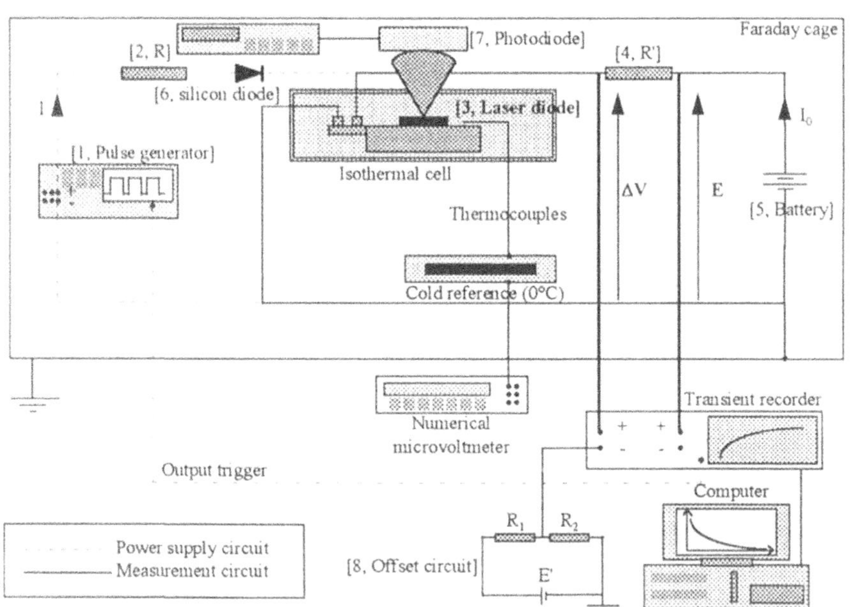

Figure 8 : Experimental arrangement

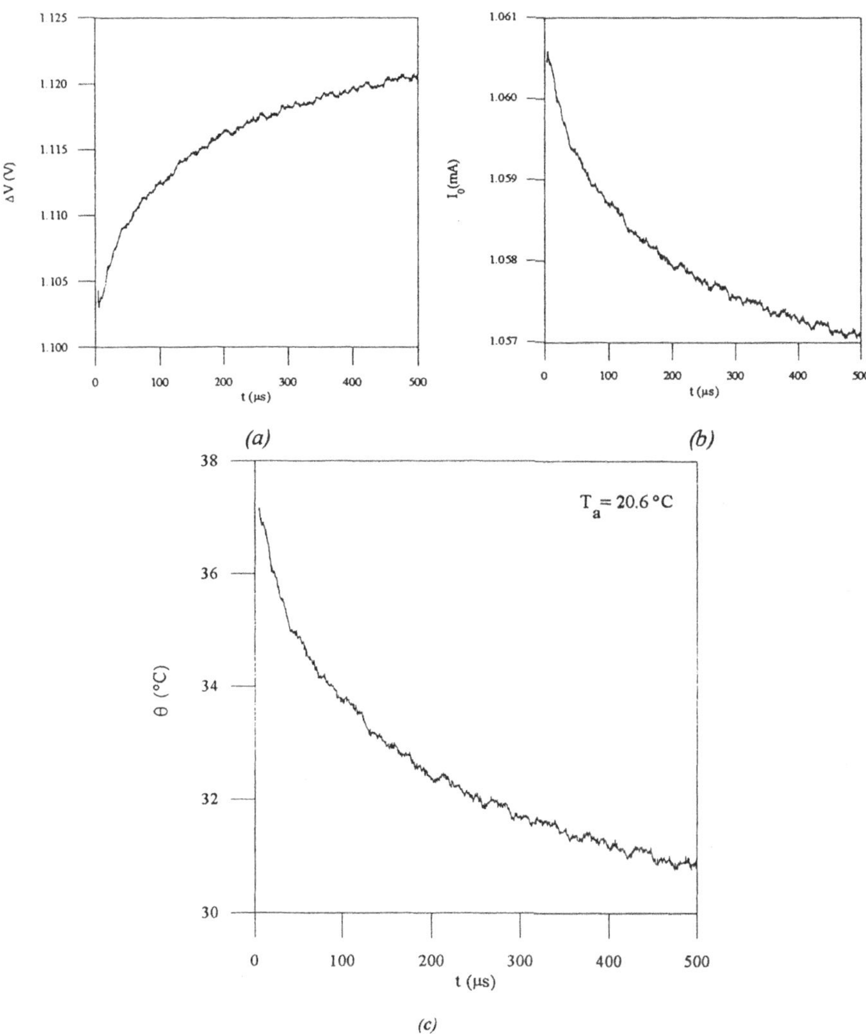

Figure 9 : Experimental results (ambient temperature T_a = 20.6 °C)

(a) : Voltage measurement - (b) : Current evolution - (c) : Junction temperature decreasing

APPLICATION OF THERMAL TERRITORIES FOR AIR-COOLED CIRCUIT BOARD DESIGN

ÅKE MÄLHAMMAR PH.D.
ELLEMTEL
Box 1505
S-125 25 Älvsjö
Sweden

Abstract

Multilayer glass epoxy circuit boards are good thermal conductors and can, for air-cooled devices, serve as heat sinks of considerable capacity. The best use of this property is achieved when the components on a circuit board are placed so that they have a conveniently sized "free" circuit board surface in all directions. A simple way to ensure this is to associate a virtual cooling area, here called a thermal territory, with each component. Although thermal territories are primarily intended for layout purposes, they are also very useful on the individual component level, particularly in the early stages of the design process. Experiences have shown that it is possible to avoid many mistakes if thermal territory estimates are made as soon as approximate data is available. Application experiences and calculation methods are discussed, and some rules of thumb are given.

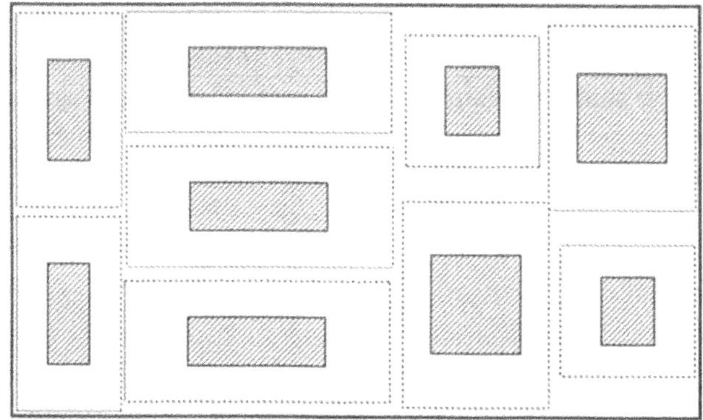

Figure 1. Thermal territories as they can be used in the layout process.

E. Beyne et al. (eds.), Thermal Management of Electronic Systems II, 179-186.

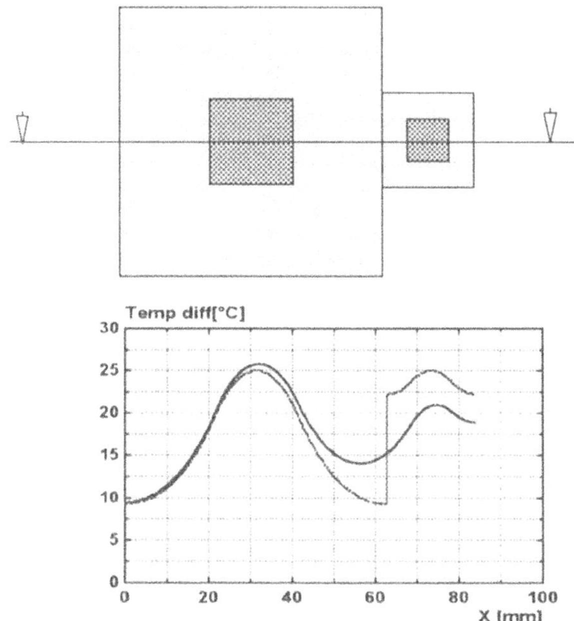

Figure 2. Temperature profile for a small and a large thermal territory, when connected and unconnected.

1 Layout Applications

A thermal territory is defined as the minimum rectangular area of a circuit board that a component requires for its cooling. The basic idea is to assign a thermal territory to each component on a circuit board and then, when the layout is made, ensure that the thermal territories do not overlap, see figure 1.

In general, it is impossible to completely avoid overlapping. When overlapping does occur, it can to some extent be compensated for by providing a "free" area in another direction. Alternative approaches are to use unsymmetrical thermal territories or thermal territories defined for groups of components. Although these approaches are theoretically possible, they would complicate the method considerably.

It is evident that it is impossible to succeed with a layout where the sum of all the thermal territories exceeds the surface area of the circuit board. A rule of thumb, based on experiences from signal processing applications, is that a success cannot be ensured unless the cover ratio is as low as 50% and that cover ratios above 70% are very difficult to deal with. These rules are, however, not directly applicable to layouts with an extreme hot spot character, such as power processing applications.

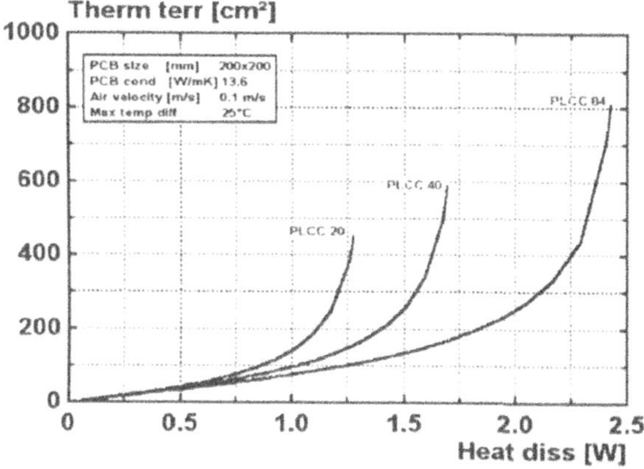

Figure 3. Thermal territory sizes for some members of the PLCC package family.

It should be noted that the thermal territory layout method is approximate. The main cause of error is non-matching temperatures at the thermal territory interfaces. The temperature profiles in figure 2 are an example of this. Here, the two thermal territories have been calculated for two uniform flat heat sources with the same maximum temperature limit, but with different sizes and source strengths. The temperature levels at their outer boundaries are therefore different, and the result is a step in the temperature profile. This is not physically possible on a circuit board. As indicated in figure 2, the temperature of the small heat source will be lower than expected, and the temperature of the large heat source will be higher than expected.

The same type of error will appear if the thermal territories are of the same size, but have different maximum temperature limits. It can, therefore, be concluded that the thermal territory layout method functions best when the thermal territories are of approximately the same size, and have the same maximum temperature limit. The method is also useful for other cases, but then the difference between a layout optimized with the thermal territory method and a layout optimized with a more accurate method, will be larger.

2 Single Component Applications

It is evident that it is more difficult to use a component that requires a large thermal territory than a component that requires a small thermal territory. Component comparisons is, therefore, another obvious application for thermal territories. One difficulty when using thermal territories on the component level is that the size varies not only with the proper-

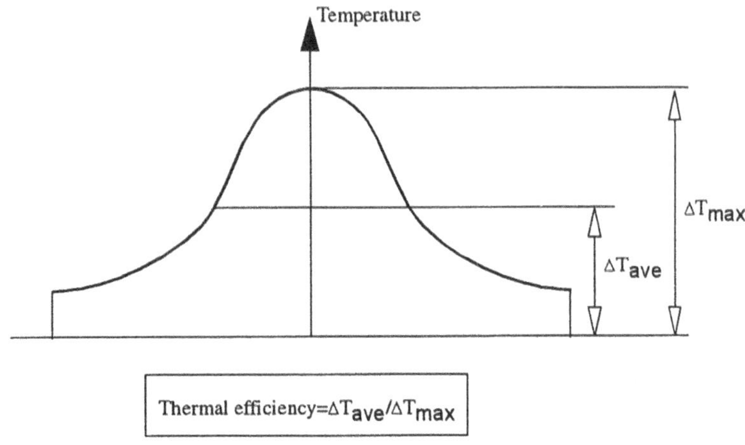

Figure 4. The thermal efficiency definition for a thermal territory.

ties of the component itself but also with the properties of its environment, for example, the thermal conductivity of the circuit board and the air velocity. Accurate thermal territory calculations can, therefore, only be made if the values of these parameters are well known. Figure 3 shows an example for some members of the PLCC package family.

A widely-used way to describe the thermal properties of a component is by the junction to ambient thermal resistance. The difficulty with this type of description is that the value, to a large extent, depends on the properties of the circuit board where the component is mounted. Thermal territories are useful for clarifying this aspect of the problem. The straightforward method is to assume a thermal territory size, create a thermal model of the type shown in figure 6 and then reduce it to a junction to ambient model. Although this method is adequate for many cases, it is sometimes difficult to know if the thermal territory size assumed is resonable.

An alternative method is to base the thermal territory size assumption on the temperature conditions of the thermal territory. The thermal efficiency definition in figure 4 is useful for this purpose. A large thermal territory will result in a low thermal efficiency, and vice versa. In the former case the cooling capacity of the circuit board will be poorly used. In the latter case the heat dissipated will be small. A reasonable size for the thermal territory must be a compromise between these two extremes. The major advantage with basing the thermal territory size assumption on a certain thermal efficiency is that a fair uniformity is achieved for all components involved in a comparison.

Figure 5 shows examples of thermal efficiencies for some members of the PLCC package family. For thermally critical designs, the rule of thumb is that the thermal efficien-

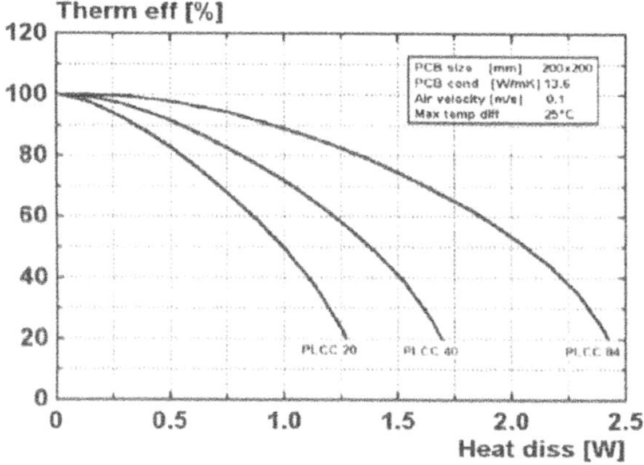

Figure 5. Thermal efficiency for some members of the PLCC package family.

cies should be larger than 75% and only exceptionally as small as 50%.

3 The Component Interface

The temperature difference between an arbitrary point in a thermal territory and the ambient air is always proportional to the heat dissipated from the territory. A thermal territory can, therefore, simply be modelled as a thermal resistance. Figure 6 shows how a thermal resistance model for a component can be interfaced with a thermal territory model. T_j is the junction temperature, T_p is the pin temperature and T_a is the ambient air temperature. The required thermal resistance for the thermal territory, R_t, can be calculated provided that the component model, the heat dissipation and either of T_p or T_j are given values.

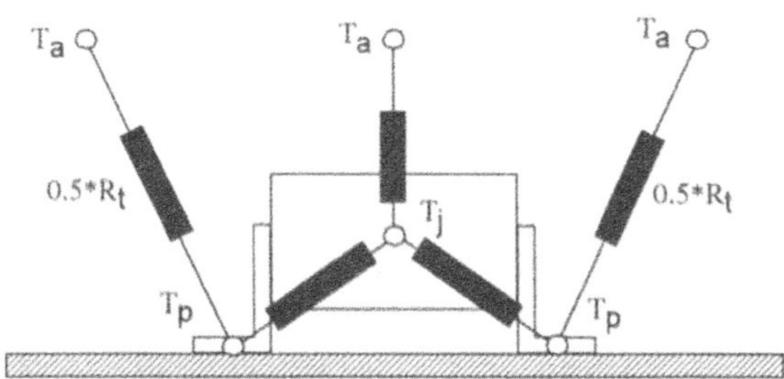

Figure 6. A thermal resistance model for a component with a thermal territory.

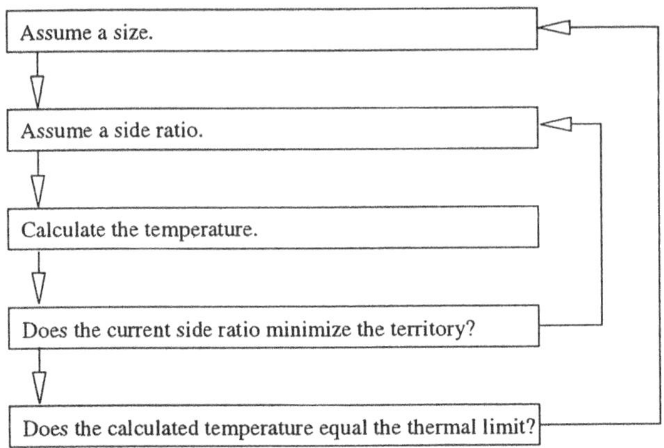

Figure 7. Iterative calculation procedure for thermal territory calculations.

The part of the circuit board that is below the component always requires a special consideration. The main reason for this is that it is cooled only from one side, whereas the rest of the circuit board is usually cooled from two sides. One way to get around this problem is to include the circuit board below the component in the thermal resistance model for the component. The thermal resistance for the thermal territory will then only represent the part of the thermal territory that is located outside the component. The same principle can be used for component models that have more than one interface to the circuit board.

4 Calculation Methods

A thermal territory can be seen as a small circuit board with a single component. All procedures that can be used to calculate temperatures on circuit boards can, therefore, also be used to calculate thermal territories. The main difference is that thermal territory calculations are made to determine the size of the circuit board, and not its temperature.

A thermal territory is defined as the smallest rectangular area of a circuit board that a component requires for its cooling. Two parameters, the area and the side ratio of the rectangle, are therefore needed to specify a thermal territory. Figure 7 shows a principle for an iterative calculation process that in its core uses a temperature calculation procedure. Two embedded iteration loops are needed, which makes the calculation very slow.

To achieve a reasonable calculation time, it is evident that iterative calculations of the type shown in figure 7 must be based on a very fast temperature calculation method. Numerical methods, such as the finite difference or the finite element method, are in this

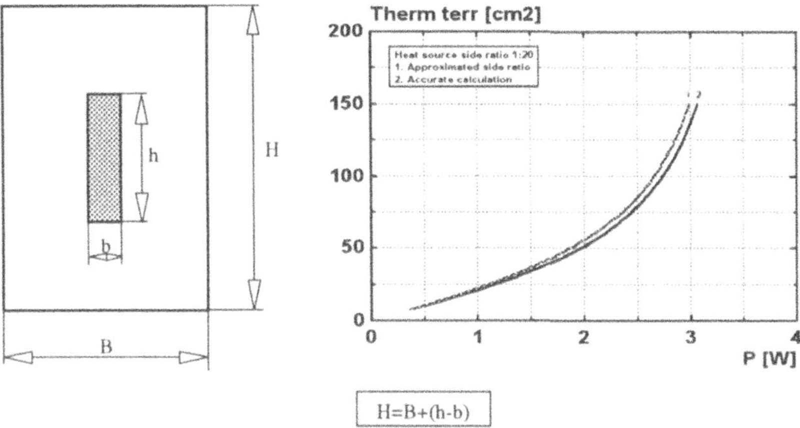

Figure 8. Impact of the side ratio approximation in the formula above.

context inconveniently slow.

Temperature calculation methods based on analytical solutions to the Fourier equation is an alternative. The analytical solution to the Fourier equation is for a rectilinear coordinate system based on a two embedded Fourier series and for a polar coordinate system on Bessler functions. Both these solutions work reasonably well for thermal territories of moderate size. For very small or very large thermal territories, however, there are problems: the Fourier series solution converges very slowly for large territories. The Bessler solution has the same problem for small territories. In both cases, there is also a problem for thermal territories that are smaller than the component. Another issue that must be addressed, is how to interface to the component.

One way to fasten the calculation time considerably, is to use an approximate approach for the side ratio calculation. For small territories the optimum side ratio can be expected to be the same as for the component. For large territories the optimum side ratio can be expected to approach 1.0. A very simple correlation that fulfils both of these tendencies and also results in a small error, is shown in figure 8.

Analytical solutions can provide a reasonably fast and accurate way to deal with the calculation problem. There are, however, cases when simpler and faster methods would be an advantage. An example of such a method is the circular approximation method, shown in figure 9. Although it may seem to be a large step to approximate a rectangle with a circle, this has with some success been tried on similar problems. When making this approximation there are two distinct possibilities: to give the circle either the same area or the same circumference as the rectangle. The former alternative results in thermal

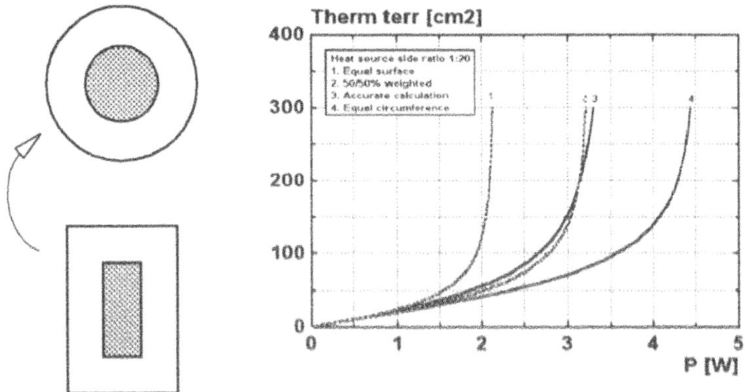

Figure 9. Circular approximations for a flat heat source.

territories that are too large and the second alternative results in thermal territories that are too small. A 50/50% weighted average between these two extremes is, however, a surprisingly good approximation.

The circular approximation method has the advantage that it is fast and can be addressed both numerically and analytically. The main disadvantage is that it is very difficult to find a good solution to the component interface problem. For cases where the thermal territory is much larger than the component, this problem is, however, of secondary importance.

From the discussion above it can be concluded that there is no perfect method for calculating thermal territories. The size of the thermal territory and the component interface problem are the major factors on which the choice of calculation method should be based.

5 References

1. Ottavy N. Méthod de superposition de maillages en éléments finis. Thèse no 493. Universtité de Poitier, ENSMA, pp. 802-803(1989).

2. Malhammar A. Heat dissipation limits for components cooled by the PCB surface, IEPS 1991, pp. 304-311.

THERMAL IMPEDANCE EVALUATION FOR INDUSTRIAL POWER PROFILES

An experimental-analytical approach

VALTER MOTTA CLAUDIO M. VILLA TIAO ZHOU
Sgs-Thomson Microelectronics
Agrate B.za Italy - Carrollton Texas USA

1. Introduction

Thermal behaviour of integrated circuit is increasingly important as the total power increases.
In some applications transient thermal behaviour is important.

This is the case of multiple power pulses, typically in the motor driving applications, as in the case of disk drives in the start-up phase.

In the industry the junction temperature in steady state condition can be estimated by measuring the temperature of the package case. The same approach is not practical for transient operation as the thermal gradient from the active device to the package surface can be very high and strongly dependent on package type and environmental conditions.

Often direct junction temperature measurement is not possible.

How to estimate the junction thermal profile quickly when the power dissipation on chip is a function of time? To what extend can measurement and simulation activities be optimised?

A combined experimental and mathematical approach was developed in order to evaluate the junction temperature profile.
Starting from simple experimental measurement, by means of a general mathematical model, a generic power pulse can be studied.

Hereafter an example of this semi-empirical method is presented. It is applied to a new package developed in SGS-Thomson: a TQFP (10x10x1.4 mm) with an internal drop in copper slug intended to improve the thermal dissipation.

E. Beyne et al. (eds.), Thermal Management of Electronic Systems II, 187-194.

2. Experimental characterisation

By first, thermal impedance of Pw TQFP was measured by means of the device P658, developed by SGS-Thomson, according to SEMI and JEDEC guidelines.
This test chip has two large resistors as heating and a calibrated diode as temperature sensor.

Thermal impedance measurement is performed by recording the voltage drop on the diode along one single pulse; the length t_o of the pulse increases from 10 milliseconds to 1000 seconds.
With a calibrated factor K = 0.49 mV / °C the thermal impedance is obtained. The Figure 1 shows the increase of the junction temperature versus time from 1 millisecond to 1000 seconds.
Resolution of the data is better than 5%.

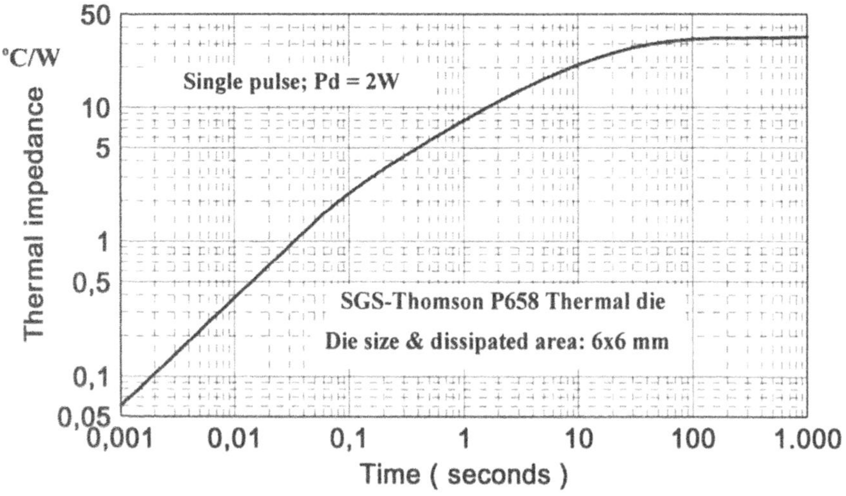

Figure 1. Thermal impedance [°C/W] for Pw TQFP mounded on a typical board

The range of thermal impedance and time is very big: the time varies 6 order of magnitude, the impedance value varies three orders of magnitude from the beginning to the end of the curve.

For this reason, a double logarithmic scale is more convenient in data presentation.

3. Polynomial interpolation of the experimental values

The typical time range for this kind of application is below one minute.
During such short time the air convection is secondary effect and can be partially ignored so the thermal impedance reported above can be used for any reasonable power level.

In order to simplify the data processing, the thermal profile obtained in this way can be interpolated mathematically with a polynomial interpolation; the estimation of thermal impedance versus the time can be easily obtained.

On the basis of our experience, the polynomial able to fit the experimental curve for each package type is a seven degree's equation:

$$Y = a_0 + a_1 \cdot x + a_2 \cdot x^2 + a_3 \cdot x^3 + a_4 \cdot x^4 + a_5 \cdot x^5 + a_6 \cdot x^6 + a_7 \cdot x^7 \qquad (1)$$

Where a_0 ... a_7 are the coefficient.
As the curve is in a log-log scale, for the polynomial interpolation for input and output can be use a logarithmic form.
Therefore:

$$x = \log T \qquad \text{where } T \text{ [milliseconds] is time} \qquad (2)$$

$$Y = \log Z \qquad \text{where } Z \text{ [°C/W] is the thermal impedance} \qquad (3)$$

As the coefficients are obtained by interpolation, their result agrees with the experimental data, with a level of confidence higher than 5%; due to the use of a logarithmic scale, this very high resolution can be guarantee only for times higher than 10 milliseconds and lower than 500 seconds.

TABLE 1. Polynomial coefficient for the thermal impedance of the Pw TQFP .

Coefficient	Power TQFP 10x10x1.4
a_0	-1.142382077
a_1	0.158208952
a_2	1.116577978
a_3	-0.779410215
a_4	0.260900561
a_5	-4.6158007396E-02
a_6	4.0314134895E-03
a_7	-1.3356200461E-04

Practically all the pulses in the motor control are longer than 100 ms and shorter of few minutes, in case of repeated start up and down, this means that all the problems are in a reasonable time range.

When the time is long - about minutes - board, ambient, environmental and system configuration affect the value of thermal impedance in some extends; so in this case a specific analysis is requested.
For the case of Pw 10x10x1.4 mm TQFP the coefficients a_j are reported in the Table 1. These data are obtained from the experimental point using a software developed in SGS-Thomson.

For example to determine the temperature increase in the Pw 10x10x1.4 mm TQFP after $4*10^3$ milliseconds due to a power pulse of 5.2 Watts, the polynomial described above is applied in the following way:

$$x = \log(T) = \log(4 \cdot 10^3) = 3.602 \tag{4}$$

By substituting this value and the coefficients a_0 ... a_7 in (1) it gives Y = 1.17486 ; this means that the thermal impedance at 4 second is (3):

$$Z = \text{Invlog}(Y) = 14.96 \ \text{°C/W}. \tag{5}$$

At the end, if the total power dissipated is 5.2 Watts, the temperature increase can be calculated by

$$\Delta T = Z \cdot P_d = 14.96 \ \text{°C/W} \cdot 5.2 \ \text{W} = 77.8 \ \text{°C} \tag{6}$$

The result obtained with a direct measurement gives a thermal impedance of 15.3 °C/W and a temperature increase of 79.5 °C.

The difference between the measured and the extrapolated data (1.7 °C) is about 2% which is in the range of the experimental error.
In the same way, point by point, it is possible to reproduce a complete power step or a complex power profile.

4. Thermal profile estimation by superposition

The goal is to determine the thermal behaviour in case of a power profile consists of multiple pulses as described in the Figure 2.

The following method can be used to estimate the maximum temperature reached.
Clearly the hottest point takes place at the end of the last peak.
It is possible, as a rough approximation, to split the profile in two components: the first one is the average power obtained by multiplying the power peak (equal to 2.2 W) by

the Duty Cycle (equal to 60%).
The second component is the remained power of the last peak (the 40% of the total power, equal to the missing part of the last peak).

For a first and simple estimation of the peak temperature, we can imagine to add this power peak at the end of the average power. Even if this approximation is not completely correct, in fact it can overestimate the peak temperature, easily and quickly we can obtain, within an acceptable error, a first indication of the max. temperature.
In this way, the maximum temperature increasing, at the end of the profile (19 seconds), will be the sum of the temperature after 19 seconds due to the average power, plus the temperature due to 1.5 seconds of the remained peak.

Power profile

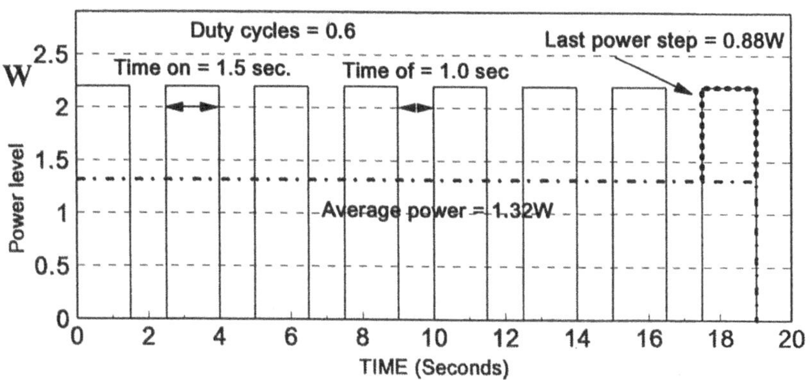

Figure 2 . Power profile under evaluation.

The elapsed time and power levels of the two components are:

T_1 = 19 seconds $\quad$ P_{avg} = 60% of total power = 1.32 Watts
T_2 = 1.5 seconds $\quad$ P_{peak} = 40% of total power = 0.88 Watts

The temperature increase is obtained by:

$$\Delta T_{max} = Z_{avg}(T_1) \cdot P_{avg} + Z_{peak}(T_2) \cdot P_{peak} \tag{7}$$

Namely:

$$\Delta T_{max} = 25.33\ ^{\circ}\mathrm{C/W} \cdot 1.32\ \mathrm{W} + 9.68\ ^{\circ}\mathrm{C/W} \cdot 0.88\ \mathrm{W} = 41.9\ ^{\circ}\mathrm{C} \tag{8}$$

To verify the accuracy of these assumptions, measurement of the thermal profile was performed for this power profile.

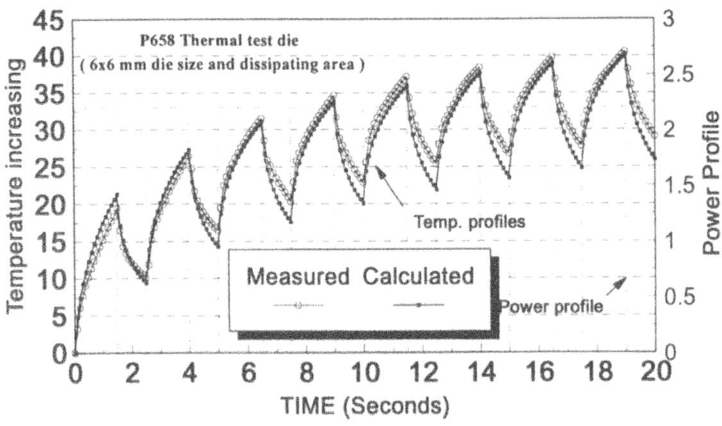

Figure 3. Measured and calculated thermal behaviour for the given power profile

As shown in Figure 3, the maximum temperature increasing measured is 40.6 °C. So the difference from the calculated data is less than 1.5 °C (about the 3% of the total value). In this case the accuracy is very good; usually with this method the accuracy is in the range of 5 - 10 % good enough for the majority of the applications.

As also shown in the previous figure it is possible also to follow the whole temperature profile using a well-defined sequence of equations; in this case the approximations can be also reduced and the accuracy will increase.

The profile above used can be expressed as a sum of 8 power steps starting at 0, 2.5, 5, 7.5, 10, 12.5, 15 and 17.5 seconds, with a power level of 2.2 W, and an other set of 8 power steps starting at 1.5, 4, 6.5, 9, 11.5, 14, 16.5 and 19 seconds, with a negative power level of -2.2 W.
Practically it is:

$$\Delta Temp._{(T)} = P_d \cdot [\, f_{(T)} - f_{(T-1.5)} + f_{(T-2.5)} - f_{(T-4)} + f_{(T-5)} - f_{(T-6.5)} \cdots\cdots f_{(T-17.5)} - f_{(T-19)} \,] \qquad (9)$$

Were: $f(T-T_i)$ are function obtained from (1) (2) and (3) with a time shift equal to T_i.

By applying the formula (9) it is possible to follow the temperature profile in the time with good accuracy.
The maximum error, comparing the measured profile and the estimated temperature, in this case occurs in the coolest peaks where the estimation gives an underevaluation of about 10% (4 °C in this case).

The error for the hottest points is under one degree (the 2% of the total temperature profile.

It should be noticed that when estimating the maximum temperature reached for mathematical approach the errors induced are different.

For this reason now the maximum temperature calculated is equal to 40.2 °C instead of 41.9 °C as obtained with the first fast computation, but this error is always in an acceptable range, under the 5%.

5. Industrial thermal profile estimation.

A general power profile can be described using a sum of power steps with different amplitude and duration.
The consequent approximation depends on the number of steps used; higher numbers of steps is required to reach better accuracy.

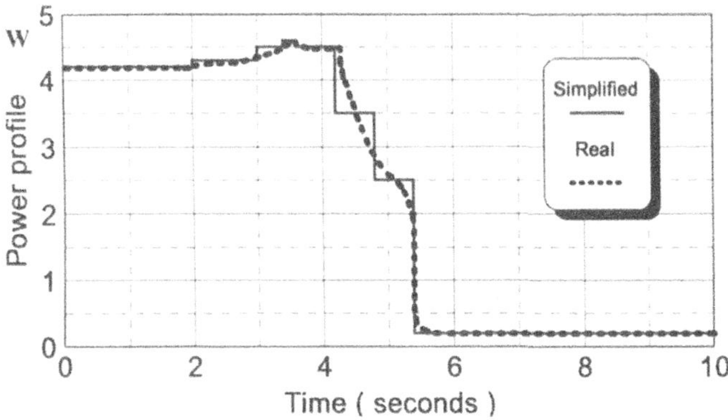

Figure 4. Example of actual and simplified power profile

In this way a general power profile can be replaced a summation of different steps and the thermal response of each package can be obtained by a linear combination of the responses of a simple power step.
This means that the thermal profile generated by a general power profile can be interpolated mathematically with a polynomial expression:

$$\Delta \text{Temp.}_{(T)} = \Sigma_i \ P_i \cdot f_{(T-T_i)} \qquad (10)$$

where P_i is the power level (positive or negative) and $f_{(T-Ti)}$ is a function equal to zero for $T < T_i$ and is calculated from equation (1) for $T > T_i$ with $x = \log(T-T_i)$.

Figure 4 shows an example of approximating a generic power profile into a step-made profile. In figure 5 the temperature that is the result of this approximation is plotted.
An experimental control of the calculated profile was done and it presents less than 10% difference from the measurements.

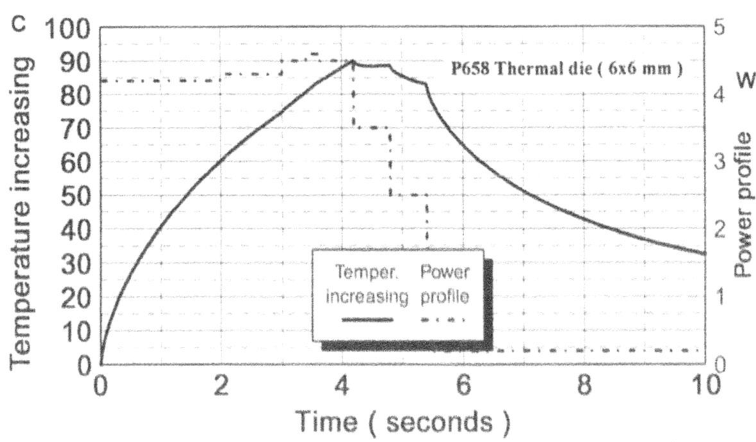

Figure 5. Calculated temperature profile

6. Limits and conclusions

As the thermal impedance is strongly dependent on the dissipating area, mainly for short time range, the prediction of junction temperature is effective only if the real dissipating area is similar to the dissipating area of the thermal test dice.

If the dissipating area on the commercial device is comparable to the test pattern, an effective predictive method to determine the junction temperature profile was presented.

In this way, without any measurement, it is possible to foresee the peak but also the complete temperature profile without performing any direct measurement.

THERMAL CHARACTERISATION AND MODELLING OF MULTICHIP MODULES USING STANDARD ELECTRONIC PACKAGES

Areski BELACHE, Thierry GAUTIER, Jean Claude BOISDE, Nicolas LEVEAU and Guy PAULET
THOMSON-CSF, Department of Technology
BP 56, L'Orée de Corbeville
91401 ORSAY, FRANCE
Email : areski.belache@thomson.sctf.fr

1. Introduction

The increase in speed and the reduction in size and weight of electronic components have resulted in the development of Multi-Chip-Module (MCM) technology in recent years. However, the increasing power density in such modules led to the need for thermal models of such electronic components. The Department of Technology has developed a Thermal-Model-Generator (TMG) for MCMs to provide thermal models. The numerical method implemented in the MCM-TMG software is based on physical representations of thermal flows and calculation of the associated thermal resistance.

2. Topology of the model

Based on previous work in the SCTF[1, 2] since 1990, a methodology and a numerical method initially designed to build thermal models for standard single-chip packages, were extended to enable the thermal modelling of multi-chip substrates using plastic (PQFP) and ceramic (PGA) packages.

TMG (Thermal Model Generator) is based on physical representation of thermal flow transmission from the silicon die junction to the ambient. Then, a thermal resistance can be associated with each significant heat flow path. The model topology used for plastic packages is shown in figure n°1.

The first step of the study was to choose representative multichip module technologies which have been subjected to thermal testing in our laboratory for different cooling modes. In a second step, thermal testing completed by additional 3D CFD simulations were carried out. In a final step, the development of our TMG software has been implemented to provide thermal model used in simulation cards.

3. Test vehicle description and thermal testing

Typically, a MCM accommodates a number of chips and interconnects them through several layers of conductive paths. The MCM used for this study contains four chips. Analysis of two types of interconnection technologies was carried out during this study to build the model generator TMG-MCM.

E. Beyne et al. (eds.), Thermal Management of Electronic Systems II, 195-201.

The technologies implemented for packaged MCMs are those used in production for reasons of reproducibility. The packaging of the MCMs is carried out by SGS-Thomson at Saint Egrève for the Plastic Quad Flat Pack (PQFP with 160 ferro-nickel pins, pitch 0.65 mm) and by Thomson-CSF/TCS also at Saint Egrève for the Pin Grid Arrays (PGA co-fired Al_2O_3 with 224 pinouts, pitch 2.54 mm). Each test vehicle is mounted on the printed circuit board.

The moulded plastic package is produced by mounting the alumina interconnection substrate, supporting the bonded dies on the lead frame. This technology is referred to as the MCM-D, the "D" indicating "Deposit material" (figure n° 2).

The silicon dies are housed and mounted on the bottom of the alumina cavity-up before lid sealing. This is referred to as MCM-C (Ceramic) technology.

The dissipating source (P638 thermal test die from SGS-Thomson) housed in each package consists of 2 power transistors and a measurement diode positioned between them. After calibration, diodes measure the junction temperature for different test configurations (dissipating or non-dissipating sources).

Experiments were set up to measure to measure the characteristics of the two types of test vehicles for different cooling modes (natural convection in the vertical position and forced convection in a wind tunnel) see figure n°3.

For each package, additional 3D simulations were carried out by means of the FLOTHERM software utility (FLOMERICS). The aims of the CFD simulations, validated by the test are :

- to analyse the specific thermal behaviour of the packaged multichip modules (coupling effects) ;
- to establish the values of the different thermal flows inside and outside the multi-chip module.

Figure n° 4 shows an example of isothermal curves and velocity vectors for the packaged MCM-C type PGA in natural convection.

Comparisons of junction to ambient temperatures, simulated on the 3D model, with those measured during test reveal in the worse case that the hot point temperatures of the MCMs are overestimated by 20 % in CFD simulations (figure n°5). This difference is substantially the same for the two types of MCMs. This difference is small considering that the values of the thermal conductivities of materials used in the simulations are the values indicated by the MCM manufacturers.

4. General Multi-Chip Module model and assumptions

Analysis of the thermal behaviour results obtained during experiments and simulations carried on MCMs under a wide range of cooling conditions, allowed us to avoid creating a new thermal model topology to describe multi-chip technology different from mono-chip packages. Moreover, the model had to be compatible with available thermal simulation cards.

A reference isotherm on the grid wedge is the most interesting regarding to the analytical equations and the tolerance of results - better than 20 % -. Each source is coupled with the other sources through an equivalent thermal resistance.

Then, we constructed the new multichip model from the mono-chip package models. The die to die attach resistance has completely disappeared and a thermal resistance network of the interconnection substrate housing the several dies has been added (figure n° 1).

Thermal resistance networks for MCM interconnection substrates are calculated using the superposition principle already published [3]. This method has been applied by Aghazadeh et al. [4] to obtain the thermal resistance for 68 lead ceramic pin grid array (CPGA) housing two chips.

5. Thermal model generator for MCM

Finally, we compared the results from the MCM-TMG generator model, with results obtained from experimental and numerical procedures. The comparison shows that the lowest accuracy is obtained with very small chips (few mm side) dissipating high power compared with the other heat sources (as noticed with simulations). This is because the very local heat transfer coefficient is difficult to take into account in our analytical method. But, usually, we consider that a variation of 20 % between predicted and test results the reality can be obtained.

6. MCM-TMG software features

The software can be implemented on a standard personal computer. MCM-TMG generates thermal models in less than 1 hour (µP 80386). A gateway has been implemented in order to simulate the chip behaviour on an electrical simulator.

7. Conclusions

The first part of the study consisted in producing test specimens that are representative of the plastic and ceramic MCM substrate technologies. The next step was the laboratory testing using the MCM-PQFP-160 and the MCM-PGA-224 packaging in a controlled cooling environment.

In natural convection and forced convection, differences between test and simulation results are overestimated by 20%, at most for the junction to ambient temperatures measured at the dissipating sources (hot points at the MCM level). These results constitute the input data for the next step with a view to producing the MCM-TMG for our multi-chip module test vehicles.

Final results from thermal GMT-MCM software are 20 percent superior to experiments and CFD simulations. This accuracy is, in most cases, sufficient and gives a good approximation of cooling requirements in the practical design of electronic systems.

References

[1] Th. Gautier, "Construction and Validation of Thermal Models of Packages", 7th IEEE SEMI-THERM Symposium, Phoenix AZ, Feb. 12-14 Feb., 1991.

[2] Th. Gautier et al, "Thermal Model Generator (TMG)", I-THERM III Symposium, Austin, TX, Feb. 5-8 Feb., 1992.

[3] Arvizu et al, "The use of superposition in calculation cooling requirements for circuit board mounted electronic components", Proceeding Electronic Components Conference, 32nd, 1982.

[4] Aghazadeh et al, "Thermal Performance of Multi-chip Modules", 7th International Electronics Packaging Conference, Boston, MA, 1987.

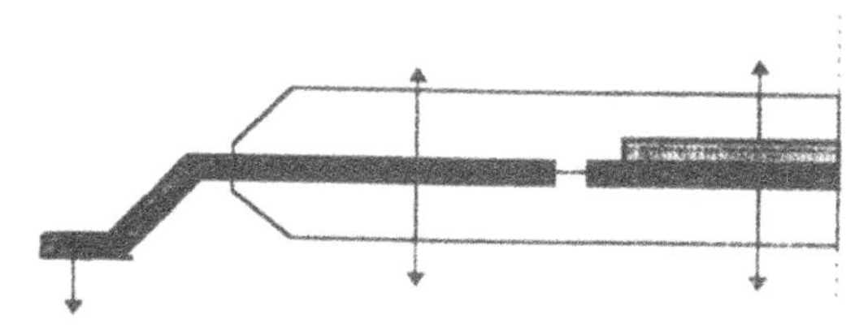

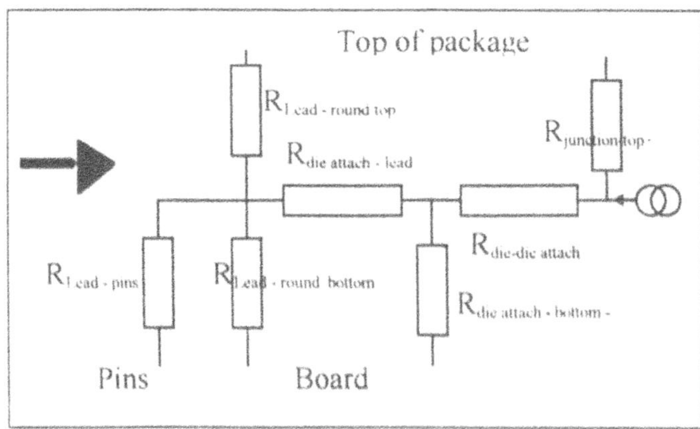

Figure 1. Physical model for plastic packages

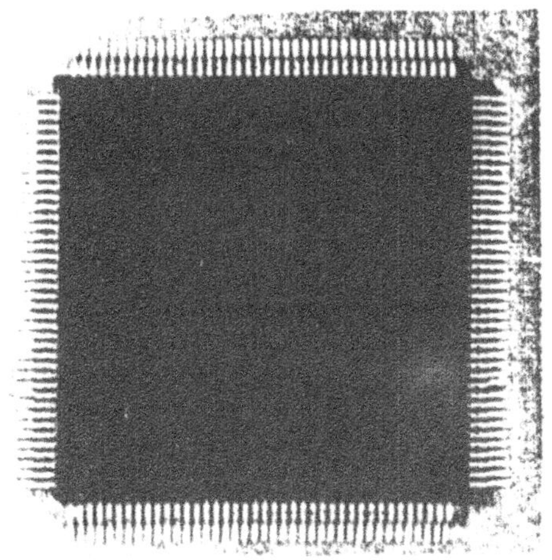

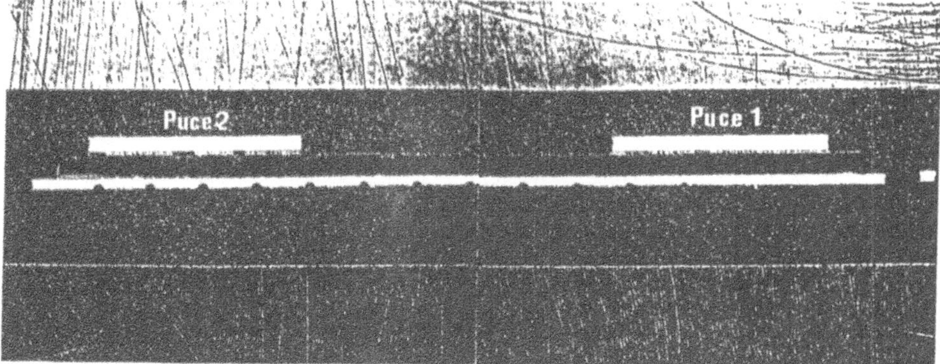

Figure 2. SEM Close-up view of micro-section showing the internal structure of the plastic packaged MCM-D type PQFP-160 leads with its alumina substrate.

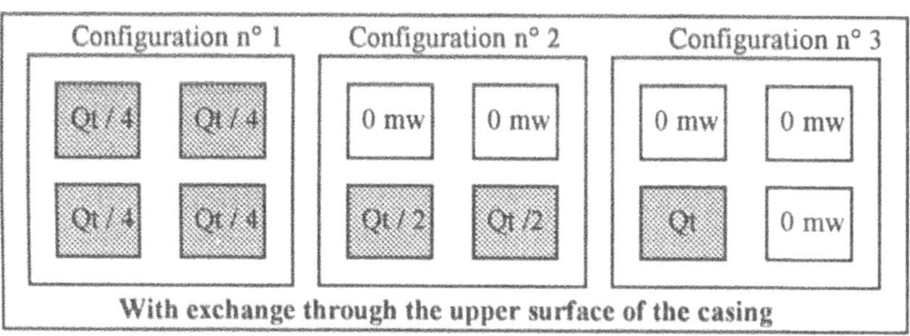

Figure 3. Test configurations (dissipating sources shaded). The total dissipated power of the MCM-PQFP is constant and equal to 0.8 watt for all configurations. This is raised to 1.6 watts for the MCM-PGA technology (The purpose here is to establish the values of inter-chip thermal flows):

- *4 dissipating sources (4 * Qtotal / 4 : configuration 1) ;*
- *2 dissipating sources (2 * Qtotal / 2 : configuration 2) ;*
- *1 dissipating source (1 * Qtotal / 1 : configuration 3).*

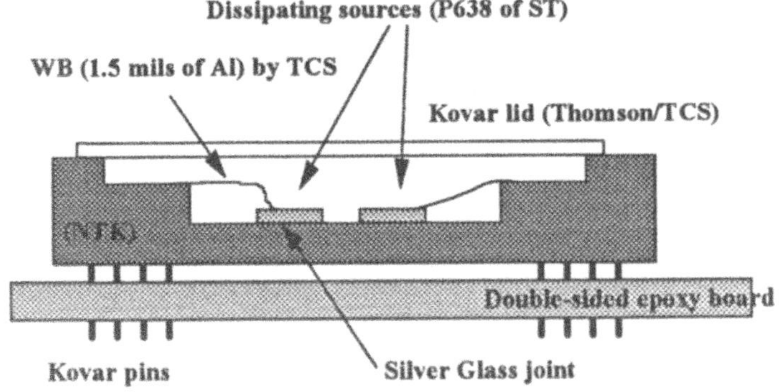

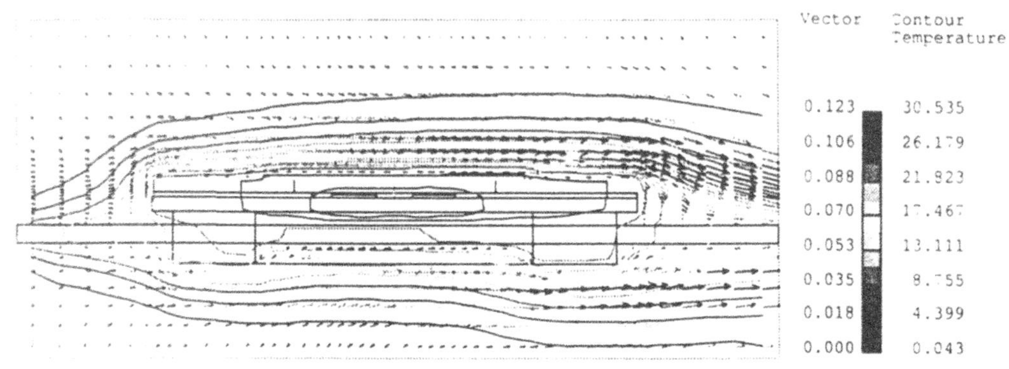

Figure 4. Description of the technological building blocks used in the production of the ceramic packaged MCM (The outer dimensions of the ceramic package are 44 x 44 mm) and Isothermal curves and velocity vectors in a cross-section of the component passing through the centre of the package . Gravity is defined in the Oy direction (natural convection for the packaged MCM-C type PGA package).

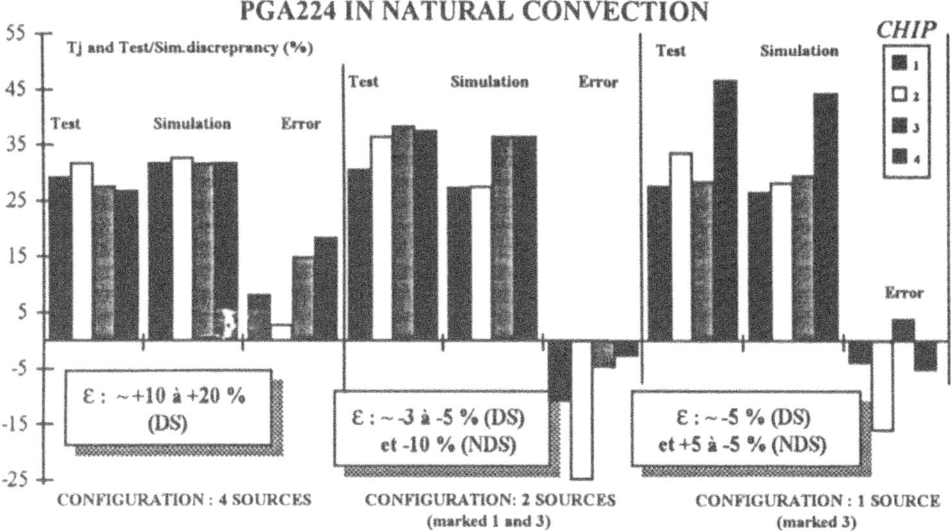

Figure 5. Junction to ambient temperatures (dissipating or non-dissipating sources) and test/simulation differences in natural convection of the ceramic packaged MCM type PGA 224 for the 3 dissipated configurations (Qtotal = 1.6 W). The board is in the vertical position. DS means dissipating sources and NDS means non-dissipating sources.

MEASUREMENT OF PRACTICAL THERMAL RESISTANCE VALUES FOR MULTI CAVITIES POWER HYBRIDS (MCPH) IN A VACUUM ENVIRONMENT

D.PETITJEAN, P.LYBAERT
Faculté Polytechnique de Mons, Service de Thermique,
rue de l'Epargne, 56, B-7000 Mons (Belgium)
E.FILIPPI, A.STURBOIS
Alcatel Bell SDT, Structural/Thermal Analysis Department,
rue Chapelle Beaussart, 101, B-6032 Mont-sur-Marchienne (Belgium)

1. Introduction

Multi cavities power hybrids (MCPH, see figure 1) are extensively used in power systems for space applications. In these multichip modules, the dies are glued on a ceramic substrate which is attached to an aluminium or magnesium structure which acts as a heat sink. This structure is screwed to the system box.

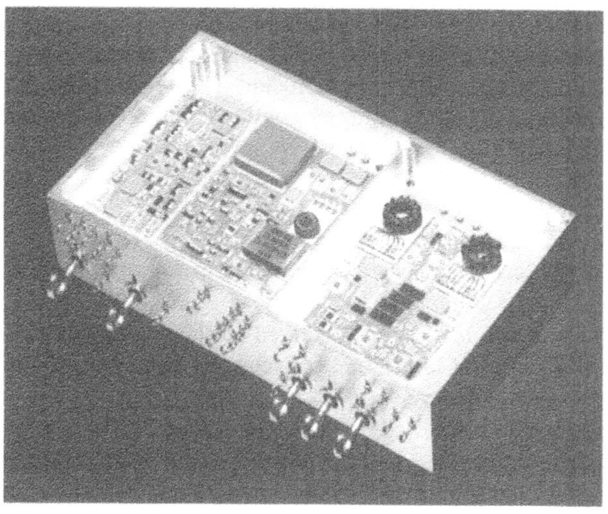

Figure 1. Multi cavities power hybrid

E. Beyne et al. (eds.), Thermal Management of Electronic Systems II, 203-212.

In vacuum environment, heat dissipated by the dies is transferred to the chassis by conduction. Thermal analysis of MCPH based power systems requires the knowledge of the thermal conductance across the contact between the hybrid base plate and the supporting chassis.

Contact heat transfer depends on many parameters, e.g. contact area, surface finish (surface roughness and flatness), contact pressure, thermal conductivity of the materials, surface treatment (coating), ambient pressure, ... Furthermore, contact conductances are often improved by using interstitial materials. The effects of these factors have been investigated by a lot of researchers and are well documented in the literature (see, for instance, Madhusudana and Fletcher, 1984 and Fletcher, 1988). However, most of the results are based on experiments performed on « ideal » contacts, in which a uniform contact pressure is obtained by pressing together small diameter - i.e. less than 25 mm - cylindrical samples. Hence, these results cannot be used to predict thermal resistance values for real contact applications, involving large contact areas and non-uniform contact pressures.

This work is devoted to the experimental determination of practical thermal conductance values for MCPH's. The effects of the following factors have been investigated: coating of the contacts surfaces, number of screws and tightening torque, use of different interstitial materials.

2. Experimental Procedure

2.1. VACUUM TEST FACILITY

The thermal resistances have been measured in a small vacuum test facility. The test chamber has an internal diameter of 300 mm and a height of about 350 mm. High vacuum conditions ($<10^{-5}$ mm Hg) can be obtained by a two-stage pumping system, which uses a mechanical pump and a turbomolecular pump. The system contains a circular copper cold plate of 200 mm diameter. The temperature of the cold plate is controlled by a thermostat. Using water as cooling fluid, cold plate temperature can be fixed between 10 and 90°C.

2.2. TEST SPECIMENS

The geometry of the test specimens is represented in figure 2. Each specimen consists of two parts :

- the support plate, 8 mm thick, which is fixed on the cooling plate of the vacuum test facility. A PAPYEX sheet is inserted between the surfaces in order to improve the contact with the cold plate.
- the hybrid case, which is screwed on the support plate by 4 or 6 screws.

Heat dissipation is provided by five screen printed resistances (about 8 Ω each), fixed on beryllium oxyde supports which are glued on the hybrid base plate.

As the interface area is relatively large (about 36 cm^2), the multipoint temperature averaging technique (Eid and Antonetti, 1987) has been used to measure the temperature difference across the interface. Five small diameter (0.2 mm) thermocouples are inserted into 0.6 mm holes arranged on each side of the contact surface, at 1 mm distance of the interface, straight below the resistances.

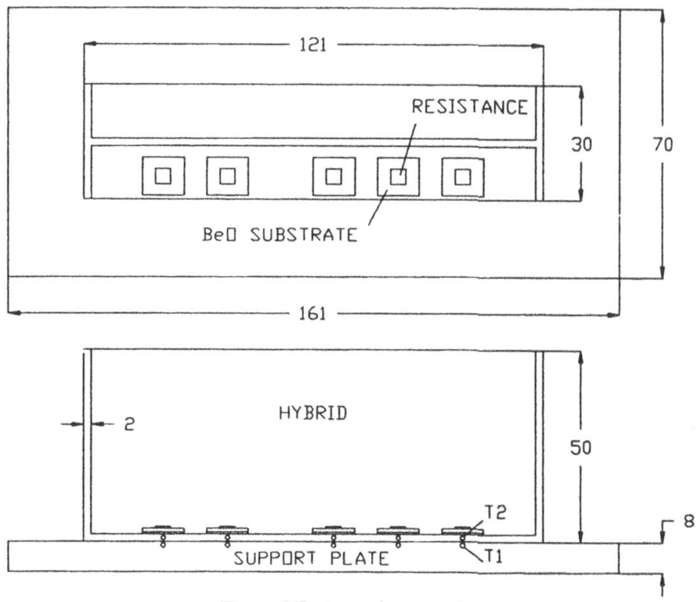

Figure 2 Test samples geometry

2.3. MEASUREMENT METHOD

In the original multipoint temperature averaging technique of Eid and Antonetti (1987), an overall contact resistance is calculated by dividing the measured average temperature difference by the heat flow rate. The mean contact resistance is then obtained by substracting from the overall resistance the thermal resistances due to conduction heat transfer between the contact surface and the plane where the temperatures are measured. This method can be used for uniform heating and cooling conditions

In our system, heating is far from being uniform. The mean contact conductance is then obtained by comparing the measured average temperature difference to the temperature difference computed by a 3-D finite element (ANSYS) model of the actual test geometry. The finite element grid is represented in figure 3. The total number of nodes is 10,035, with some nodes coinciding with the measurement points.

In the model, the interface is modelled by a uniform thermal conductance h_c . Simulations are then performed for different values of the conductance, ranging from 50 to 10000 $W.m^{-2}.K^{-1}$ and the resulting average temperature differences accross the interface (at the locations of the thermocouples) are computed. This procedure should be repeated for each combination of materials in contact. However, as the distance between the thermocouples and the contact surface is small and the thermal conductivities of the materials are high, a single « calibration » curve can be used for all the tests, the error on the computed temperature differences being less than 4 percent. This curve is given in figure 4.

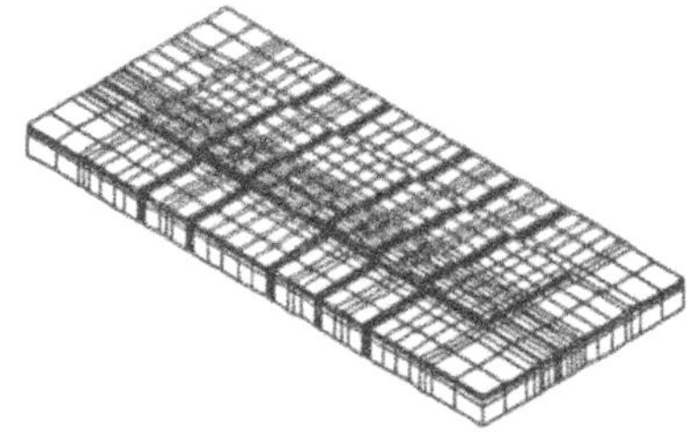

Figure 3. Finite element model of the test specimen

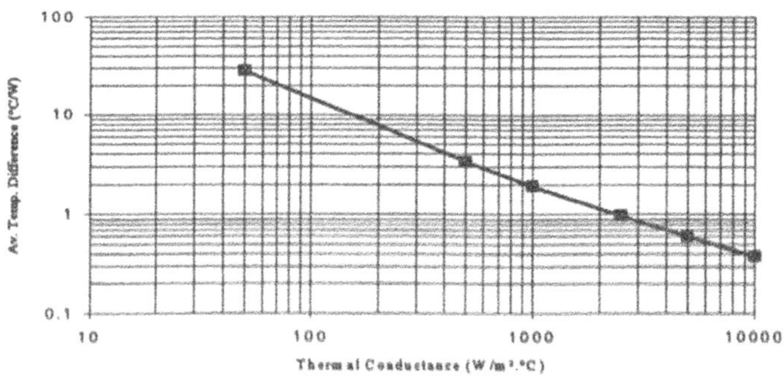

Figure 4. Determination of the mean contact conductance - Calibration curve

The procedure to find the conductance values is then the following:

(1) from the measured temperatures $T_{1,i}$ and $T_{2,i}$, , and the power dissipated by each resistance P_R , compute the measured average temperature difference by

$$\Delta T_{R,av} = \frac{\sum_{i=1}^{5}(T_{2,i} - T_{1,i})}{5.P_R} \tag{1}$$

(2) using the calibration curve, compute the conductance value.

2.4. UNCERTAINTY ANALYSIS

The error associated with the conductance measurement mainly results from the following errors:

(1) the error on the measured temperature difference (about 0.1 °C for each temperature measurement i.e. 0.2 °C on the difference),
(2) the error on the power dissipated (about 3 %),
(3) the radiation heat losses of the specimen,
(4) the error resulting from the use of a single calibration curve.

The most significant contributions are due to the temperature measurement error and the radiation heat losses. Both contributions depend on the value of the thermal conductance. The measured temperature differences are higher for lower conductance values, and the relative error on the average temperature drop is lower. But, for a fixed cold plate temperature, the temperature level of the specimen and the radiation heat losses are higher. The maximum error on the conductance is estimated to be about 10 percent for a conductance value of about 100 $W.m^{-2}.K^{-i}$ and 15 percent for conductances of 10,000 $W.m^{-2}.K^{-1}$.

3. Experimental Conditions

The following factors have been investigated:

- *material and coating of the contact surfaces :*
  1. MCPH material: the most usual material (reference case) is gold plated Al6061 (aluminium, k=152 $W.m^{-1}.K^{-1}$), tests have also been performed with uncoated and chromated Al6061.
  2. support plate material: the usual material (reference case) is chromated Al5754 (aluminium, k=132 $W.m^{-1}.K^{-1}$) but when the weight of the system is a critical concern, as for high orbit applications, AZ31b magnesium (k=78 $W.m^{-1}.K^{-1}$), with DOW 17 (galvanic protection) or DOW 19 (corrosion protection) surface treatments, is often used. Gold plated Al6061 has also been tested.

  Surface roughness of the contact surfaces is less than 1.6 μm, except when DOW 17 treatment is applied. In this case, surface roughness is about 9 μm.
- *number of screws and tightening torque :* the MCPH can be bolted to the support plate by 4 (reference case) or 6 stainless steel screws. The diameter of the screws is 3 mm (M3 screws). Three values of the tightening torque have been selected, 4.8 kg.cm, 8 kg.cm and 11 kg.cm.
- *interstitial materials :* interstitial materials (fillers) are used to improve the contact conductance. According to the application, electrical conducors are allowed or prohibited. Some of the most usual space qualified materials have been tested. Their properties are summarized in table 1 and 2, respectively for electrical conductors and insulators. Most of the fillers are available in the form of thin polymer or metal foils of fixed thickness t. The other fillers (MAPSIL and RTVS 691) are pasty materials which are applied on the MCPH. The contact surfaces are then pressed together by tightening the screws with the desired torque and the excess material is removed. For insulating contacts, we also tested anodised aluminium foils (electrically insulating) sandwiched between two conducting fillers.

TABLE 1. Properties of the interstitial materials - Electrical conductors

Name	*Description*	*t (μm)*	*k (W.m^{-1}.K^{-1})*
SIGRAFLEX	stratified graphite foil	250	5
PAPYEX N	stratified graphite foil	200	4
MAPSIL	silicone grease	-	0.5
RTVS 691	silicone compound - pealable	-	0.23
In	indium foil	150	82
Al	aluminium foil	15 or 70	204
Sn	tin foil	50	65
THERMSTRATE 1100	Al/paraffin sandwich foil	51 + 2*6	?

TABLE 2. Properties of the interstitial materials - Electrical insulators

Name	*Description*	*t (μm)*	*k (W.m^{-1}.K^{-1})*
CHO-TERM R1671	glass fiber reinforced silicone	380	4
CHO-TERM T500	glass fiber reinforced silicone	250	4
ISOSTRATE	kapton/paraffin sandwich foil	51 + 2*6	?
SIL-PAD 400	glass fiber reinforced silicone	230	0.9
POLY-PAD K10	kapton/polyester composite	120	1.3
TRANSIL 80-300	glass fiber reinforced silicone	230	2.22
TRANSIL 83-300	reinforced polyimide	51	2.22
Anodized Al	anodized (2 x 40 μm) Al foil	130	?

4. Results and Discussion

The measured thermal contact conductances are given in tables 3-6. The values range from a few hundreds W.m^{-2}.K^{-1}, when the surfaces are in dry contact, up to 10,000 to 15,000 W.m^{-2}.K^{-1}, with gold plated aluminium combined with the best interstitial materials.

4.1. DRY CONTACT

The values obtained without filler are given in table 3. They range from 100 to about 1500 W.m^{-2}.K^{-1}, depending on the materials and the coating of the surfaces in contact, the number of screws and the tightening torque. For a given hybrid material, higher values are obtained when the support plate material is gold plated aluminium, the lowest values are obtained for AZ31b magnesium with DOW17 surface treatment. The best results are obtained when both surfaces are gold plated aluminium.

Increasing the tightening torque from 4.8 to 11 kg.cm improves the conductance value by up to 50 percent. The effect is more significant when the conductance is low and one of the surface is chromated aluminium or DOW17 treated magnesium. Gold plating seems to give a very hard coating layer which do not deform when the tightening torque (or the contact pressure) is increased.

TABLE 3. Measured conductance values ($W.m^{-2}.K^{-1}$) - Dry contact
Effect of number of screws and tightening torque

Hybrid	*Support*	*4x4.8 kg.cm*	*4x8 kg.cm*	*4x11 kg.cm*	*6x8 kg.cm*
Al6061 gold	Al6061 gold	1005	1125	950	1490
Al6061 gold	Al5754 chrom.	305	420	450	795
Al6061 gold	AZ31b DOW17	100	135	155	250
Al6061 gold	AZ31b DOW19	525	680	600	970
Al6061	Al6061 gold	605	640	655	-
Al6061	Al5754 chrom.	410	495	515	850
Al6061	AZ31b DOW17	185	225	245	310
Al6061 chr.	Al5754 chrom.	340	400	425	695
Al6061 chr.	AZ31b DOW17	180	220	240	310

Using six screws instead of four improves the conductance by a factor of 1.4 to 1.9. This is due to a more uniform contact pressure, particularly in the central part of the interface.

4.2. USE OF INTERSTITIAL MATERIALS

Much higher contact conductances are obtained by using interstitial fillers. The values measured with the different filler materials are compared in table 4. All conductance values given in this table have been obtained with four screws and a tightening torque of 8 kg.cm.

TABLE 4. Measured conductance values ($W.m^{-2}.K^{-1}$) - Interstitial materials - Screws : 4 x 8 kg.cm

Filler	*Al6061 gold Al6061 gold*	*Al6061 gold Al5754 chr.*	*Al6061 gold AZ DOW17*	*Al6061 gold AZ DOW19*
Dry contact	1125	420	135	680
SIGRAFLEX	-	4465	2580	-
PAPYEX N	7770	5350	1380	10330
MAPSIL	16095	-	10275	-
In	17980	5475	1990	15150
Al 15 μm	-	575	-	-
Al 70 μm	-	495	-	-
Sn	-	670	-	-
THERMSTRATE	-	1495	-	-
CHOTERM 1671	1995	1760	1305	2405
CHOTERM 500	2330	1945	1195	2595
ISOSTRATE	-	985	-	-
SILPAD	-	855	-	-
POLYPAD	-	450	-	-
TRANSIL 80-300	-	625	-	-
TRANSIL 83-300	-	1050	-	-

For the reference situation (gold plated Al6061/chromated Al5754), the indium foil, PAPYEX and SIGRAFLEX are the best filler materials: they improve the contact conductance by a factor of more than 10 with respect to dry contact. These fillers are also the best ones for the other combinations of materials in contact. MAPSIL also gives very high contact conductances, but this material can migrate under high vacuum conditions. Dismantling the hybrid is also difficult when MAPSIL has been applied on the contact surfaces. Apart from indium, the other metal foils (aluminium and tin) do not lead to significant improvements. The table also shows that the electrical conductors are better than the insulating fillers.

The effects of the number of screws and the tightening torque on the conductance values obtained with the different fillers are given in table 5.

TABLE 5. Measured conductance values ($W.m^{-2}.K^{-1}$) - Hybrid material: Al6061 gold
Effect of support material, interstitial material, number of screws and tightening torque

Support	*Filler*	*4 screws 4.8 kg.cm*	*4 screws 8 kg.cm*	*4 screws 11 kg.cm*	*6 screws 8 kg.cm*
Al6061 gold	CHOTERM T500	2290	2330	2295	2845
	CHOTERM R1671	1915	1995	2030	2490
	PAPYEX	7595	7770	7990	9460
	MAPSIL	15100	16095	16495	-
Al5754 chrom.	CHOTERM T500	1670	1945	2295	2620
	CHOTERM R1671	-	1760	2265	2545
	PAPYEX	-	5350	-	7545
	SIGRAFLEX	-	4465	-	6680
AZ31b DOW17	CHOTERM T500	1115	1195	1225	1755
	CHOTERM R1671	1200	1305	1385	1790
	PAPYEX	1220	1380	1490	2330
	SIGRAFLEX	2325	2580	2805	3860
	MAPSIL	9630	10275	10555	-

This table shows that increasing the tightening torque has but a small influence - about 10 to 20 percent - on the contact conductance. Using six screws instead of four improves the conductance by a factor of 1.2 to 1.5, the improvement being higher for the lower conductance values.

Some tests have also been performed with RTVS 691. This filler is a pealable epoxy compound which allows dismantling of the contact surfaces. The measured conductance values are compared with those obtained with PAPYEX in table 6.

TABLE 6. Thermal conductance values obtained with RTVS 691 - 6 screws, 8 kg.cm

Hybrid/Support	*PAPYEX*	*RTVS691*
Al gold/Al gold	7545	10530
Al gold/AZ31 DOW17	2330	8110

The table shows that RTVS691 is a very efficient filler, which improves the contact conductance for a magnesium support by a factor of about 3.5 with respect to PAPYEX.

4.4. ASSESSMENT OF ANODIZED ALUMINIUM BASED SANDWICH FILLERS

As mentioned before, commercially available electrically insulating fillers do not lead to very high contact conductances. Higher values could be obtained however by using « sandwich » fillers, in which an insulating filler is sandwiched between two layers of electrically conducting interstitial materials. Some tests have been performed by combining an anodized aluminium foil (total thickness 130 µm, both sides anodized 40 µm deep) with some of the best conducting fillers (In foil, PAPYEX, MAPSIL). The results are given in table 7.

TABLE 7. Measured thermal conductances ($W.m^{-2}.K^{-1}$) - Sandwich Filler 1 / Anodized Al / Filler 2 Al6061 gold / Al6061 gold - 4 screws, 8 kg.cm

Filler 1	*Filler 2*	h_c ($W.m^{-2}.K^{-1}$)
In foil	In foil	2906
PAPYEX	PAPYEX	3400
MAPSIL	In foil	3380
MAPSIL	PAPYEX	4820
MAPSIL	MAPSIL	6705

High conductance values can be obtained, from about 3000 to 7000 $W.m^{-2}.K^{-1}$, i.e. of the same order of magnitude as conducting fillers. From the practical point of vue, the best combination is MAPSIL/Anodized Al/PAPYEX which allows dismantling (the interface between the anodized aluminium and the support plate remains « clean »). In order to avoid migration of the MAPSIL, the interface between the hybrid and the aluminium foil has to be sealed by a silicone joint.

5. Conclusions

Practical thermal conductance values have been determined for multi cavities power hybrids, when the contact surfaces are in dry contact as well as when using different interfacial fillers.

The measured contact conductances range from a few hundreds W/(m^2°C), when the surfaces are in dry contact, up to 10,000 to 15,000 W/(m^2°C), with gold plated aluminium combined with the best interstitial materials.

The following conclusions can also be drawn from the results:

- for dry contacts (i.e. without interstitial material), the highest conductance values are obtained with gold plated aluminium;
- increasing the tightening torque significantly improves the thermal conductance of dry contacts, the improvement is marginal when interstitial materials are used;

- higher contact conductances are obtained with 6 screws instead of 4, the improvement (about 30 to 80 percent) being higher for dry contacts.

High thermal conductance electrically insulating interfaces can be obtained by using an anodized aluminium foil (electrical insulator) sandwiched between two conducting fillers.

6. Acknowledgments

The authors would like to acknowledge the support of the Walloon Regional Government.

7. References

Madhusudana, C.V. and Fletcher, L.S. (1984) Contact Heat Transfer - The Last Decade, AIAA Journal, Vol. 24, nr 3, 521-523.

Fletcher, L.S. (1988), Recent Developments in Contact Conductance Heat Transfer, ASME Journal of Heat Transfer, Vol. 110, 1059-1070.

Eid, J.C. and Antonetti, V.W. (1987) Thermal contact resistance measurement techniques for microelectronic packages, in ASME Heat Transfer Division, Vol. 89, 31-36.

4. SINGLE AND MULTI-PHASE CONVECTIVE COOLING

FLOW BOILING AND FILM CONDENSATION HEAT TRANSFER IN NARROW CHANNELS OF THERMOSIPHONS FOR COOLING OF ELECTRONIC COMPONENTS

N. TENGBLAD, B. PALM
Department of Energy Technology -
Division of Applied Thermodynamics and Refrigeration
Kungliga Tekniska Högskolan, S-100 44, Stockholm, Sweden

Abstract - A closed loop, two-phase thermosiphon system is an efficient method for dissipating high heat fluxes from a hot spot to a secondary coolant with a small temperature drop. At KTH in Stockholm, several prototypes of such systems have been built and tested over the past years. In this paper heat transfer coefficients for boiling and condensation with the refrigerants R142b, R134a, R22 and propane (R290) are presented, analysed and compared to correlations from the literature. The influence of channel diameter on boiling heat transfer is also examined.

1. Introduction

Closed loop two-phase thermosiphons resemble heat pipes in that they transport heat as heat of vaporisation. An advantage with thermosiphons vs. traditional heat pipes is the wide freedom of choice to design and enlarge the internal surfaces of the evaporator and condenser. This freedom makes it possible to achieve low temperature differences for boiling and condensation as well as for the one phase flow on the secondary coolant side.

Tests at KTH (Tengblad and Palm, 1995) have shown that with R142b as refrigerant a heat flux of 150 W/cm^2 (from the simulated component) can be dissipated with a temperature difference of less than 12°C over the thermosiphon, i.e. between the evaporator front wall and the condenser wall. The low temperature differences that were measured for boiling and condensation correspond to very high heat transfer coefficients.

In spite of the large number of studies made on flow boiling phenomena there is no generally accepted method of predicting boiling heat transfer coefficients. In the tests reported here, the aim has been to gain a better understanding of

E. Beyne et al. (eds.), Thermal Management of Electronic Systems II, 215-226.

boiling and condensation characteristics in *narrow* channels, and to get a base for developing heat transfer correlations for boiling and condensation in these channels. The ultimate goal is to incorporate such correlations into an existing fluid flow model of the thermosiphon.

An advantage with an external thermosiphon as opposed to immersion cooling is that the refrigerant may be chosen more freely as the hot component is not in contact with the fluid. The refrigerants tested here (halocarbons and propane) have higher saturation and reduced pressure at a given temperature than the FC fluids often used in immersion cooling. Higher pressure is beneficial in boiling (VDI Heat Atlas, 1993 and Cooper, 1984). The hysteresis, or temperature overshoot, is also expected to be smaller with the tested fluids. Halocarbons and propane also have higher liquid thermal conductivity, which is beneficial in both boiling and condensation heat transfer.

2. Thermosiphon design, Experimental set-up

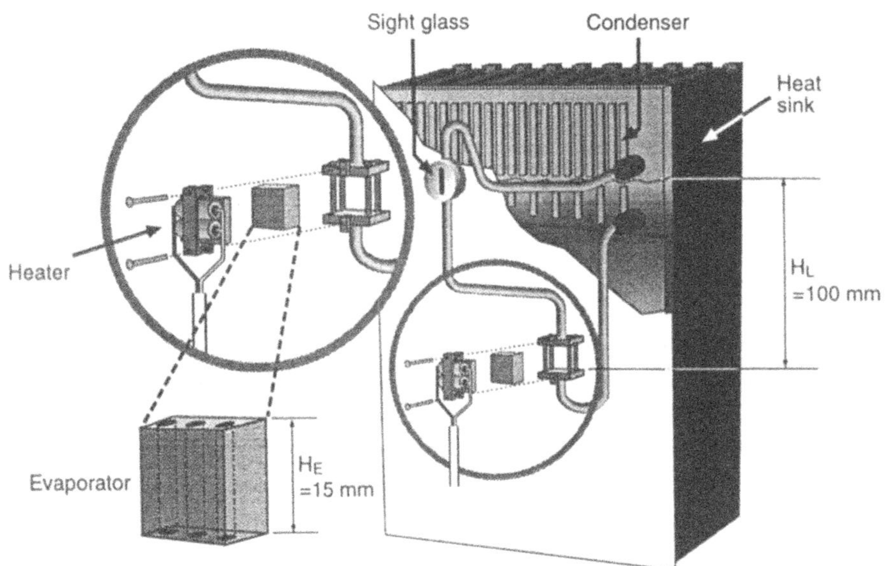

Figure 1. Design of tested thermosiphon system.

The thermosiphon system is designed to cool an individual electronic component on a vertical PCB placed in a sub rack. It consists of an evaporator, a condenser and connecting tubing (Fig. 1). The thermosiphon contains about 10 g of refrigerant in liquid and vapour phase. The evaporator is placed in contact with the component (heater), from which heat is absorbed as heat of vaporisation as the liquid boils. In the condenser, which is cooled by ambient air in free convection, the heat is given off as the vapour condenses back to liquid. The refrigerant is circulated by the density difference between the vapour and liquid and no external power is needed for the fluid transport.

Four evaporators with different channel diameter were tested (Fig. 2). The evaporators were made from small blocks of copper in which 2-7 vertical, channels were drilled. They were made easily exchangeable by a mechanical joint with Teflon sealings.

The condenser was made from a 4 mm copper sheet of the size of a PCB (161x265 mm). In the 55 mm top part of the sheet a condenser channel system was milled (Fig. 3). The channel system was sealed by brazing a 1 mm copper lid onto the top part of the sheet. To additionally increase the surface area in contact with ambient air, a standard aluminium heat sink was attached to the copper sheet.

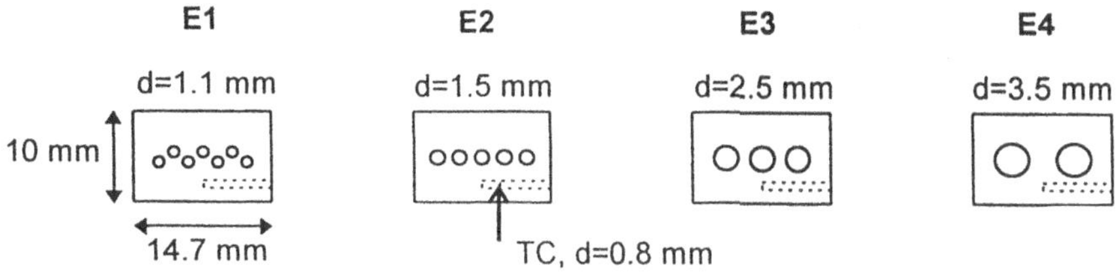

Figure 2. Top view of the tested evaporators (TC=thermocouple).

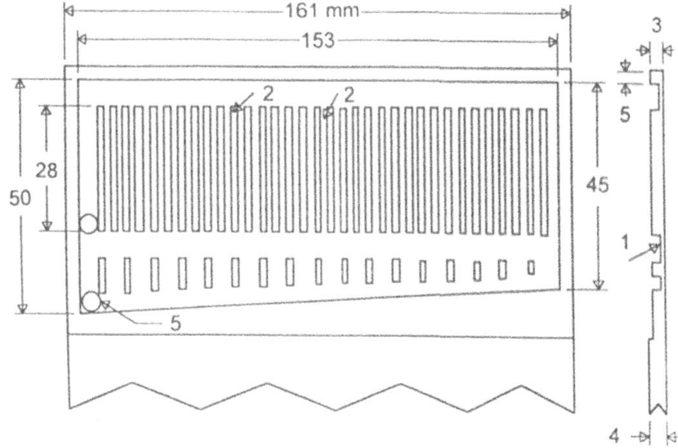

Figure 3. Front- and side view of channel system inside the top part of tested condenser sheet.

The evaporator and condenser were connected by copper tubing and all connections were brazed. The length of the connecting tubing corresponded to a specific liquid head H_L, i.e. the distance between the liquid level in the condenser and the evaporator inlet (Table 1).

TABLE 1. Geometrical data of the tested thermosiphon system (cf. fig. 1).

Liquid head, H_L (mm)	100
Evaporator height, H_E (mm)	15
Inner diameter of connecting tubing (mm)	3.0
Front area of simulated component (cm^2)	1.0
Evaporator front area (cm^2)	2.2
Evaporator inside area (cm^2)	3.3-3.6 (cf. fig.2)
Condenser inside area (cm^2)	108
Condenser outside area (cm^2)	5400

The hot component was simulated by a piece of copper with a front area of 1 cm^2, heated from the back side by an electric heater. The heater, evaporator and connecting tubing were well insulated and all supplied heat was assumed to be transferred through the thermosiphon. The heat input was obtained from the electric voltage and current, measured by a DDM (Fluke, 45) with an estimated accuracy of ± 0.7%.

The data obtained from the thermocouples (type T) and absolute pressure transducer (Druck, PCDR 960) were recorded by a micrologger (Campbell, 21X). One thermocouple was located in the evaporator front wall, three in the top part of the condenser sheet and one in the air close to the experimental set-up. The estimated accuracy in *absolute* temperatures were ± 0.2°C and in measured temperature *differences* ± 0,05°C. Saturation temperatures were calculated from the corresponding pressures inside the system, using Refprop (NIST, 1994). The heat transfer coefficients were calculated as $h = \dot{q} / |\Delta t|$ where $\dot{q}$ is the heat flux and Δt is the temperature difference between the wall and saturation temperatures. To achieve a high accuracy in the differences between saturation temperature and wall temperatures, the calculation of saturation temperatures was calibrated before each test to give zero temperature difference at zero heat flux. The estimated accuracy in this temperature difference was ± 0.05°C at small differences and up to ± 0.2°C at high differences.

Before each test the thermosiphon system was opened to ambient air and cooled down to about 0°C. The system was then evacuated before it was completely refilled. To achieve a fixed liquid level in the condenser, the system was partially emptied while the liquid level was visually controlled in a sight glass.

The test runs were carried out by first increasing the supplied electrical heat from zero to the maximum heat input (60 W). After a stabilisation period of 90 minutes the applied power was *decreased* in steps every 30 minutes.

3. Experimental results and discussion

3.1 GENERAL CHARACTERISTICS OF THE THERMOSIPHON

Figures 4a-b show how the wall temperatures and saturation pressure of the thermosiphon increase with heat input. The condenser was cooled by free convection air with an inlet temperature of 26°C. The increase in condenser wall temperature with increasing heat input is due to that the temperature difference at the air-side rises substantially with increasing power. The saturation temperature was very close to the condenser wall temperature with all tested refrigerants. Note that propane (and also R134a and R22) has a considerably higher pressure than R142b.

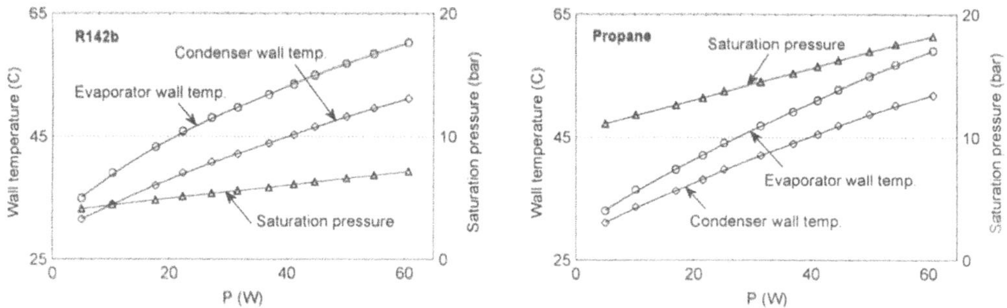

Figures 4a and 4b. Wall temperatures and saturation pressure of the thermosiphon at different heat input. The condenser was cooled by free convection air at 26°C (evaporator E2).

3.2 CONDENSATION WITH DIFFERENT REFRIGERANTS

In the test runs for measuring condensation heat transfer coefficients, the liquid level in the condenser was kept low, so as to free most of its internal surface area for condensation and also to minimise the splashing of liquid drops on to the condenser walls. The calculated heat transfer coefficients are based on the total internal surface area of the condenser, i.e. 108 cm^2.

Figures 5a-d show experimental average heat transfer coefficients for condensation in comparison with Nusselt's correlation for film condensation (Nusselt, 1916). As is seen, all tested refrigerants give very high heat transfer coefficients, especially R142b which gives about twice as high values as the other refrigerants (15000 - 40000 $W/(m^2 \cdot K)$). It should be mentioned that at heat fluxes *below* 2000 W/m^2 the temperature differences are very small, 0.1°C or less. This is within the estimated uncertainty of the temperature difference measurements.

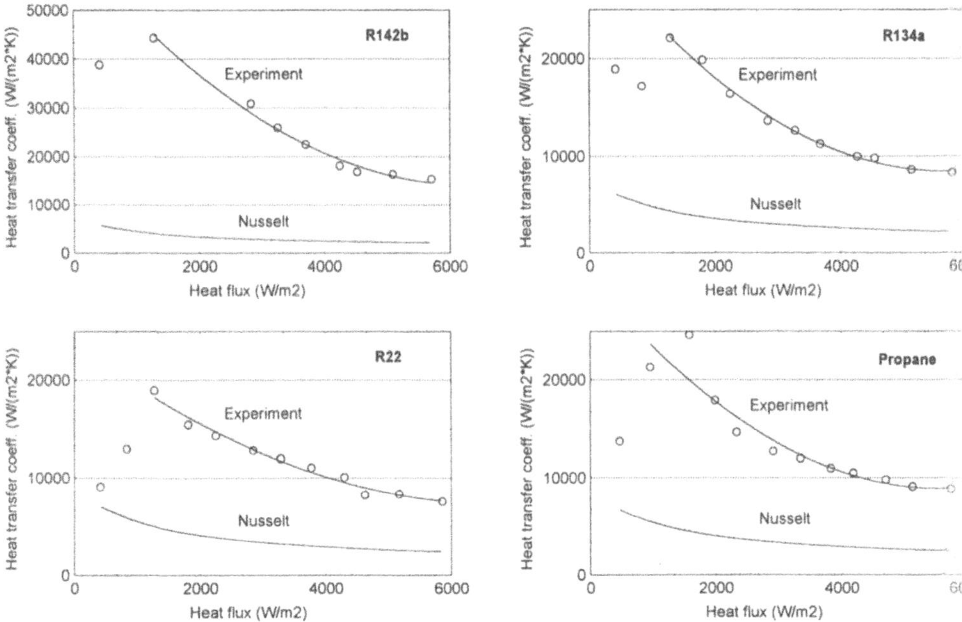

Figures 5a - d. Measured heat transfer coefficients for condensation with R142b, R134a, R22 and propane vs. heat flux, compared to Nusselt´s correlation with wall height = 45 mm.

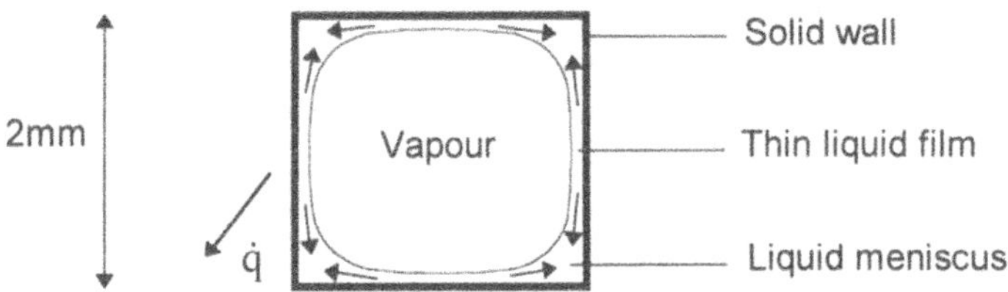

Figure 6. Horizontal cut view of the square and vertical condenser channel with thin liquid films in between cooled walls and vapour phase.

For both low and high heat fluxes, all refrigerants give heat transfer coefficients that are considerably higher than predicted by Nusselt´s correlation. The high experimental values are probably due to rapid drainage of the condensed liquid film by *capillary forces*, induced by the sharp-edged corners of the square and vertical channels inside the condenser (Fig. 6). It may also be that an *attractive surface force* (Israelachvili, 1991) operates during the film condensation and further increase the drainage rate. At steady state, a finite thickness of the liquid film will result due to a balance between the viscous, capillary, surface and gravitational forces. This finite thickness is smaller than what would be expected for a flat wall or circular channel.

3.3 BOILING WITH DIFFERENT REFRIGERANTS

Figures 7a-d show experimental average boiling heat transfer coefficients for the four different fluids at heat fluxes up to 18 W/cm^2. All tested refrigerants give very high heat transfer coefficients, especially R134a and R22 which give values up to about 45000 W/(m^2·K) (Figs. 7b and 7c). For comparison the correlation of VDI Heat Atlas (1993) for flow boiling (upward flow) is plotted.

The most probable explanation for the lower heat transfer coefficients with R142b compared to the other fluids is its lower reduced pressure. For propane the reduced pressure is close to that of R22 and R134a. The reason for the slightly lower heat transfer coefficients in this case is probably to be found in the thermodynamic and transport properties of propane.

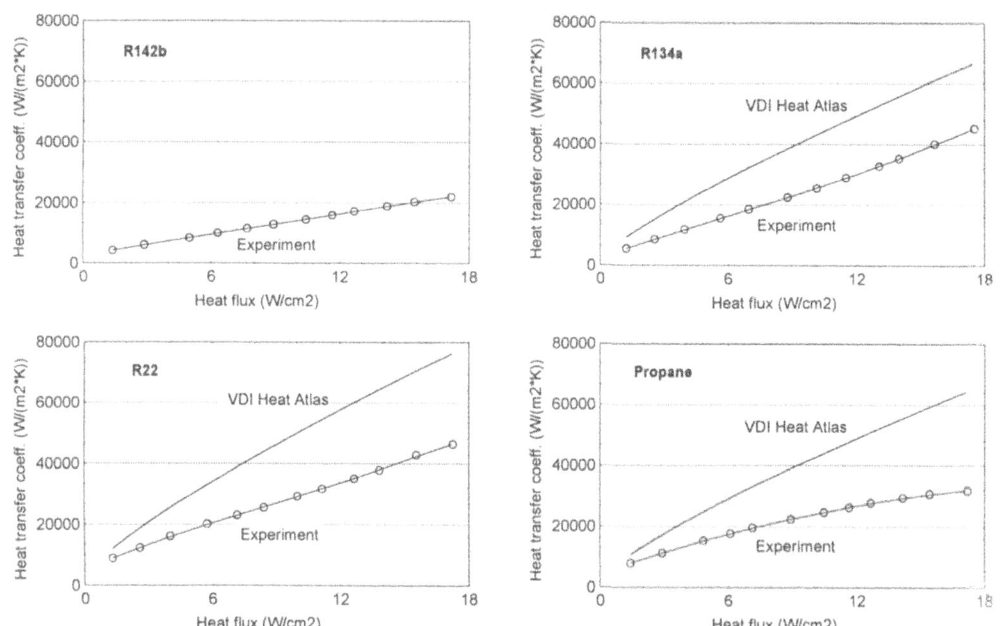

Figures 7a - d. Measured heat transfer coefficients for flow boiling (upward flow) with R142b, R134a, R22 and propane vs. the heat flux inside the evaporator (evaporator E2), compared to the correlation from VDI Heat Atlas.

3.4 DIAMETER EFFECTS ON BOILING

Four evaporators with different channel diameters were tested with R142b (cf. Fig. 2). The internal surface areas of the evaporators were held approximately constant by varying the number of channels. The mass flow rate was also approximately constant as only a minor part of the fluid flow resistance was caused by the evaporator channels.

For the three smallest channels there is a small decrease in the heat transfer coefficients with increasing diameter (Fig. 8). The largest diameter (3.5 mm) gives heat transfer coefficients as high as, or higher than, the smallest diameter (1.1 mm). This deviating behaviour is not understood and need further investigation. The difference in heat transfer coefficients for all four diameters is relatively small, less than 25%. This is less than predicted by the VDI Heat Atlas-correlation, which predicts an increase in the heat transfer coefficient by 60% as the diameter is decreased from 3.5 to 1.1 mm, or 40% in going from 2.5 to 1.1 mm.

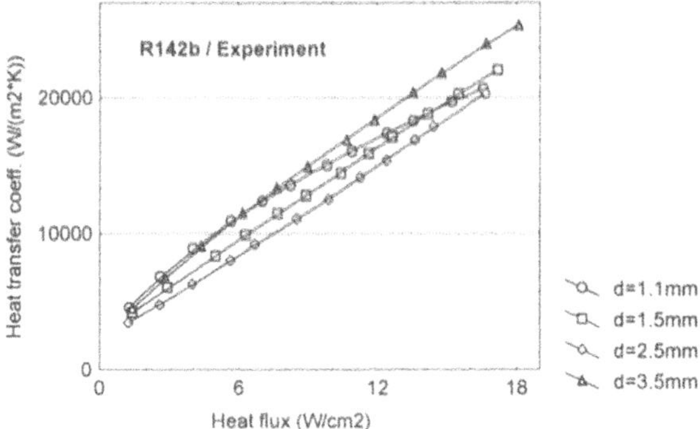

Figure 8. Measured boiling heat transfer coefficients vs. heat flux with different channel diameter.

3.5 DISCUSSION ON THE RESULTS FOR BOILING

For boiling in narrow channels, fluid visualisation indicates that, the bubbles fill the channel from side to side or at least partially attach to the wall. This leads to a situation similar to that in condensation (cf. Section 3.2), with thin liquid films in between the channel wall and vapour phase (Fig. 9). Thus, in boiling, the primary heat transfer mode is most likely *film evaporation* and the heat transfer is mainly limited by the heat resistance from *conductance* through the thin liquid film. With these assumptions the experimental heat transfer coefficients correspond to *average* film thicknesses* of about 1-3 μm for R22, R134a and propane, and 3-6 μm for R142b. These thicknesses are in accordance with statements by Mesler (1979) regarding the liquid film region beneath a growing vapour bubble in *pool boiling*.

* *The surface area that is active for film evaporation is approximately proportional to the average void fraction α in the evaporator. Thus the average film thickness δ is given as $\delta=\alpha\cdot(\lambda/h)$, where λ is the thermal conductivity of the liquid film. The average void fraction was calculated with a homogenous fluid flow model.*

Strong forces are required to effectuate refilling of the evaporating liquid film. Stephan and Hammer (1994) have recently shown the importance of capillary forces for nucleate boiling. *Repulsive surface forces* might also in the end contribute to the transport of liquid into the thin film part.

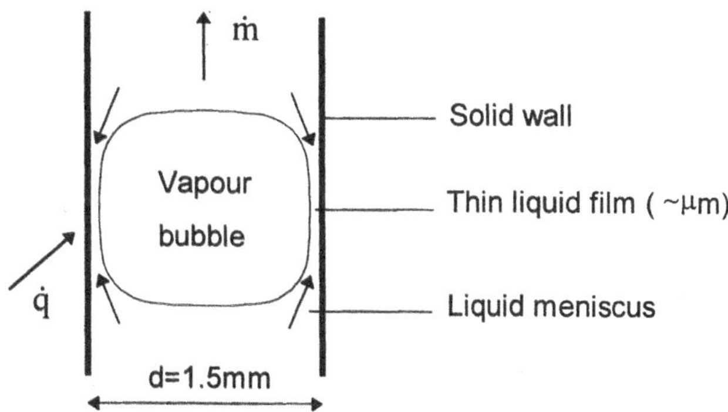

Figure 9. Vertical cut view of circular and vertical evaporator channel with thin liquid films in between heated wall and a vapour bubble.

It is clearly seen in Figures 7b-d that the VDI Heat Atlas-correlation over predicts the heat transfer coefficients in the short and narrow channels of the thermosiphon evaporator, though the tendency is quite good. The local heat transfer coefficient $h(z)_B$ for flow boiling, by VDI Heat Atlas, is given as

$$\frac{h(z)_B}{h_0} = C_F \cdot \left(\frac{\dot{q}}{\dot{q}_o}\right)^{n(p^*)} \cdot F(p^*) \cdot \left(\frac{d_o}{d}\right)^{0.4} \cdot \left(\frac{R_a}{R_{ao}}\right)^{0.133} \tag{1}$$

where

h_0 = reference heat transfer coefficient for pool boiling at $p_o^* = p/p_c = 0.1$, $\dot{q}_o = 20000$ W/m^2, $R_{ao} = 1\mu$m, $d_o = 0.01$ m.

and

$$C_F = 0.36 \cdot M^{0.27}; \quad n(p^*) = 0.8 - 0.1 \cdot 10^{(0.76 p^*)}$$

$$F(p^*) = 2.816 \cdot (p^*)^{0.45} + \left(3.4 + \frac{1.7}{1-(p^*)^7}\right) \cdot (p^*)^{3.7}$$

The range of validity of Equation (1) is: $1 \le d \le 32$ (mm); $0.05 \le R_a \le 5$ (μm); $G \le 4500$ kg/(m^2·s); $0 \le x \le 0.8$; $0.001 \le p^* \le 0.985$. As the roughness height of the tested channels is unknown, $R_a = 1$ μm was chosen for calculation. In Table 2 some properties and reference heat transfer coefficients of the tested refrigerants are given.

TABLE 2. Reference heat transfer coefficients h_0 at $p_o^* = 0.1$ and $R_{ao} = 1 \cdot 10^{-6}$ m, according to VDI Heat Atlas. The underlined value of C_F is given in VDI Heat Atlas.

Refrigerant	p_c (bar)	C_F	M	$\dot{q}_o$ (W/m^2)	h_0 (W/(m^2·K))
R142b	41.20	1.250	100.49	20000	------
R22	49.9	1.2	86.47	20000	3930
R134a	40.67	1.255	102.03	20000	3500
Propane (R290)	42.47	0.991	44.10	20000	4000

One possible explanation to the higher heat transfer coefficients with the correlation compared to our experimental data, may be that the diameter-parameter, at the right-hand side of Equation (1), is not generally valid for narrow channels. For small diameters (d=1-5 mm), the correlation is based on just a few experiments with boiling helium I. As our measured heat transfer coefficients show a weak dependence of diameter (d=1.1-3.5 mm, cf. Fig. 8), a slight modification of the VDI Heat Atlas-correlation for small diameters is proposed. By exchanging the original diameter-parameter

$$\left(\frac{d_o}{d}\right)^{0.4} \quad \text{with} \quad \left(\frac{d_o}{0.005}\right)^{0.4} \quad \text{, when } 1 \leq d \leq 5 \text{ mm} \qquad (2)$$

the experimental data, in Figures 7b-d, are predicted with a maximum error of approximately ± 10% for R134a and R22 and ± 25% for propane.

Summarily, it is interesting to note that with Equation (2) in (1), this modified VDI Heat Atlas-correlation satisfactoraly represents the experimental data without explicitly including capillary or surface forces.

4. Conclusions

Square (2x2 mm) and vertical condenser channels give very high condensation heat transfer coefficients with all tested refrigerants (halocarbons and propane), especially with R142b which gives values of 15000 - 40000 W/(m^2·K).

Circular (d=1.1-3.5 mm) and vertical evaporator channels give very high boiling heat transfer coefficients, especially with R134a and R22 which give values up to 45000 W/(m^2·K).

In both boiling and condensation heat transfer in narrow channels, *capillary* and presumably also *surface forces* have fundamental influence on the liquid and heat transport in the thin liquid films (~μm) that develop in between solid wall and vapour phase.

In narrow channels (d=1.1-3.5 mm), the diameter seem to have a smaller influence on boiling heat transfer coefficients than is expected for larger tubes. With a slightly modified version of the VDI Heat Atlas-correlation for flow boiling (upward flow), the experimental heat transfer coefficients are predicted with a maximum error of approximately ± 10% for R134a and R22 and ± 25% for propane.

Nomenclature

α	void fraction
δ	liquid film thickness, m
d	diameter, m
d_o	reference diameter, m
G	mass flux, $kg/(m^2 \cdot s)$
h	heat transfer coefficient, $W/(m^2 \cdot K)$
$h(z)_B$	local boiling heat transfer coefficient, $W/(m^2 \cdot K)$
h_0	reference heat transfer coefficient, $W/(m^2 \cdot K)$
H_E	evaporator height, m
H_L	liquid head, m
λ	liquid heat conductivity, $W/(m \cdot K)$
$\dot{m}$	mass flow rate, kg/s
M	molecular weight
p	pressure, bar
p_c	thermodynamic critical pressure, bar
p^*	reduced pressure
p_o^*	reference reduced pressure
$\dot{q}$	heat flux, W/m^2
$\dot{q}_o$	reference heat flux, W/m^2
R_a	roughness height, m
R_{ao}	reference roughness height, m
x	vapour quality
z	Cartesian coordinate

Acknowledgements

This research project is sponsored by Ericsson Company and the Swedish National Board for Industrial and Technical Development. The design of the thermosiphon system is the result of fruitful discussions with dr. Åke Mälhammar of Ericsson Co. and professor Eric Granryd at the Department of Energy technology, KTH.

Conclusions regarding the influence of capillary and surface forces in boiling and condensation heat transfer have been discussed with professor Jan Christer Eriksson at the Department of Physical chemistry, KTH.

References

Cooper, M.G., (1984), *Heat Flow Rates in Saturated Pool Boiling - A Wide Ranging Examination Using Reduced Properties*, in Advances in Heat Transfer, Vol. 16, ed. Harnett, J.P., and Irvine, T.F.Jr., Academic Press, Orlando, Florida, USA.

Israelachvili, J., (1991), *Intermolecular & Surface Forces*, Second Edition, Academic Press.

Mesler, R., (1979), *Boiling Phenomena*, ed. van Stralen, S., and Cole, R., McGraw-Hill, Vol. 1, pp. 813-815.

NIST, (1994), refrigerant properties program "Refprop" version 4.01.

Nusselt, W., (1916), *Surface Condensation of Water*, Z. Ver. Ing., Vol. 60, pp. 569-575 and pp. 541-546.

Stephan, P., and Hammer, J., (1994), *A new Model for Nucleate Boiling Heat Transfer*, Wärme- und Stoffübertragung, Vol. 30, pp. 119-125.

Tengblad, N., and Palm, B., (1995), *Two-Phase Thermosiphon for Cooling of Electronic Components*, Advances in Electronic Packaging, Vol. 2, pp. 969-974.

VDI Heat Atlas, (1993), VDI-Verlag GmbH, Düsseldorf, Germany, pp. Hbb 1-33.

APPLICATION OF PHASE CHANGE MATERIALS (PCMS) TO THE PASSIVE THERMAL CONTROL OF A PLASTIC QUAD FLAT PACKAGE: EFFECT OF ORIENTATION OF THE PACKAGE

D. Pal and Y. K. Joshi
CALCE Electronic Packaging Research Center
University of Maryland.
College Park, Maryland
U.S.A

ABSTRACT. A transient three-dimensional computational study is performed for passive thermal control of plastic quad flat packages (PQFP) using organic phase change material (PCM) placed in a heat sink under the printed wiring board (PWB). Governing conservation equations for mass, momentum and energy are solved using a finite volume technique. The effects of phase change are handled by a single-domain enthalpy method. It is found that the use of organic PCMs can stabilize the package temperature for transient periods, without substantial weight penalty. The effect of horizontal and vertical package orientations on the thermal response is studied. It is found that the vertical orientation results in convection dominated melting of the PCM, and a slightly improved thermal performance. For the horizontal orientation, melting is conduction dominated. It is found that incorporation of fins in the PCM results in a slight improvement in thermal performance. Results are presented as time-wise variations of maximum package and substrate temperatures, heat transfer histories, velocity vectors, isotherms and melt shapes.

1. Introduction

Miniaturization of integrated circuits in the recent years has enabled a continuing reduction in the system sizes. Increasing levels of functional hardware is being included in single mono-chip packages (Rosten and Viswanath, 1994). On a system level, this has resulted in a fewer packages with higher heat dissipation. Due to their high heat flux levels, efficient thermal management for the packages is required for a reliable operation of the system. Forced convection cooling is often used in such situations by flowing air over the components and PWB. Such cooling schemes can not be implemented in portable systems, such as lap-top or notebook computers, where fans can not be used and the enclosures are sealed. Cooling schemes for such systems typically utilize conduction in solids, and natural convection in air, to transfer the dissipated heat from the packages to the ambient. The limits of passive thermal management can be pushed further by efficient component placement, and with the use of heat sinks, high conductivity substrates or heat pipes (Oktay, 1994). However, such cooling

E. Beyne et al. (eds.), Thermal Management of Electronic Systems II, 227-242.

schemes pose weight penalties on the portable systems, and also will not be sufficient for higher heat dissipation levels. For portable systems, operation is often transient or intermittent, and passive cooling using solid-liquid phase change materials (PCM) can be an attractive option.

Solid-liquid PCMs have a high latent heat of melting. Dissipated heat from the components can be efficiently stored in PCM heat sinks, as the PCM changes phase from solid to liquid. The PCM heat sink can be of a very simple configuration, such as a metal cavity filled with PCM. Molten PCM can re-solidify by dissipating heat in the surroundings, once the equipment is idle. Applications of PCMs in thermal management of electronics have been investigated in the past. Organic paraffins have been considered for transient cooling of avionics (Duffy, 1970; Witzman et al. 1983). A two dimensional computational modeling is performed by Snyder (1991) for use of organic PCM for thermal control of a radar module. Ishizuka and Fukuoka (1991) have performed experiments with metallic PCMs for transient cooling of high density packages. They also developed a simple network model using the concept of thermal resistances and capacitances.

The modeling studies available in the literature have studied simplified packaging configurations often using simplistic numerical models. The internal details of the packages, such as die, paddle, lead-frame and leads were ignored. Phase change is assumed to be conduction driven, and effects of buoyancy were neglected. While such simplifications can reduce the computational effort significantly, they do not provide information about the dominant heat transfer paths from the packages, which is essential to the design of PCM heat sinks. For example, for a package, where the primary heat transfer path is through the leads, it will be beneficial to include the PCM heat sink under the printed wiring board (PWB). The present modeling study is done in a three-dimensional domain, and the internal details of the package are modelled. Melting is modelled using an implicit enthalpy technique, and the effect of natural convection in the melt are considered.

2. Analysis

The geometry of the enclosure is shown in Figure 1. The package (28 mm x 28 mm x 3 mm) is mounted on a 5 cm x 5 cm ceramic substrate which is 1 mm thick. An organic paraffin PCM, n-Eicosene, with a melting point of 37^0C is contained underneath the substrate. The PCM laminate is 10 mm thick. The enclosure is 5 cm x 5 cm x 5 cm (XL x YL x ZL). During the operation, dissipated heat from the package is conducted through the leads to the substrate and is absorbed by the PCM. Part of the dissipated heat from the package is lost by convection and radiation from the top and the sides of the package. The contribution of radiation heat transfer is neglected in the present study.

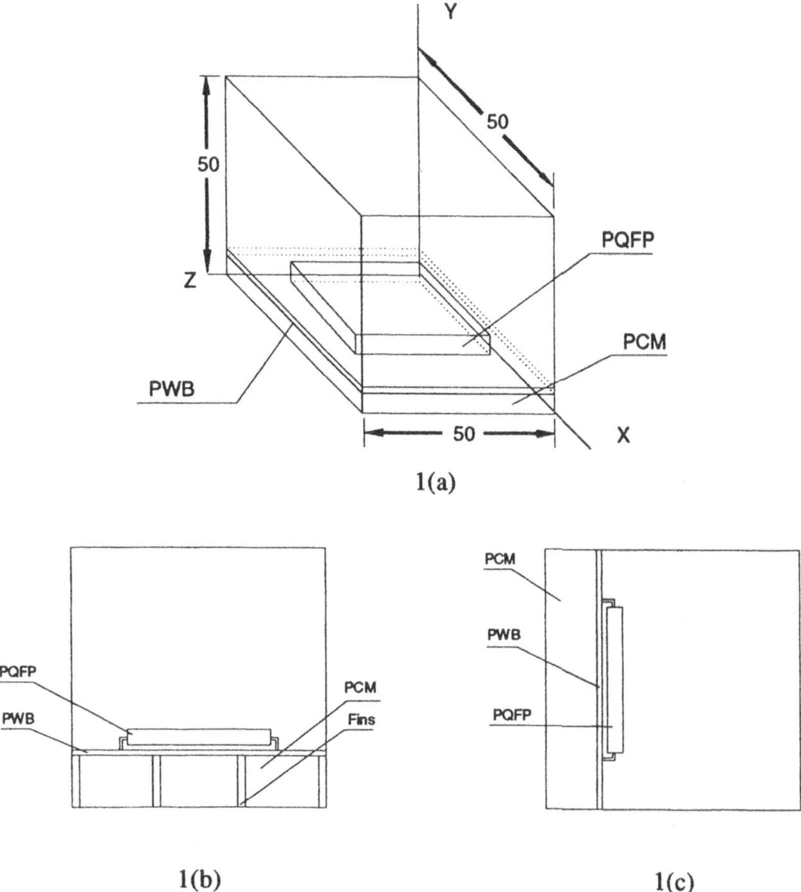

Figure 1. Geometry of the enclosure with the package and the PCM laminate. Both horizontal, Figures 1(a), 1(b) and vertical, Figure 1(c) configurations have been studied. Horizontal configurations have been considered both with fins, Figure 1(b), as well as without fins, Figure 1(a). All dimensions are in mm.

The plastic quad flat pack is placed on a ceramic printed wiring board. PCM is contained in an enclosure and is placed underneath the substrate. The package assembly and the PCM are placed inside an enclosure. The internal details of the package such as the leads, lead-frame, paddle, die and plastic encapsulant have been modelled. The contact resistance and the influence of the solder layer between the package and the PWB are neglected. An earlier study (Pal and Joshi, 1995) was done with the identical geometry but with the package oriented in vertical direction as seen in Figure 1(c). The present study considers the horizontal geometry, Figure 1(a,b) to simulate low aspect ratio enclosures, such as portable computers, with horizontal placement of packages. Comparison is done in terms of thermal performance

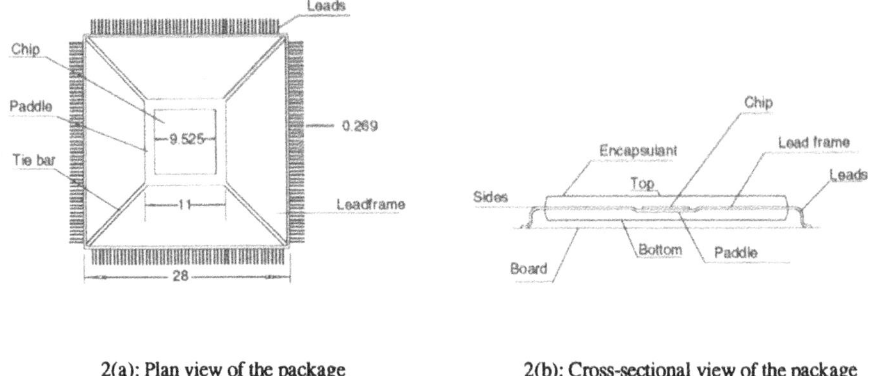

2(a): Plan view of the package

2(b): Cross-sectional view of the package

Figure 2. Internal details of the plastic quad flat package (PQFP). All dimensions are in mm.

between the horizontal and vertical orientations. One disadvantage of organic PCMs is their poor thermal conductivity. Use of fins inside the PCM laminate can augment the heat transfer, and may result in a better thermal performance. The effect of fins in the PCM is also investigated in the present study.

The details of the package are shown in Figure 2. The Silicon die is placed on the paddle which is connected to the lead-frame. The lead-frame connects to the leads outside of the package. The package considered consisted of 208 leads. For ease of modelling, each 5.2 leads are lumped as one. This reduces the total number of leads per side as 10, with the same blockage area as the actual package. The lead-frame and encapsulant layer inside the package are treated as a mixture and equivalent properties are used. The computational method used can handle conjugate conduction and convection in the domain and anisotropy of thermophysical properties in the solids. This method does not require a prescription of heat transfer coefficient on the solid-fluid interfaces, and a single solver can be used to solve the governing equations for the whole domain.

2.1. GOVERNING EQUATIONS

The fluid flow of air is assumed as laminar. For the organic PCM and other solids, the thermophysical properties are assumed to be independent of temperature except density. The density of the molten PCM and air are assumed functions of temperature and Boussinesq approximation is used for solving natural convection flow in these media. The time dependent governing equations are written as:

continuity:

$$\frac{\partial \rho}{\partial t}+\frac{\partial}{\partial x}(\rho u)+\frac{\partial}{\partial y}(\rho v)+\frac{\partial}{\partial z}(\rho w)=0 \tag{1}$$

x momentum:

$$\frac{\partial}{\partial t}(\rho u)+\frac{\partial}{\partial x}(\rho uu)+\frac{\partial}{\partial y}(\rho vu)+\frac{\partial}{\partial z}(\rho wu) =$$

$$\frac{\partial}{\partial x}(\mu\frac{\partial u}{\partial x})+\frac{\partial}{\partial y}(\mu\frac{\partial u}{\partial y})+\frac{\partial}{\partial z}(\mu\frac{\partial u}{\partial z})-\frac{\partial p}{\partial x}+S_x \tag{2}$$

y momentum:

$$\frac{\partial}{\partial t}(\rho v)+\frac{\partial}{\partial x}(\rho uv)+\frac{\partial}{\partial y}(\rho vv)+\frac{\partial}{\partial z}(\rho wv) =$$

$$\frac{\partial}{\partial x}(\mu\frac{\partial v}{\partial x})+\frac{\partial}{\partial y}(\mu\frac{\partial v}{\partial y})+\frac{\partial}{\partial z}(\mu\frac{\partial v}{\partial z})-\frac{\partial p}{\partial y}+S_y \tag{3}$$

z momentum:

$$\frac{\partial}{\partial t}(\rho w)+\frac{\partial}{\partial x}(\rho uw)+\frac{\partial}{\partial y}(\rho vw)+\frac{\partial}{\partial z}(\rho ww) =$$

$$\frac{\partial}{\partial x}(\mu\frac{\partial w}{\partial x})+\frac{\partial}{\partial y}(\mu\frac{\partial w}{\partial y})+\frac{\partial}{\partial z}(\mu\frac{\partial w}{\partial z})-\frac{\partial p}{\partial z}+S_z \tag{4}$$

Energy:

$$\frac{\partial}{\partial t}(\rho c_p T)+\frac{\partial}{\partial x}(\rho u c_p T)+\frac{\partial}{\partial y}(\rho v c_p T)+\frac{\partial}{\partial z}(\rho w c_p T) =$$

$$\frac{\partial}{\partial x}(k\frac{\partial T}{\partial x})+\frac{\partial}{\partial y}(k\frac{\partial T}{\partial y})+\frac{\partial}{\partial z}(k\frac{\partial T}{\partial z})+S_h \tag{5}$$

The same sets of equations apply to all the zones. However, for the solids, a very high value of viscosity is set. The values of various coefficients and source terms for various materials are presented in Table 1. The thermophysical properties of various materials are listed in Table 2. Heat dissipation within the silicon die is assumed to be volumetrically uniform.

2.2. BOUNDARY CONDITIONS

The boundary conditions for the six bounding surfaces of the computational domain are as follows.

$$x = 0,\ XL;\ \frac{\partial T}{\partial x} = 0,\ u = v = w = 0$$

$$y = 0;\ \frac{\partial T}{\partial y} = 0,\ u = v = w = 0$$

$$y = YL;\ T = 25^0C,\ u = v = w = 0$$

$$z = 0,\ ZL;\ \frac{\partial T}{\partial z} = 0,\ u = v = w = 0$$

At the dissimilar material interfaces, additional set of boundary conditions is satisfied using the harmonic mean formulation (Patankar, 1980).

TABLE 1. Source terms for various materials

	S_x	S_y	S_z	S_h
Air	0	$\rho g \beta (T-T_0)$	0	0
PCM	$\frac{C(1-\epsilon)^2}{\epsilon^3+b}u$	$\frac{C(1-\epsilon)^2}{\epsilon^3+b}v$ $+\rho g \beta (T-T_m)$	$\frac{C(1-\epsilon)^2}{\epsilon^3+b}w$	$-\rho \frac{\partial(\Delta H)}{\partial t}$
Silicon	0	0	0	Q'''
Other solids	0	0	0	0

2.3. MODELING OF PHASE CHANGE

The process of phase change is modelled using an implicit enthalpy technique. Details of this technique can be found elsewhere (Brent et al. 1988) and will not be discussed in detail. This method uses a source/sink term S_h in the energy equation for the absorption of latent heat

TABLE 2. Thermophysical properties used for computation

Materials	Thermal conductivity (W/m^0 C)	Density (kg/m^3)	Specific heat (kJ/kg-K)	Dynamic viscosity (kg-m/s)
Air	0.0261	1.177	1005	1.85×10^{-5}
FR-4	0.35	1938	1600	10^{30}
Ceramic (C-786)	18.00	3875	840	10^{30}
Fins (Aluminum)	237	2702	903	10^{30}
Leads and Paddle (Copper)	385	8933	385	10^{30}
Encapsulant	0.31	1070	100	10^{30}
Leadframe Encapsulant mixture	154.17	4215.2	214	10^{30}
Chip (Silicon)	$154.86(\frac{300}{T})^{\frac{4}{3}}$	691	2330	10^{30}
PCM (n-Eicosene)	0.23	795	2050	3.57×10^{-3}
Melting point of PCM: 37^0 C		Latent heat per unit mass : 241 kJ/kg		

during phase change. In the numerical calculation, the latent heat content of each control volume in the PCM is updated after each cycle of calculations. Depending on the latent heat content, a control volume either contains solid or liquid PCM or a mixture of both. An effective porosity ($\epsilon = \Delta H/L$) is used for the control volumes. For control volumes containing solid PCM (ϵ=0), a high momentum source term S_n (n=x,y,z) is used to treat them as solid. For control volumes in fully molten state (ϵ=1), the momentum source terms S_n are zero. Control volumes, which are not fully molten, are treated as mushy zone. The momentum source terms for these are set as that for a Darcian porous medium. The source terms S_n thus cause a smooth transition from solid to liquid, and this helps the numerical convergence.

2.4. NUMERICAL METHOD AND VALIDATION

The governing equations are solved numerically using a finite volume technique. SIMPLER algorithm (Patankar, 1980) with power law scheme for convection diffusion terms is used for numerical solution. Conjugate conduction in the solids is handled by harmonic mean formulation of thermal conductivity. The algorithm is fully implicit in time and solution is obtained at each time step using an iterative solution procedure. Due to the difference in the time scales in various heat transfer processes involved, a control of time step is necessary to assure convergence at a time step. During the conduction dominated stages, a large time step is used (about 10 sec). However, at the onset of natural convection, time steps are reduced to 2 sec. This step is further reduced by a factor of 2, when the PCM layer started melting. The maximum change in the temperature and a global energy balance between two successive iterations are chosen as convergence criteria. Convergence is said to be achieved, when the change in maximum temperature between two successive iterations is below 0.0001 times the maximum temperature at that time step and the global energy balance residual is 0.0005 times the heat generation rate.

In order to validate the code for solving phase change problems, the benchmark problem of melting of pure Gallium in a rectangular cavity is considered. This problem considers a two dimensional rectangular cavity with the top and bottom walls insulated and isothermal hot and cold left and right walls. The solutions are compared with the results of Brent et al. (1988) and a maximum difference of only 1% in the temperature are obtained.

For the current problem, a grid size of 39 x 34 x 39 is used. Local grid refinement is done to model the details of the package. Local grid refinement is also done near the solid surfaces to capture the boundary layers.

3. Results and Discussions

For the numerical simulations a total of four cases are considered. These four cases, described in Table 3, consider a power dissipation of 1 W. For case A, no phase change is considered. Case B uses a PCM laminate and the package in horizontal orientation (Figure 1(a)). Case C considers the effect of fins in the PCM layer. The fins are longitudinal and placed parallel to the z direction. Four 1.5 mm thick Aluminum fins are used (Figure 1(b)). The fins occupy about 12% of the heat sink volume. Case D considers the package and the board oriented in a vertical configuration (Figure 1(c)). Computations are performed for cases A-D for up to 1800 s following the start of power dissipation. Results are presented in the form of time-wise variation of maximum component and substrate temperatures, and heat transfer from various surfaces. Velocity vectors, isotherms and melt shapes for the vertical mid-plane ($Z=ZL/2$) are also presented and discussed.

TABLE 3. Summary of parameters for various cases

Case #	Power(watts)	Geometry	Computation
A	1	No fins, horizontal	Without phase change
B	1	No fins, horizontal	With phase change
C	1	Fins, horizontal	With phase change
D	1	No fins, vertical	With phase change

3.1. TIME-WISE VARIATION OF TEMPERATURE AND FLOW FIELD

Figure 3 shows the time-wise variation of maximum package and substrate temperatures for cases A and B. During the initial phase, there is a rapid increase in the package temperature. This rate slows down as the natural convection in air develops. The substrate temperature shows a lower rate of rise during the initial phase. Temperature variation in the x and z direction of the substrate is found to be very small. The

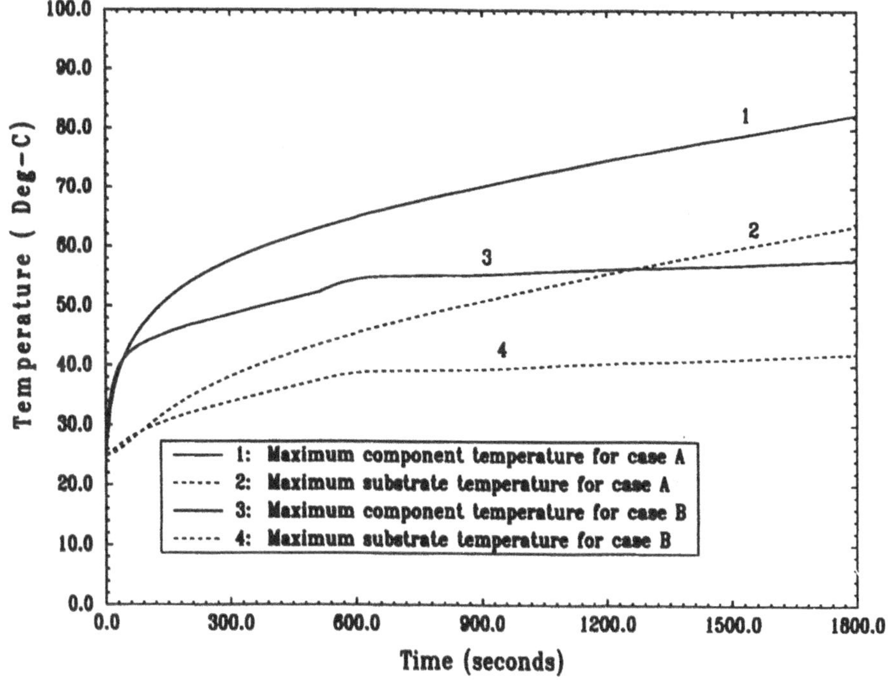

Figure 3. Time-wise variation of maximum package and substrate temperature with and without phase change (cases A and B) for a package power dissipation of 1 W.

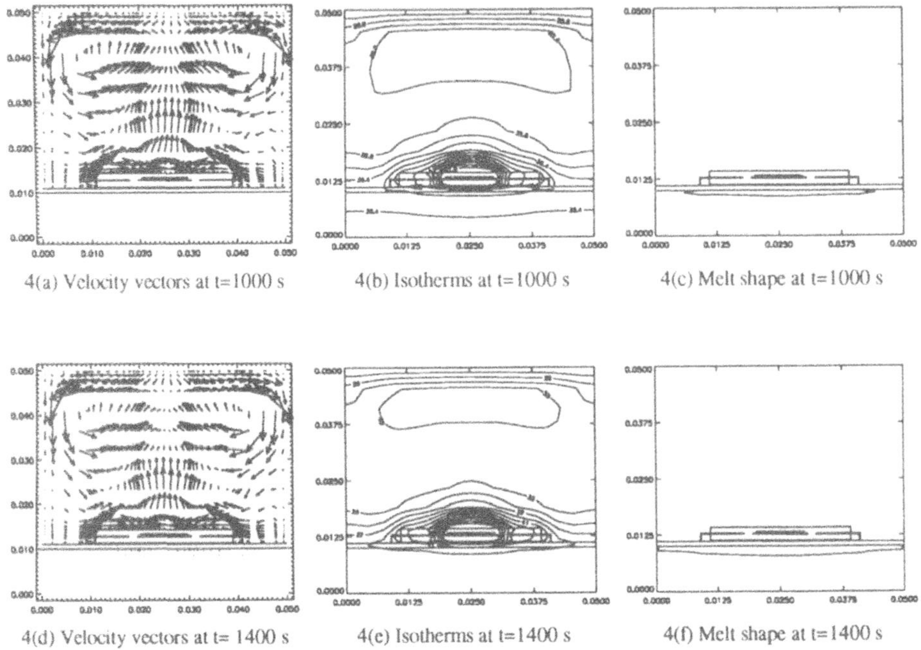

Figure 4. Velocity vectors, isotherms and melt shapes at the vertical mid-plane (Z=ZL/2) for case B

interface between the PCM and the substrate reaches the PCM melting point and melting is initiated at t=480 s. As melting continues, the package temperature stabilizes, and at this point, the time-wise variations for case B departs from that of case A. For case A, which does not consider phase change, temperature increase continues. At 1800 s, The maximum temperature of the package was stabilized at 57°C for case B, and was about 82.5°C for case A.

The velocity vectors, isotherms and the melt shapes in the vertical mid-plane are presented in Figure 4. The velocity field in air results in a vertical plume above the package. This plume impinges on the cold isothermal top wall, where heat transfer takes place from the hot air to the wall. Cooler air then comes down along the sides of the enclosure and moves parallel to the substrate and the top of the package, as it heats up and becomes lighter and moves up. The Rayleigh number for a power level of 1 W is about 8.66 x 10^6. Velocity field exhibits symmetry though still evolving in time.

Temperature contours in Figure 4(b) and 4(c) are found to be closely spaced on and near the package, where there are temperature gradients due to conduction from the package to the air. The temperature variation within the PCM is smaller than in the air. Melt front shapes are

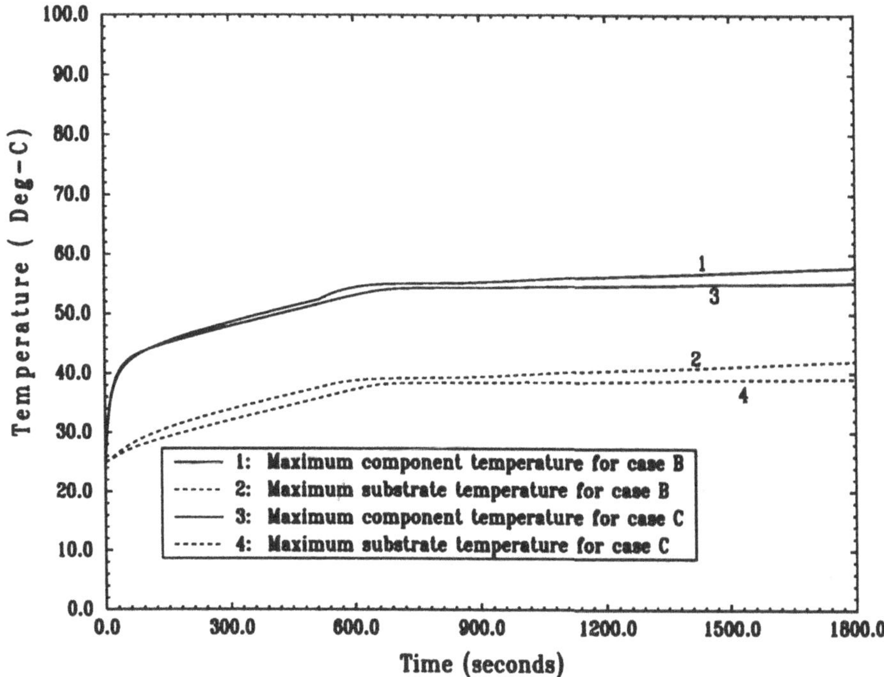

Figure 5. Time-wise variation of maximum package and substrate temperature with and without fins (cases B and C) for a package power dissipation of 1 W

found to be planar, and parallel to the PWB. At 1000 s, the melt front is found localized under the package footprint. At 1400 s, however, the melt front is spread along the length and the width of the substrate. The planar nature of the melt shape indicates that the melting is mainly conduction dominated. For the molten layer, temperature at the top (PWB-molten PCM interface) is higher than that at the bottom (molten PCM-solid PCM interface). Due to this fact, natural convection within the melt does not occur up to 1800 s.

3.2. EFFECT OF FINS

Fins placed inside the PCM laminate increase its effective thermal conductivity, and thus the rate of heat transfer into the PCM. Figure 5 presents the time-wise variation of the maximum temperature for the cases with and without fins. For both cases, the maximum package temperatures are quite close before the initiation of melting. Melting is initiated at about 480 s for case B, and somewhat later (about 720 s) for case C due to the efficient heat spreading. As melting continues, the maximum package temperature for case B increases by about 3.5^{0}C. However, for case C, with fins, the

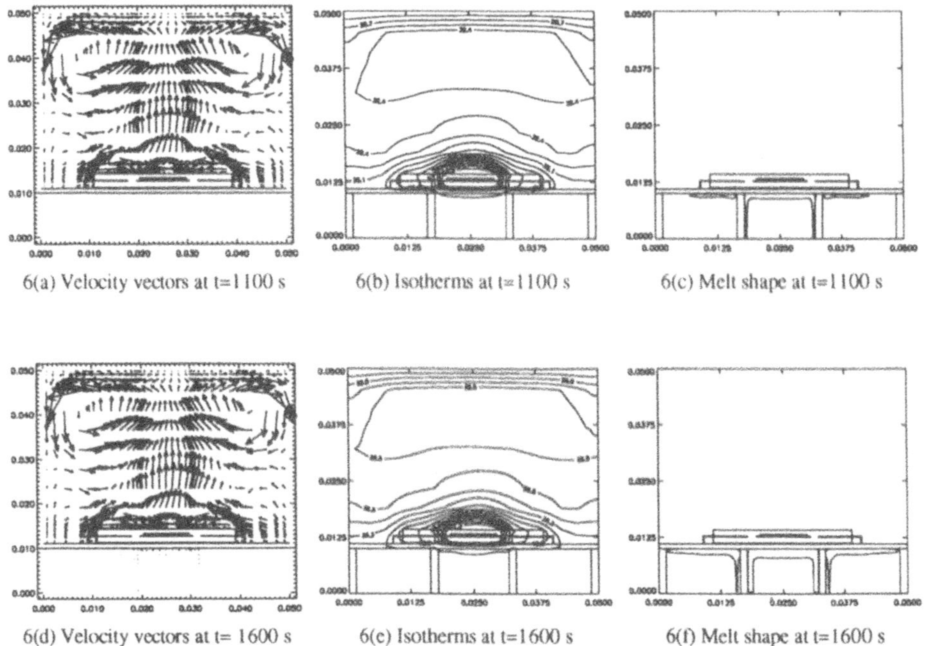

6(a) Velocity vectors at t=1100 s 6(b) Isotherms at t=1100 s 6(c) Melt shape at t=1100 s

6(d) Velocity vectors at t= 1600 s 6(e) Isotherms at t=1600 s 6(f) Melt shape at t=1600 s

Figure 6. Velocity vectors, isotherms and melt shapes at the vertical mid-plane (Z=ZL/2) for case C

existence of fins arrests the temperature of the package within 1°C for up to t=1800 s.

Velocity vectors (Figures 6(a), 6(d)) and isotherms (Figures 6(b), 6(c)) in the air exhibit similar nature to that of case B (Figure 4(a), 4(d)). Interesting features are however observed for the melt shapes. At t=1100 s, the melt front is a thin layer under the substrate and largely localized under the package footprint, with a thin layer along the inner edges of the two fins under the package. This is due to the fact that the straight fins provide a conduction path from the PWB to the PCM. At a later time (t=1600 s), both the horizontal melt layer and those along the fins increase in thickness. Melt layers also develop along the outer edges of the two fins under the package. For the power level of 1 W, overall impact of the fins on the thermal performance in terms of the maximum package temperature is found to be about 2.5°C, after t=1800 s. However fins can provide a significant improvement over laminate PCMs at higher power levels.

3.3. EFFECT OF ORIENTATION OF THE PACKAGE

Figure 7 compares the maximum temperature for the package oriented in horizontal and vertical

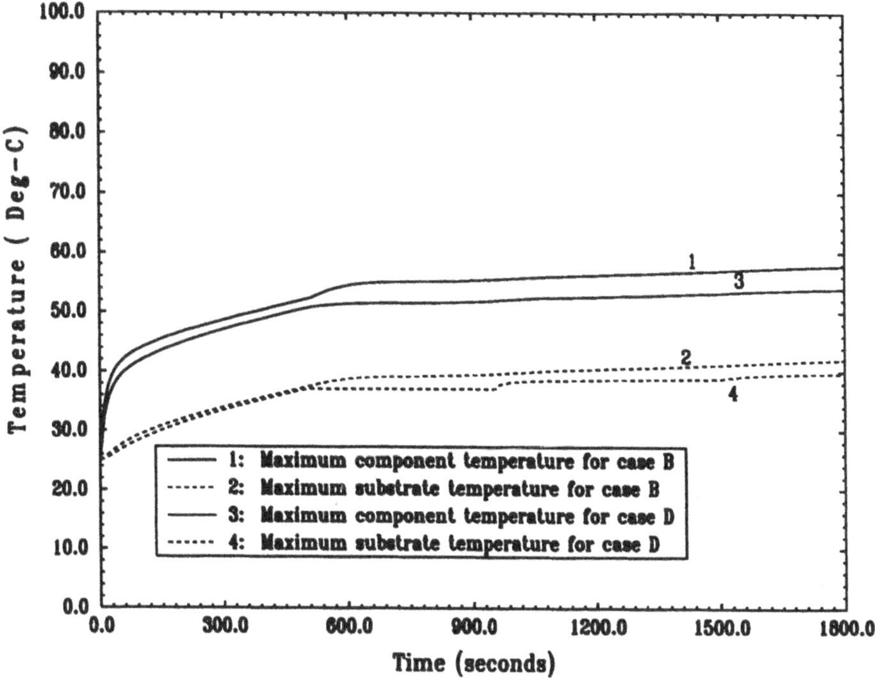

Figure 7. Time-wise variation of maximum package and substrate temperature for horizontal and vertical orientations (cases B and D) for a package power dissipation of 1 W

configurations. For all times, the package temperature for the case B is found to be higher than that for case D. The fluid flow in the air for the vertical orientation is similar to the side heated cavity (Pal and Joshi, 1996). Hot air rises up along the package and PWB surface and impinges on the top wall. This hot air then moves to the right wall, where it loses heat in contact with the cold isothermal wall and comes down. This gives rise to a single clock-wise rotating cell. Following melting, natural convection in the melt is found to occur for case D. At 1800 s, the maximum package temperature difference between the two configurations was found to be 3^0C.

3.4. HEAT TRANSFER RATES

Heat transfer rate history from various package surfaces for case B are presented in Figure 8. From the package, conduction through the leads was found to be the dominant heat transfer path. About 81% of the dissipated heat is conducted through the leads, 14% is removed by conduction in the air gap from the bottom of the package, and about 5% is dissipated from the top of the package. Heat transfer to the PCM laminate is about 68% of the total dissipated heat

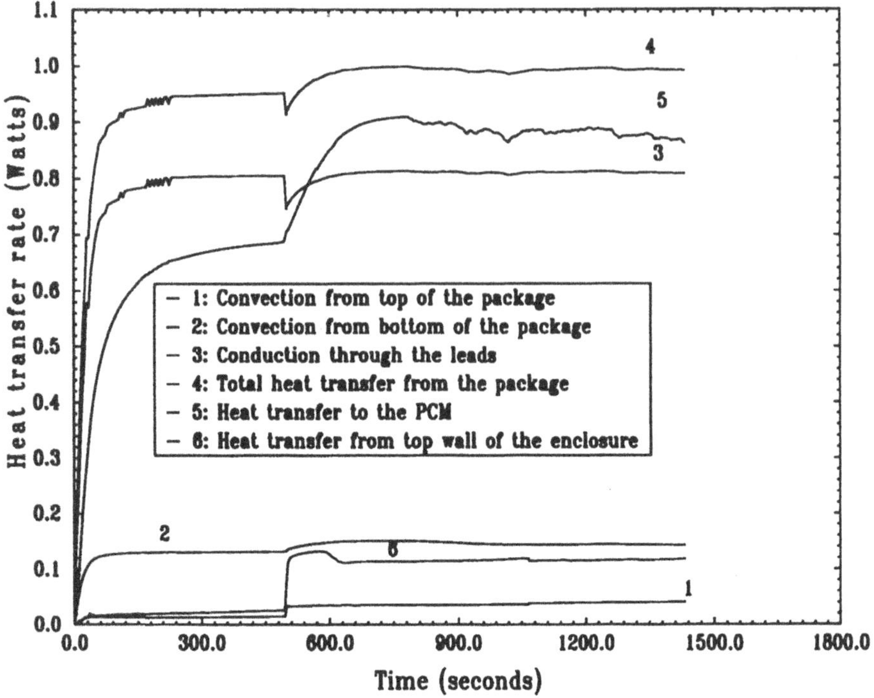

Figure 8. Time-wise variation of heat transfer rates for case B for a package power dissipation of 1 W

from the package before the beginning of melting, and increases to about 86% following the initiation of melting. The rest is lost through the cold isothermal top wall. For case C, the heat absorption by the PCM heat sink during the melting was found to be about 91% for the case with fins, compared to 86% for the case without fins. The heat transfer rates show slight fluctuations around mean values. This may be caused by unsteadiness of the flow structure in air.

4. Conclusions

1. For a package power dissipation of 1 W, the maximum package temperature is found to be stabilized at 57°C with phase change, as opposed to 82.5°C without phase change.

2. The effect of fins inside the PCM resulted in higher rate of melting, and in terms of thermal performance, this is found favorable. The incorporation of fins resulted in stabilization of package temperature within less than 1°C, as opposed to about 3.5°C for the case without fins.

3. Vertical orientation of the package resulted in convection dominated melting of PCM, which resulted in a reduction of about 4°C in the maximum package temperature compared to the case

with horizontal package orientation.

Acknowledgement

The authors are thankful to the members of the CALCE Electronic Packaging Research Center for their support of this research.

Nomenclature

b	Constant in the porosity source term
β	Coefficient of volumetric expansion (1/K)
C	Morphological constant
c_p	Specific Heat (kJ/kg-K)
ΔH	Latent heat component of enthalpy (kJ/kg)
g	Gravitational acceleration (m/s^2)
h	Heat transfer coefficient (W/m^2-K)
L	Latent heat of fusion (J/kg-K)
μ	Dynamic viscosity (kg-m/s)
p	Pressure (N/m^2)
Q'''	Volumetric heat generation rate (W/m^3)
ρ	Density (kg/m^3)
S	Source term (W/m^3)
T	Temperature (°C)
u	Velocity in x direction (m/s)
v	Velocity in y direction (m/s)
w	Velocity in z direction (m/s)

Subscripts

0	Ambient
a	Air
H	Enthalpy
m	Melting point
p	Phase Change Material

References

Brent, A. D., Voller, V. R., and Reid, K. J., 1988, "Enthalpy-Porosity Technique for Modeling Convection-Diffusion Phase Change: Application to the Melting of a Pure Metal, *Numerical Heat Transfer*, Vol 13, pp. 297-318.

Duffy, V., 1970, "Thermal Control Through Fusible Materials", *Electronic Packaging and Production*, July, pp. 45-53.

Ishizuka, M., and Fukuoka, Y., 1991, "Development of a New High Density Package Cooling Technique Using Low Melting Point Alloys," *Proceedings, ASME/JSME Thermal Engineering Joint Conference*, Vol 2, pp. 375-380.

Oktay, S., 1994, "Beyond Thermal Limits in Computer Systems Cooling", *ASI Proceedings on Cooling of Electronic Systems*, Dordrecht, The Netherlands. pp. 47-70.

Pal, D and Joshi, Y., 1996, "Application of Phase Change Materials for Passive Thermal Control of Plastic Quad Flat Packages (PQFP): A Computational Study", *Numerical Heat Transfer, Part A*, Vol. 30, pp. 19-34.

Patankar, S., 1980, "Numerical Heat Transfer and Fluid Flow," 1980, Hemisphere Publishing Corporation, Washington D. C.

Rosten, H. and Viswanath, R., 1994, "Thermal Modelling of the Pentium™ Processor Package," *Proceedings of Electronic Component and Technology Conference, Thermal Simulation and Characterization of Electronic Packages.*

Snyder, K. W., 1991, "An Investigation of Using a Phase-Change Material to Improve the Heat Transfer in a Small Electronic Module for an Airborne Radar Application," *Proceeding, International Electronics Packaging Conference*, San Diego, CA, Vol 1, pp. 276-303.

Witzman, S., Shitzer, A., and Zvirin, Y., 1983, "Simplified Calculation Procedure of a Latent Heat Reservoir for Stabilizing the Temperature of Electronic Devices," *Proceedings, The Winter Annual Meeting of the ASME*, ed., S. Oktay et al, Boston, Mass., HTD Vol 28, pp. 29-34.

HIGH-PERFORMANCE AIR-COOLED HEAT SINKS FOR POWER PACKAGES

A. ARANYOSI[1,2], L. BOLLE[1] and H. BUYSE[2]
[1] *Unité TERM*, [2] *Unité LEI*
Faculté des Sciences Appliquées
Université Catholique de Louvain
1348 Louvain-la-Neuve, Belgium

ABSTRACT. Compact heat exchangers have been designed for spot-cooling of high-power electronic components (power transistors of TO-2xx type and the like) by extending the limits of air-cooling through the use of novel heat sink configurations and impingement flow arrangement with attention given to meet noise requirements. Experiments have been conducted to test and compare the thermal performance and pressure drop characteristics of two narrow channel heat sinks, a woven wire screen and a porous metal fibre structure. Thermal resistance measurements were made both for steady-state and transient conditions using the MOSFET's source-gate voltage as TSEP to obtain the peak junction temperature. The effects of heat sink structure, mass-flow rate of air and component power dissipation have been investigated. Measured heat sink-to-ambient thermal resistances varied from 0.53 to 0.20 K/W for air velocities between 4 and 21 m/s, yielding average areal and volumetric heat transfer coefficients from 150 to 280 W/m²K (based on the total surface area) and from 0.14 to 0.53 W/cm³K, respectively, corresponding to maximum removable base plate heat flux of 12 W/cm². Obtained results demonstrate that confined ducting of high-velocity laminar air-flows through compact structures of these types, combined with central feeding of air, provides greatly improved thermal performance compared to conventional forced air cooling schemes taking up considerably larger volumes.

1. Introduction

The demands of compact packaging of today's advanced microelectronic devices and power control equipment dissipating ever increasing heat-fluxes, have strained the role of traditional heat sinking, requiring innovative solutions to the thermal problems. While the successful removal of heat densities encountered in high-end systems necessitates some form of liquid cooling (immersion, jet impingement, heat pipe etc.) or the use of Peltier elements, in many applications, because of advantages related to simplicity of design, low installation and operating costs as well as ease of

E. Beyne et al. (eds.), Thermal Management of Electronic Systems II, 243-252.

maintenance, air cooling is still preferred. Traditional forced air-cooling of current practice (fan-cooled extruded heat sinks), however, is limited to maximum removable base plate heat-fluxes of around 1 W/cm². Thus, if air is to be retained as the coolant, in order to satisfy the stringent requirements for efficient thermal and space management, one has to overcome the limitations stemming from the poor thermal properties of air. This goal can be realized by applying appropriate passive and active heat transfer enhancement techniques (Bergles, 1988). This paper presents the experimental work conducted in the development of compact air-cooled heat sinks tailored for spot-cooling of power packages by extending the limits of air cooling through the use of novel heat sink configurations and high-velocity laminar flows. Covered are: review of relevant literature, main design considerations, heat sink description, experimental apparatus and TSEP technique used in the thermal resistance measurements, evaluation of heat sink efficacy, and potential applications.

2. State-of-the-Art

To push the limits of air cooling, one can turn to the use of novel heat sink structures, flow configurations, high velocity and/or unsteady flows, and alternate gases (Incropera, 1988). Although upper bounds will ultimately be imposed by available manufacturing technologies and excessive pressure drop/operating noise, there is still room for improvement in air cooling. Good examples of advanced air-cooled systems can be found in the literature. Aluminum pin fins receiving impinging air maintained the chip junction temperatures below prescribed limits in the IBM 4381 (Biskeborn et al.,1984) and in the VAX-9000 (Pei et al.,1990) MCMs, both structures being capable of dissipating 2.5 W/cm² at the module level. Gabuzda (1988) used die-cast radial fins to take out the waste heat from encumbered IC packages. Employing turbulence promoting fins and high-velocity (50 m/s) closed-cycle gas flow, Kishimoto et al. (1984) reported the removal of 3.6 W/cm². Lorenzetti (1987) applied convoluted fins for cooling of high-power semiconductors and achieved excellent thermal performance. Inspired by the landmark work of Tuckerman and Pease (1981) on water-cooled microchannel heat sinks, many investigators (e.g. Goldberg, 1984, Mahalingam and Andrews, 1988) carried out experiments for similar microstructures using gases, and succeeded in extending the limits up to 6.3 W/cm². Hilbert et al. (1990) developed a novel geometry, microchannel heat exchanger and could accomodate heat-fluxes of as high as 16 W/cm². More recently Gromoll (1994) attained heat transfer coefficients over 1000 W/m²K, corresponding to foot print area heat flux of 15 W/cm², with extremely high velocity (200 m/s) air, forced through microstructures manufactured by silicon etching technology. Although the foregoing studies indicate that significant progress has been made in the enhancement of forced air-cooling, in many cases the improved thermal performance was achieved at the expense of unacceptably high pressure drops and accompanying acoustic noise. For this reason, the present study, while keeping the favourable features of cited works, places more emphasis on the reduction of these restrictive factors to get workable air-cooling system.

3. Main Design Considerations and Heat Sink Description

In this work three types of structures (narrow channel, woven wire screen and porous metal fibre) have been chosen to form heat sinks, each offering large surface area to volume ratio (25-30 cm^2/cm^3) and small hydraulic diameters (0.5-0.8 mm), thus, high heat transfer coefficients. To provide sufficient volumetric air-flow rate, while maintaining low pressure drop and acoustic noise, high-velocity laminar flows and impinging air, as well as short streamwise structure lengths were employed. Central feeding of air has the additional advantage of lowering the temperature gradients across the heat sink. Further factors considered in the tradeoff studies included manufacturability, servicability, cost-effectiveness and the pressure flow characteristics of available air sources. The heat sink geometries (Figure 1.) were designed with the intent of cooling a discrete power-MOSFET package (type APT 5025 BN, frequently used in power supply and motor control applications) dissipating 140 W, while its junction temperature had to be maintained below 110 °C.

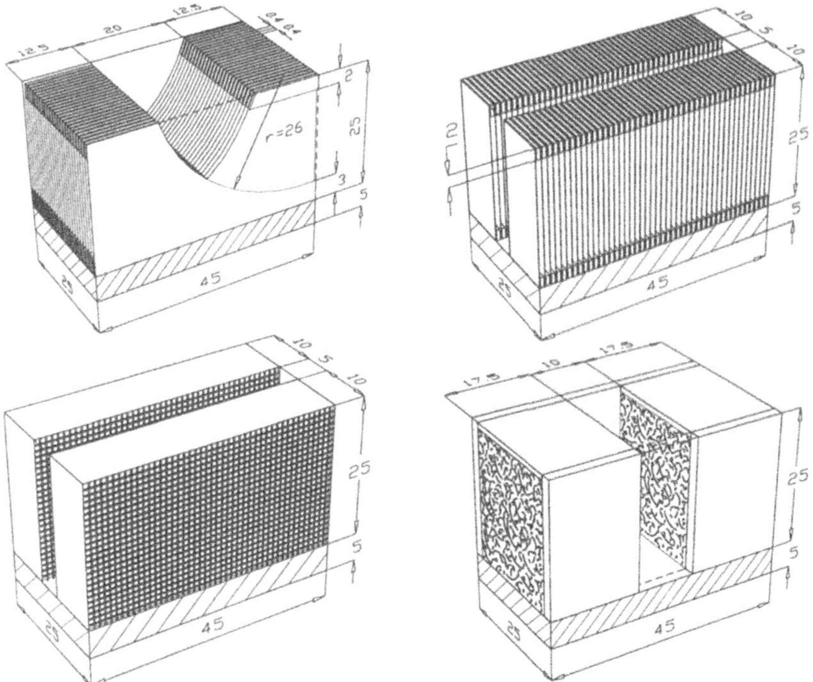

Figure 1. Tested heat sink configurations (clockwise from left top): curved-, rectangular-profile narrow channel, porous metal fibre, woven wire screen structure

To maximize the thermal performance, each heat exchanger was made of copper. The first narrow channel heat sink (henceforth HS #1) is composed of individually cut and alternately positioned curved-profile plates facing in opposite directions. Folded fin tips, while acting as spacers, close the structure at the top. The other narrow channel heat sink (HS #2) was fabricated by two-directional folding of copper foils. The third heat sink (HS #3) is made up of woven wire screens of 0.28 mm wire diameter, with 0.63 mm aperture width (Haver&Boecker, 1990). HS #4 is a sintered porous fibre structure of 80 % porosity, constructed from 0.5 mm diameter wires of 4 cm length. Details of manufacturing can be found in Colin et al. (1993). All structures were brazed to a base plate serving to mount the MOSFET. In the design of HS #1 and #2 we applied the simplified analytical method of Tuckerman and Pease (1981), while HS #3 and #4 were made to examine the feasibility of using these types of structures as heat sinks for electronics cooling.

4. Experimental Apparatus and Procedures

To be able to evaluate the cooling capabilities of the heat sinks, an experimental setup was built (Figure 2.), consisting of a metered air supply, a test section, an electronic switch (with power supplies) and a data acquisition/control system.

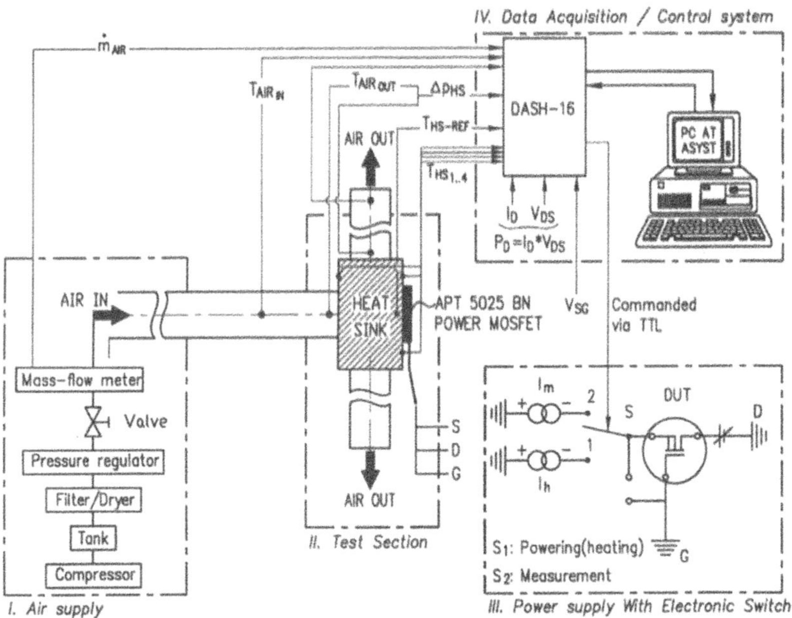

Figure 2. Diagram of the experimental apparatus and its instrumentation

The apparatus was designed to allow determination of thermal resistances/impedances of the coupled SDHS (Semiconductor Device - Heat Sink) system, as well as average heat transfer coefficients for the tested heat sinks. The compressed air, after being filtered, regulated and metered, via the inlet pipe, entered the top of the heat sink and, picking up the heat dissipated by the MOSFET, exited at the two sides. The flow-rate of air was measured by a thermal mass-flow meter (Hastings) with ± 1 % accuracy. Type K (chromel-alumel) thermocouples of 0.2 mm dia (calibrated in a constant temperature bath to within 0.1 °C) were used to measure the air temperature rise, providing means to perform energy-balance calculations. The inlet and exit pipes as well as the heat sinks were insulated to minimize heat losses. The pressure drop through the heat sink was measured by a differential pressure transducer (± 1 Pa). The generated acoustic noise was measured by means of a hand held sound level meter at a distance of 1 m. In all cases, to reduce the thermal contact resistance, the MOSFET was bolted to the heat sink base with a screw torque of around 1 Nm and with high thermal conductivity (k = 2.4 W/mK) metal-filled thermal grease between the case and the sink. The tested heat sinks were instrumented with thermocouples (type K), which were glued with conductive epoxy to the surface of the heat sink at six locations. Four were placed on the outer surface of the heat sink base to measure the temperature drop along its axes, and two others at the fin tip to give information about the 'fin' efficiency. An additional thermocouple was inserted - via a small drilled hole, which was then filled with thermal paste - into the heat sink base, positioned right under the center of the heat dissipating chip, to measure the temperature at the reference point. The close-up view of the MOSFET-mounting with the identification of the thermal resistances is shown in Figure 3.

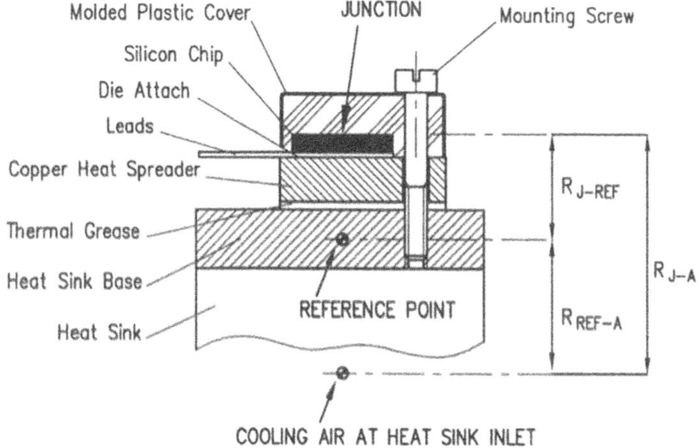

Figure 3. Detailed view of MOSFET-mounting with interpretation of thermal resistances

To obtain the MOSFET's junction temperature, its source-gate voltage was used as temperature sensitive electrical parameter (Oettinger and Blackburn, 1990). Use of a power transistor as both heater and sensor requires that the heating current be removed for a brief window during which V_{SG} is measured. This was done with a heating/switching circuit. The MOSFET was powered by varying the drain current (I_D) while the drain-source voltage (V_{DS}) was kept constant (20 V). In the measurement phase a switching transistor (which was coupled to the circuit by an optoisolator and commanded by the computer) cut out the heating current supply, and left only the measurement circuit encompassed by V_{DS} and I_D, identical to those applied during the calibration (20 V and 1 mA, respectively). Then V_{SG} was measured and traced back to power cut-off (that is, corrected for the electrical and thermal switching transients) to find the actual junction temperature at the instant of switching. Since the temperature of the power MOSFET depends only upon the power level and not upon the combination of I_D and V_{DS} used to achieve that power, this technique provides a good accuracy (to within 2 % of the IR peak temperature, Blackburn, 1982). Prior to the measurements, several MOSFETs were calibrated in an oven. Figure 4. shows a typical calibration curve.

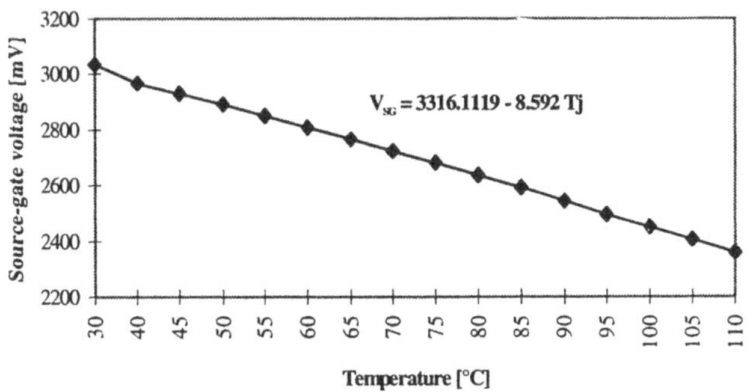

Figure 4. Calibration curve for an APT 5025 BN MOSFET

Measurements were made both for steady-state and transient conditions. During the measurements the air-flow rate and the component power dissipation were varied from 1 to 28 g/s and from 80 to 180 W, respectively. In all cases, the pressure drop and sound pressure level (less than 60 dBA) criteria determined the reasonable ranges of operation. The results are presented in terms of thermal resistances/impedances as a function of mass-flow rate of air. For practical reasons, the total thermal resistance of the SDHS system was broken into two terms as follows :

$$R_{J\text{-}A} = R_{J\text{-}REF} + R_{REF\text{-}A} \tag{1}$$

The first term consists of the internal (chip, die-attach, heat spreader) and the thermal contact resistances, as well as a portion of the conductive resistance of the heat sink base, while the second one represents the heat sink-to-ambient resistance including the effects of both conduction within and convection from the heat sink structure, as well as heating of the air. Since, however, the heat sink-to-ambient thermal resistance is not a true figure-of-merit of the heat transfer, average areal heat transfer coefficients (based on the mixed mean fluid temperature) were calculated for each heat sink structure to obtain a more realistic measure of their performance, using

$$\bar{h} = \frac{P_D}{\sum A_S(\bar{T}_S - \bar{T}_{m,A})}, \tag{2}$$

where
$$\overline{T}_S = \frac{\overline{T}_{base} + \overline{T}_{top}}{2}, \tag{3}$$

and
$$\overline{T}_{m,A} = T_{in,A} + \frac{P_D}{2c_p\dot{m}}. \tag{4}$$

All the measurement and control operations were performed by a DASH-16 data acquisition system. The experimental uncertainty for $\overline{h}$ was determined using the method proposed by Kline and McClintock (1953). With uncertainties of 0.55, 1.35, 1.00 %, 10 cm² and ± 0.1 °C associated with measurements of the drain-source voltage, heating current, mass-flow rate of air, total surface area, average heat sink base and fluid inlet temperature, the maximum uncertainty in $\overline{h}$ was found to be ± 4.1 % for HS #1 and #2. For HS #3 and #4, since their surface areas can not be measured directly, this value may attain ± 15 %.

5. Results and Discussion

5.1. STEADY-STATE

A typical variation of the thermal resistances for mass-flow rates of air between 4 and 12 g/s is shown in Figure 5. (left) with the corresponding pressure drops (right).

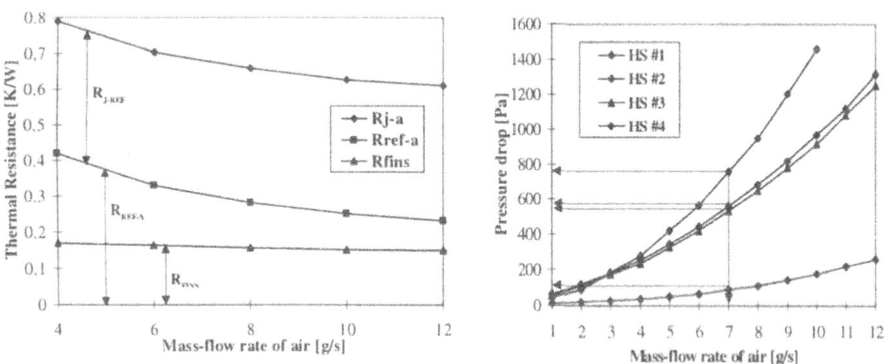

Figure 5. (left) Thermal resistances of the SDHS system for HS #1 at P_D = 140 W, (right) Measured pressure drops for the tested heat sinks

As can be seen, the heat sink-to-ambient thermal resistance (R_{REF-A}) decreases with increasing mass-flow rate and runs parallel to the (R_{J-A})-curve. R_{REF-A} was further divided into fin (R_{fins}) and heat resistances. From measured parameters (T_{REF}, $T_{in,A}$, P_D and $\dot{m}$) these quantities can be deduced. The quasi-constancy of the former (around

0.17 K/W for HS #1), confirms that the temperature profile is effectively laminar fully developed, that is, the average Nusselt number is independent of the flow velocity. At $\dot{m} = 7$ g/s the contributions of the convection and heat terms are approximately equal, yielding $R_{J\text{-}A} = 0.67$ K/W, resulting in a peak junction temperature of 112 °C, along with $\Delta P_{HS} = 560$ Pa, accompanying acoustic noise of 55 dBA and average areal heat transfer coefficient of 280 W/m²K. It is to be noted that the sum of the internal and thermal contact resistances (0.39 K/W) gives more than 50 % of the total resistance, indicating the essential role of the MOSFET-mounting, die and footprint area size. Recently, to alleviate the problem, new plastic power packages (TO-264) appeared (Artusi and Frank, 1994) with double die and package footprint area, halving in this way the mentioned resistances and allowing the removal of significantly higher powers. It is also noteworthy that for the whole range of power dissipation (from 80 to 180 W), due to the strong temperature-dependence of the thermal conductivity of silicon, $R_{J\text{-}REF}$ changed nearly 0.1 K/W. Similar heat transfer trends were observed for HS #2, with lower R_{fins} (0.12 K/W), ΔP_{HS} (88 Pa at $\dot{m} = 7$ g/s) and acoustic noise (50 dBA) along with $\bar{h} = 280$ W/m²K, coming from the larger total surface area and the double free flow area, as well as the invariable heat sink structure, respectively. HS #3 and #4 showed higher values for R_{fins} (from 0.23 to 0.19 K/W and from 0.28 to 0.24 K/W, respectively). The slight decrease is attributable to the increasing portions of developing flows. The corresponding average areal heat transfer coefficients were 150 and 185 W/m²K. The latter two structures, in spite of the higher pressure drops (535 and 758 Pa), owing to the more uniform velocity profiles, generated the same levels of acoustic noise (50-51 dBA).

5.2. TRANSIENT

The transient measurements served to identify the contributions of the individual elements of the coupled SDHS system to the total thermal resistance/impedance. Heating pulses of variable duration were used, and the TSEP (V_{SG}) of the MOSFET was measured before and after the heating pulse was applied. The cooling was permanently maintained during the whole procedure. Between successive heating pulses a waiting period of 180 s was kept to assure the cooling back of the MOSFET to the ambient temperature. Knowing the dimensions and the physical properties of the elements composing the mechanical structure, the heat capacities and time constants can be determined. Then the intersections of the vertical time-constant lines with the complete thermal impedance curve give the cumulative impedances. As can be seen in Figure 6. (left), the thermal wavefront originating at the chip and travelling through the die-attach, heat spreader and thermal grease, attains the reference point after 1000 ms, and steady-state conditions are reached at 50 s. For the thermal contact resistance the value of 0.11 K/W was obtained, corresponding to an average thermal grease thickness of around 50 μm. The above heat capacity analysis also reveals that for accidental conditions (i.e., when the air flow abruptly ceases), due to the large thermal mass of the heat sink, several tens of seconds elapse before the MOSFET's junction temperature attains a value close to the maximum allowable one. This time

span would provide a safe margin during which the cooling system failure could be sensed and the MOSFET could be switched off to prevent permanent damage.
Figure 6. (right) demonstrates the effects of package power dissipation, mounting conditions and thermal environment (heat sink structure and mass-flow rate of air) on the evolutions of the thermal impedances.

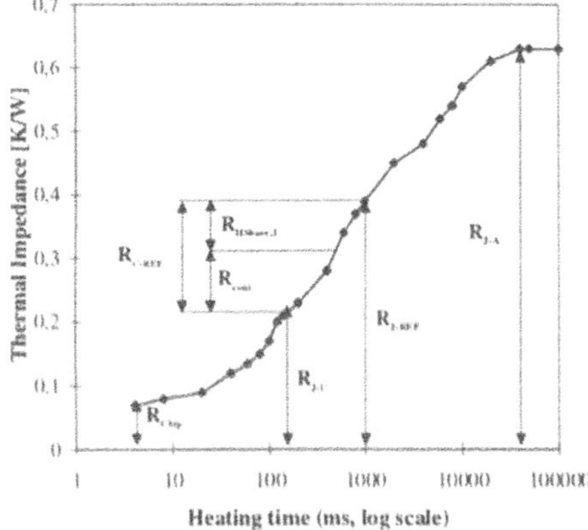

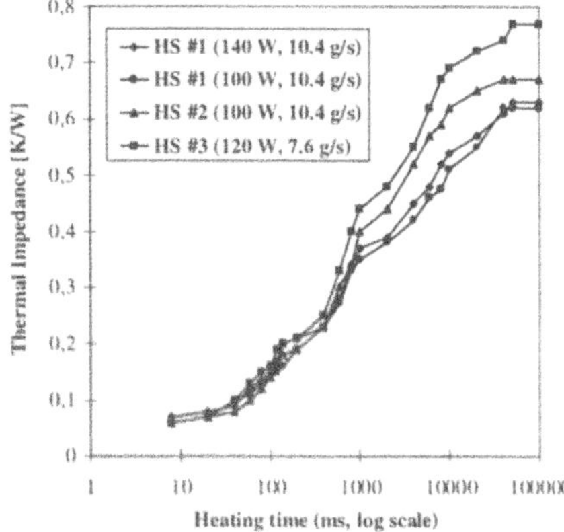

Figure 6. Transient thermal impedances of the SDHS coupled system : (left) for HS #1 at P_D = 140 W and m=10.4 g/s and (right) for different heat sink types at different P_D levels and mass-flow rates

6. Conclusions

Experimental investigation has been carried out to test and compare the thermal performance and pressure drop characteristics of compact heat exchangers of three different types developed for spot-cooling of high-power electronic components. The tested configurations included narrow channel, woven wire screen and porous fibre structures. For all heat sinks the convective term of the heat sink-to-ambient thermal resistance was found to be independent of the mass-flow rate of air and the average heat transfer coefficients of the dissipated power, manifesting constant (but heat sink dependent) average Nu numbers, proving the laminar nature of the air-flow and the consistency of the design. The tested structures were capable of accomodating base plate heat fluxes in the range of 10-15 W/cm² while staying within the limits set by the allowable acoustic noise and pressure drop. Obtained results demonstrate that confined ducting of air through compact structures of these types provides improved thermal performance compared to conventional systems taking up much larger volumes, that either allow air to pass around open heat sink geometries (and so make only partial use of the cooling potential of air) or, although also duct the air, employ extruded profiles with large fin spacing (like cooling aggregates driven by fan) providing significantly lower convective heat transfer rates. By virtue of their modularity, the presented heat sinks can be optimized to meet different system requirements. In practical applications, supplied by air via flexible pipes, they can be used to cool one or several high-power components operating in a confined space. To provide the cooling air, depending on the pressure and air flow requirements, conventional forward curved or radial wheel centrifugal blowers can be employed.

Nomenclature

A	surface area [m²]
c_p	specific heat [J/kgK]
$\bar{h}$	average heat transfer coefficient based on mixed mean fluid temperature [W/m²K]
$\dot{m}$	mass-flow rate of air [g/s]
P_D	power dissipation [W]
ΔP_{HS}	pressure drop through the heat sink [Pa]
R_{chip}	junction-to-chip thermal resistance [K/W]
R_{cont}	thermal contact resistance [K/W]
R_{C-REF}	case-to-reference point thermal resistance [K/W]
$R_{HS,basI}$	thermal resistance of the heat sink base from bottom to reference point [K/W]
R_{J-A}	junction-to-ambient thermal resistance [K/W]
R_{J-C}	junction-to-case thermal resistance [K/W]
R_{J-REF}	junction-to-reference point thermal resistance [K/W]
R_{REF-A}	reference point-to-ambient thermal resistance [K/W]
$\bar{T}_{base}$	average temperature of the heat sink base [°C]
$T_{in,A}$	air inlet temperature [°C]
$\bar{T}_{m,A}$	mixed mean air temperature [°C]
$\bar{T}_S$	average temperature of the heat sink base [°C]
$\bar{T}_{top}$	average temperature of the heat sink top [°C]

References

Artusi, D. and Frank, R. (1994) New Plastic Package Bridges the Gap Between Discrete Power Semiconductors and Power Modules, *Powerconv. Intell. Motion*, Sept, pp. 19-24.

Bergles, A.E. (1988) Some Perspectives on Enhanced Heat Transfer - Second-Generation Heat Transfer Technology, *J. Heat Transfer*, Vol. 110, pp. 1082-1096.

Biskeborn, R.G., Horvath, J.L. and Hultmark, E.B. (1984) Integral Cap Heat Sink Assembly for the IBM 4381 Processor, *Proc. 4th Annual Int. Elec. Packaging Conf.*, Anaheim, CA, pp. 468-474.

Blackburn, D.L. (1982) Power MOSFET Temperature Measurement, PESC '82 Record, *IEEE Power Elecrtonics Specialists Conference*, Boston, MA, pp. 400-407.

Colin, C., Marchal, Y., Boland, F. and Delannay, F. (1993) Stainless Steel Reinforced Aluminum Matrix Composites Processed by Squeeze Casting: Relationship Between Processing and Interfacial Microstructure, *Journal de Physique*, Nov, pp. 1749-1752.

Gabuzda, P.G. (1988) Heat Sink Device Assembly for Encumbered IC Package, *U.S. Patent No. 4,733,293*, Mar 22.

Goldberg, N. (1984) Narrow Channel Forced Air Heat Sink, *IEEE Trans. Components, Hybrids and Manufacturing Techn.*, March, pp. 154-159.

Gromoll, B. (1994) Advanced Micro Air-Cooling Systems for High Density Packaging, *Proc. 10th IEEE Semiconductor Thermal Measurement and Management Symposium*, San José, CA, pp. 53-58.

Hilbert, C., Sommerfeldt, S., Gupta, O. and Herrell, D.J. (1990) High Performance Microchannel Air Cooling, *Proc. 6th IEEE Semiconductor Thermal and Temperature Measurement Symposium*, Phoenix, AZ, pp. 108-114.

THERMAL CONTROL OF BROADCAST TRANSMITTERS BY AIR COOLED COLD-PLATES

G.CESINI, V.MORO, R.RICCI
Dipartimento di Energetica - Università di Ancona
via Brecce Bianche - Ancona (AN)
E.FUMI
Itelco S.p.A.
via dei Merciari - Orvieto (TR)

ABSTRACT

The thermofluidynamic behaviour of a cold-plate system, consisting of forced convection air rectangular channels, used for the cooling of medium power broadcast transmitters, is being examined.

Several power components are assembled on the same cold-plate in order to make the cooling system more compact.

The fluid dynamic characterization of the cold plate is carried out by using a small wind tunnel, designed and realized on purpose. Thus, pressure losses are experimentally determined and the results are compared to the most common relations existing in literature.

Then the thermal behaviour of the cold-plate is studied, by measuring its thermal resistance as a function of the flow rate variation. In particular the effect of the thermal flow localization is examined in order to assess the applicability of the usual design techniques based on uniform temperature or uniform thermal flux boundary conditions.

Finally some preliminary results, aiming at assessing the effect of thermal interaction among the various power sources on the value of thermal resistance between component and fluid thermal carrier, are shown.

1. Introduction

The cooling of power electronic components such as MOS transistors, is highly relevant in the field of electronic applications.

In particular, broadcast transmitters use power electronic components in VHF or UHF amplifiers, where, owing to the low efficiency, a large amount of heat to be dissipated by means of suitable cooling systems, is generated.

The specific thermal flow produced by MOS power transistors is ranging from 40 to 70 W/cm^2; so common air forced extruded dissipators cannot be generally used.

E. Beyne et al. (eds.), Thermal Management of Electronic Systems II, 253-261.

In addition, in order to reduce the overall dimension of the transmitter, it is very important to assemble the electronic components on heat sinks with highly compact configurations, both by increasing the exchange surface exposed to the cooling fluid and by heightening the heat transfer coefficient.

Therefore, forced convection air cold-plates are relevant in these applications, mainly because they make it possible to assemble numerous components even in a double-sided configuration. In addition tin brazed rectangular channeled cold-plates, make it possible to execute channels whose fins can be 0.5 mm. thick, without increasing pressure losses to excess.

This work aims at studying the thermofluid dynamic behaviour of a cold-plate system, as above described, which is used for the thermal control of a medium power transmitter for broadcasting television. The transmitter is a common amplification of vision and sound carriers, operating on the whole UHF band, from 470 to 860 MHz; each unit inside the rack is designed to be easily removed and checked.

The Vision/Sound Amplifier, cooled by the cold-plate dissipator, is composed of four different power stages. Each power module is composed of two transistors isolated from each other.

The first power stage works in class A and delivers 4W power. The second stage is composed of one classs A 30 W amplifier. The third stage is composed of four class A amplifiers, that can deliver 100 W. The final stage, composed of five class A/B amplifiers, delivers 600 W [1].

Literature shows a lot of works concerning the study of the cooling of electronic components in channels in direct contact with forced convection air [2,3], regarding also the evaluation of the mutual influence between the different heated sources.

Only a few papers are concerning the study of finned heat sinks [4,5], in forced convection with localized heat sources fitted on the cold plate without a direct contact with the cooling fluid. The correlations which are available or the curves peculiar to finned surfaces, tipically employed in compact exchangers, are generally used for the thermofluid dynamic sizing of these devices [6].

The reliability of these correlations is linked to the assumption of particularly simplified boundary conditions (uniform thermal flow or uniform temperature), which hardly show the real behaviour of electronic devices with localized thermal sources.

This work, after carrying out the fluid dynamic characterization of the cold-plate, experimentally studies the effect of the localization of sources and their mutual influence on the thermal resistance between the component and the cooling fluid versus the flow rate and the thermal power produced by the components.

2. Experimental Apparatus

A small sized wind tunnel, which allows to perform controlled fluid dynamic conditions through the inlet at the cold-plate, has been used to determine the thermofluid dynamic operating features of the dissipator.

The cold-plate, 150 mm. long and 500 mm.wide on the plan, is made of copper and inside shows 200 rectangular channels, 2mm. wide and 40 mm.high.

Up to 30 components can be fitted on a face, on the supporting surface of the cold-plate. They simulate the real electronic components, which can be separately fed and be located both lined-up and staggered.

Monitoring system allows to determine the pressure losses through the cold-plate and the total thermal resistance as a function of air flow rate, of component location and of thermal power.

3. Experimental Results

3.1. FLUID DYNAMIC ANALYSIS

The fluid dynamic operating features of the cold-plate have been assessed by means of suitable measurements of the volumetric flow inside the wind tunnel and of the pressure loss between the inlet and the outlet of the cold-plate itself.

In order to measure the air flow rate, it has been used a Pitot tube mounted on the suction circular duct. The measured local velocity has suitably been rectified in order to take in account the intrusive nature of the method [8].

A relation between the axial velocity and the flow rate [7] has ensued from the velocity profile and the geometric features of the circular suction channel.

To evaluate the pressure loss on the cold-plate, the overall drop in the fluid, when crossing the rectangular channels of the dissipator, from the inlet channel to the outlet one, has been measured.

The experimental results have been compared with the correlations existing in literature [6,9] for rectangular channels

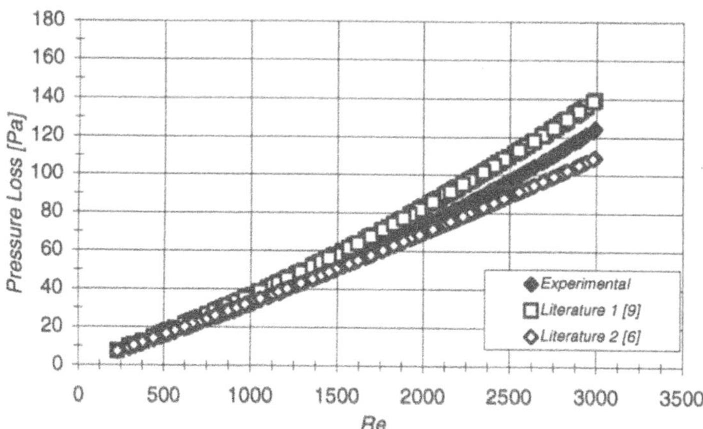

Figure 1 *Pressure loss in the cold-plate according to Reynolds number inside the rectangular channels.*

In particular, the relations on terms of Fanning's factor and of Hagenbach's factor [9]:

- *Literature 1 curve*

$$\Delta P = \left[(K_c + K_e) + f \cdot \frac{L}{r_h} \right] \cdot \frac{\rho \cdot V_m^2}{2} \tag{1}$$

and on terms of contraction and expansion factors [6]:

- *Literature 2 curve*

$$\Delta P = \left[K(\infty) + f \cdot \frac{L}{r_h} \right] \cdot \frac{\rho \cdot V_m^2}{2} \tag{2}$$

have been used.

The experimental results are reported in Fig.1 as a function of the Reynolds number, based on the hydraulic diameter of the rectangular channel: they are in good agreement with above mentioned correlations.

3.2 THERMAL ANALYSIS

The main parameter used to characterize thermofluidynamically the system, composed of the cold-plate and of the electronic components, is the "Total Thermal Resistance", defined as :

$$R_{Tot} = \frac{T_c - T_a}{P} \tag{3}$$

where it is assumed that the thermal power provided to the component is wholly carried off to the cold-plate, thus neglecting the losses due to radiation or natural convection between the component and the surrounding environment. These losses are about 1-2% of the total power.

As the thermal resistance between the component and the cold-plate is much smaller than the thermal resistance between the component itself and the environment, it ensues, quite approximately, that the case temperature of the component is very close to the electronic junction temperature.

The experimental research aims at measuring the total thermal resistance of each component as a function of the flow rate, the electric power provided, the position of the component on the cold-plate and the presence of powered components near-by.

In order to measure the case temperature of the components, two different thermographic systems have been used: an AGA Thermovision 870 working in the range 2-5 micron [7] and, on a second stage, a Long Wave (8-12 micron) AGA Thermovision 880 apparatus equipped with a processing system AGA 880-BRUT.

Measurements have been taken on a group of resistances assembled according to Fig.2. First we carried out some surveys on the thermal behaviour of electric resistances, heated one by one. Then we have investigated the mutual thermal influence among the resistances of a column of the matrix (Fig. 2).

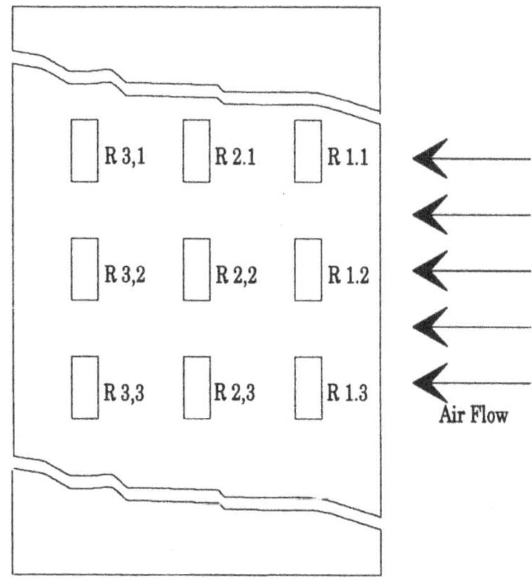

Figure 2 *Layout of the components on the cold-plate.*

Fig.3 shows the experimental findings concerning the 3 components of the central row ($R_{1,2}$-$R_{2,2}$-$R_{3,2}$), heated one by one, as a function of the volumetric flow rate, ranging from 150 to 620 mc/h, to which a Reynolds number ranging from 1000 to 3800 corresponds.

The trials have been carried out by taking into account three different power values: 50, 80, 110 watts.

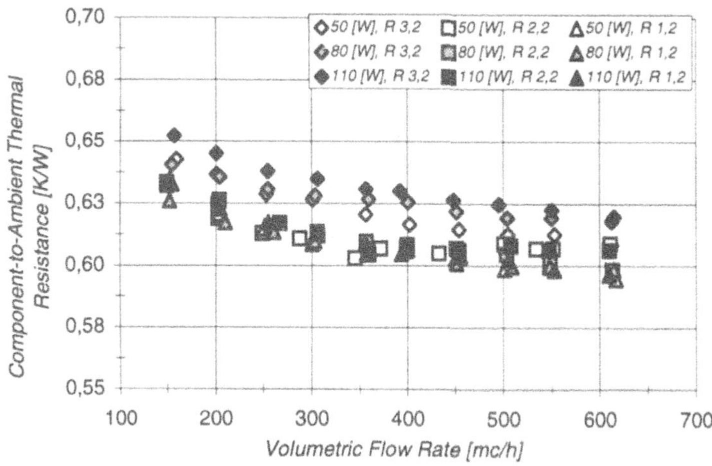

Figure 3. *Total thermal resistance (Rt) versus volumetric flow rate for different values of thermal power and for different component positions*

It ensues from the measurements that thermal power is a parameter weakly affecting the value of total thermal resistance (about 2%); it is the same with volumetric flow rate (about 4%). Such behaviour, which gets us to consider constant the total resistance value of the separately heated components, is the result of the thermal self-adjustment phenomenon of the component itself.

As a matter of fact, at a low flow rate, when the values of the convective heat transfer coefficient in the channels are smaller, conductive diffusion on the surface of the cold-plate plays a significant part, as it allows the component to work thermically on a wider area.

When the flow rate rises, the thermal flow shows a conductive diffusion on a reduced area; however it works at higher values of the convective coefficient; the combined counteracting effect of the two phenomena help keep total thermal resistance constant. The same can be said about thermal flow.

Thus we can conclude that, if a component has an available exchange area, considerably wider than its support surface, its total thermal resistance is only an inherent parameter of the component-dissipator coupling; it does not depend on the most common thermofluid dynamic parameters, such as rate and thermal flow.

The effect of mutual thermal influence between the resistances of a column has been studied by observing the temperature raising on a powered component, while the other two resistances on the column were on.

Figg.4 show the experimental results for the total thermal resistance of the 3 components, while Fig.5 shows the percentage changes of the total thermal resistance.

It is interesting to point out that total resistance is affected by the flow rate if the components are on all together; in fact the thermal boundary condition, carried out by the three powered components, is much closer to the uniform thermal flow condition and, consequently, the raising of the convective coefficient, when the flow rate increases, is much less counter-balanced by a decrease in the spreading area.

Fig.5 clearly shows how central resistance ($R_{2,2}$) is the least facilitated on account of mutual influence, as its exchange area is highly reduced by the presence of the two components near-by, whith an average raising of 25% of its resistance.

The effect of mutual influence results lower on $R_{1,2}$ where the raising of total resistance is about 16% and intermediate on $R_{3,2}$ with an average value of 20%.

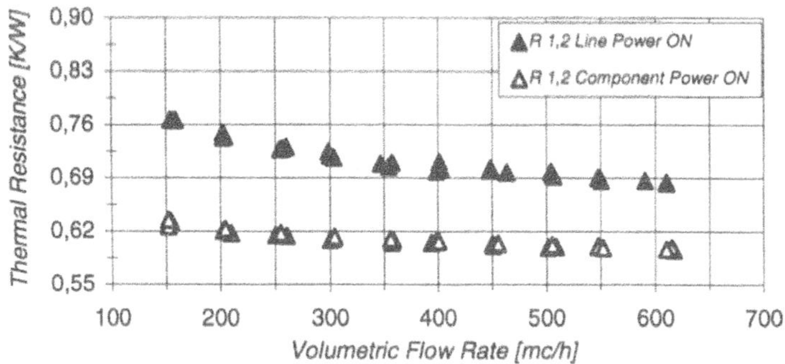

a) Total thermal resistance of component R 1,2 versus volumetric flow rate.

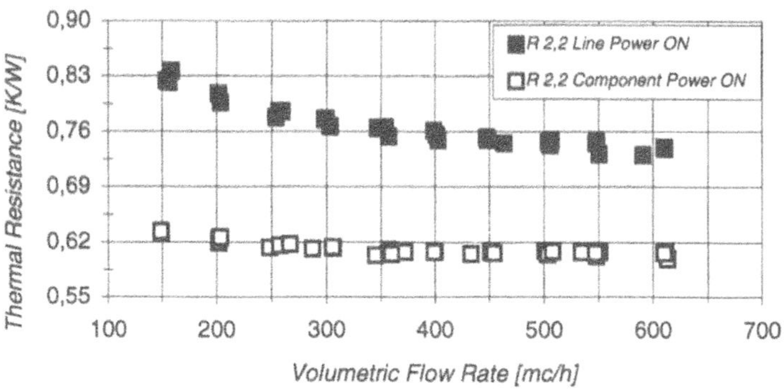

b) Total thermal resistance of component R 2,2 versus volumetric flow rate.

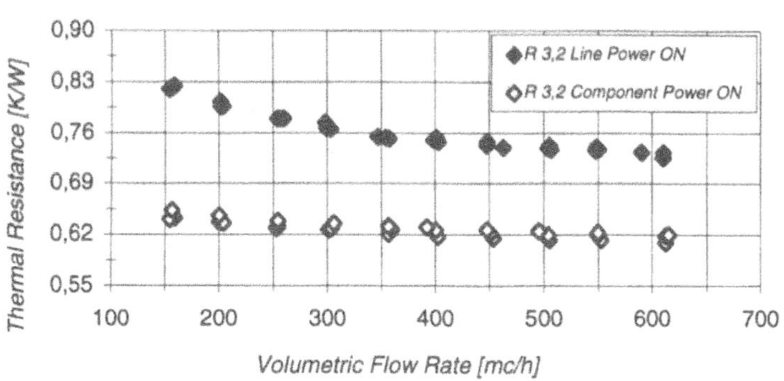

c) Total thermal resistance of component R 3,2 versus volumetric flow rate.

Figure 4. *Mutual influence on total thermal resistance versus volumetric flow rate under two working conditions: a component on and a column of components on.*

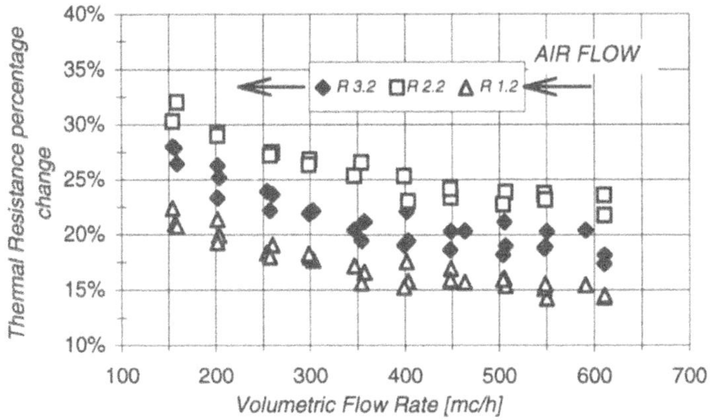

Figure 5. Percentage change in the total thermal resistance due to mutual influence.

4. Conclusions

This work has achieved preliminary results concerning the analysis of the thermofluidynamic behaviour of a rectangular channelled air cold-plate, used for the cooling of power electronic components.

Regarding to fluid dynamic operating features, experimental findings quite agree with the correlations normally used to evaluate the pressure losses.

Thermal analysis has pointed out that thermal exchange features, on localized thermal flow conditions, can be considerably different from those provided by the usual design methods based on simplified boundary conditions.

In particular when heating is highly localized, total thermal resistance does not practically depend on the flow rate in the ranges normally used for practical applications.

In addition, in the case of localized components, mounted on the cold-plate on more in-line rows, it has been checked that the position with respect to the fluid inlet edge, considerably affects thermal resistance between component and fluid and, in particular, the central position results less satisfactory.

However, the generalitation of the results concerning the influence of the component position in the cold-plate is not possible at this stage of the research owing to the presence of numerous scale factors; further investigations are being carried out to analyse the phenomenon thoroughly.

NOMENCLATURE

T_a:	inlet air temperature
T_c:	case temperature of electronic component
R_{Tot}:	total thermal resistance
P:	heat load
V:	local air velocity
V_m:	mean air velocity
D_h:	hydraulic diameter
r_h:	hydraulic radius
w_c:	rectangular channel cross section width
L:	cold-plate lenght
Q:	volumetric flow rate
ρ:	air density
Re:	Reynolds Number
K_c, K_e:	Contraction and Expansion pressure-loss Coefficients
f:	Fanning Factor
$K(\infty)$:	Hagenbach Factor
ΔP:	cold-plate pressure losses

REFERENCES

[1] ITELCO SpA (1995) UHF medium power solid state transmitters and transposers 'X' series, Sheet n. 1181950003, Orvieto (Italy).

[2] Moffat, R.J., Ortega, A., (1988) Direct air-cooling of electronic components, ch. 3 in *Advances in Thermal Modeling of Electronic Components and Systems Vol. 1*, 129-282, Bar-Cohen and Kraus eds, Hemisphere Publishing Corporation.

[3] Lasance, C.J.M., (1993) Thermal management of air-cooled electronic systems: new challenges for research, *Thermal management of electronic systems*, 3-24, Kluwer Academic Publishers, Dordrecht.

[4] Fujii, M., Seshimo, Y., Ueno, S. and Yamanaka, G. (1989) Forced air heat sink new enhanced fins, *Heat Transfer: Japanese Research* **18**, n. 6, 53-65.

[5] Boesmans, B., Christiaens, F., Berghmans, J., Beyne, E. (1993) Design of an optimal heat-sink geometry for forced convection air cooling of Multi-Chip Modules, *Thermal management of electronic systems*, 267-276, Kluwer Academic Publishers, Dordrecht.

[6] Kays, W.M. and London, A. L. (1984) *Compact Heat Exchangers*, 3rd ed., Mc Graw-Hill Book Company, 108-114, New York.

[7] Cesini, G., Fumi, E., Moro, V., Ricci, R., (1995) Cold-plates ad aria per il raffreddamento di componenti elettronici di potenza localizzati, *Proc. XIII Congr. Naz. sulla Trasmissione del Calore U.I.T.*, 109-118, Bologna (Italy).

[8] Benedict, R.P. (1976) *Fundamentals of temperature, pressure and flow measurements*, John Wiley & Sons, New York.

[9] Kakac, S., Shah, R.K., Aung, W. (1987) *Handbook of single-phase convective heat transfer*, John Wiley & Sons, New York.

This work has been carried out with the contribution of CNR, concerning the special project: *Metodi avanzati per il controllo termico dei dispositivi elettronici di potenza: raffreddamento mediante cold-plate in convezione forzata"* (fund CB CNR 94.01813.ct07)

NATURAL CONVECTION EXPERIMENTS WITH CUBOIDS AND CYLINDERS OF EQUAL AREA

R.A. VAN ES, R.M. NOORT AND C.J.M. LASANCE
Philips Centre for Manufacturing Technology
P.O. Box 218
5600 MD Eindhoven
The Netherlands.

Abstract
Heat transfer coefficients have been measured for small cuboids and cylinders of equal outer area in horizontal and vertical positions. Measurements show small, but significant differences between cuboids and cylinders.

1. Introduction

Some commercially available software packages only allow for the use of rectangular grids for reasons of userfriendliness, speed and ease of modelling. When dealing with numerical analysis of electronic systems, cylindrical objects such as electronic capacitors, power resistors and transformers, are often part of the system. Although the possibility exists to approximate a cylinder by a cuboid to an arbitrary degree of accuracy using a Cartesian grid, this solution is not recommended because of the very fine grid needed. It is to be expected that somewhere in the future local refinement techniques such as multigridding will be added to these codes, but for some time to come another approach is required.
The common way of handling a cylinder with this type of software is to replace the cylinder by a cuboid with the same outer area.
However, there are a few problems associated with this translation from one shape to another.

For one, simple mathematics show that it is not possible to retain the same outer area as well as the same length.
Secondly, when performing transient studies, the masses should be adapted in such a way as to preserve the time constant.
Thirdly, a change in shape has invariably consequences for the local pressure loss, and in the end, for the temperature, because the total pressure loss drives the resulting flow. However, this effect can be neglected as compared to the total pressure loss in an electronic system, because of the relatively low air velocities.
This topic is not further discussed in the paper.

E. Beyne et al. (eds.), Thermal Management of Electronic Systems II, 263-272.

Most importantly, the question should be answered whether the heat transfer coefficients are invariant to the change in shape. Especially the study of this aspect is the subject of this paper.

It is believed that the outcome of this research could be very valuable to a designer who wants to perform a numerical analysis of a complete electronic system, usually contains a large number of dissipating and non-dissipating cylindrical shapes. In such a case, it is much easier to replace the cylinders by cuboids.
The results can be used to answer the question if it is allowed to make this transformation, given a certain required accuracy.

Because a 3D numerical study involving all small-scale fluid phenomena that influence the heat transfer from the object to the air requires a considerable computational effort which is still beyond our capabilities, an experimental study has been chosen.
Thinking of the experimental set-up, two fundamentally different ways of tackling the problem present themselves: steady-state and transient. When performing a steady-state analysis, an accurate determination of the heat balance rules the final accuracy. Because of the rather small size of the objects (typically 7.5 mm diameter and 15 mm length), heat losses via the power leads cannot be neglected, while at the same time these losses are difficult to estimate with the required accuracy. Therefore, a transient method was considered to be the most appropriate for this research, because the power dissipation is not part of the equation, only the temperature as a function of time.

2. Experimental Set-up

Figure 1 shows the experimental set-up for measuring the heat transfer coefficient of small components. The overall heat transfer coefficient is determined from the cooling down curve of the object. Power is applied by means of a CO_2-laser and temperature is measured by an infra-red thermometer. Contactless measurement of the temperatures avoids heat losses through thermocouple wires. When the infra-red thermometer is properly calibrated and the temperature is measured perpendicular to the anodised surface of the object, temperature read out can be very accurate, say within 0.3 °C. The emissivity of the anodization layer is measured separately, using a wideband infra-red thermometer and a heating source with known temperature.
All measurements were performed in our "room-within-a-room" to minimise disturbances from large scale convection currents and changing ambient temperatures, which are usually present in standard laboratories.

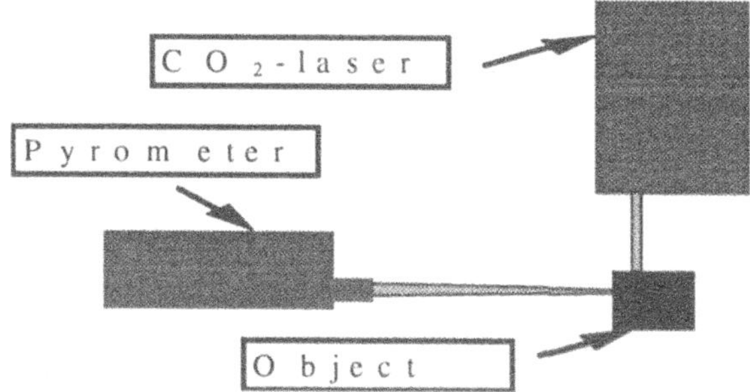

Fig. re 1. measurement set-up

The component itself is suspended on two 0.12 mm Nylon wires, which pass through two holes in the object. Conduction losses through the wires were estimated with simple fin theory as virtually absent.

Figure 2 shows a typical cooling curve for the vertical cylinder.

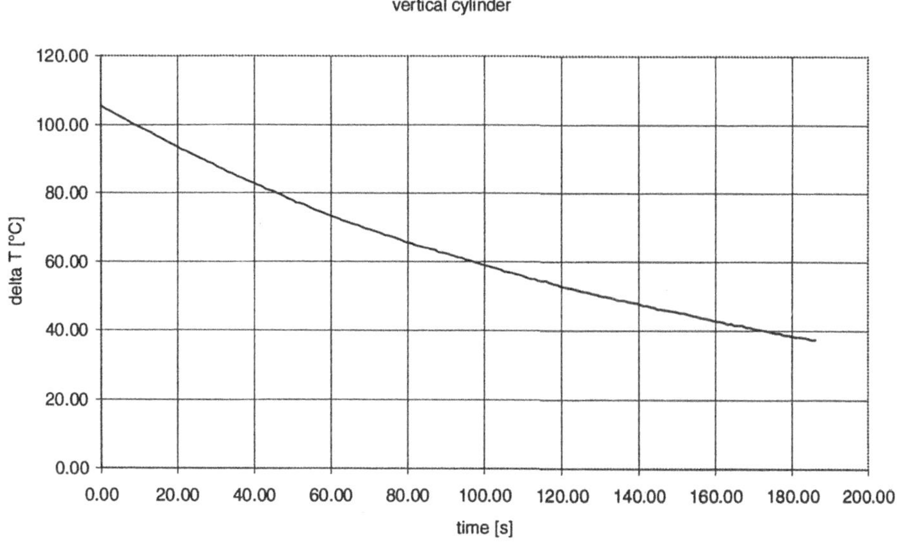

Figure 2. typical cooling curve of vertical cylinder

Figure 3 shows the natural logarithm of the temperature difference against time

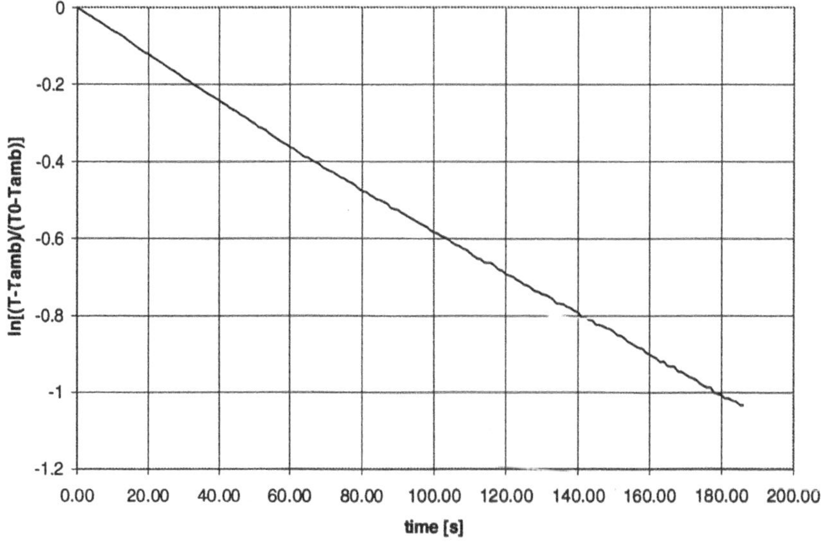

Figure 3. $\ln[(T-T_{amb})/(T_0-T_{amb})]$ versus time

The component consists of aluminium with an anodization layer. This layer has a high emissivity in order to permit proper temperature readout with an infra-red sensor. Both cylinders and cuboids have the same top and side surface area. The cylinder has a diameter of 7.5 mm and a length of 15 mm. The cuboid has a side of 6.65 mm and a length of 13.39 mm. It is not possible to retain the same volume for both objects with the above restraints. The volume of the cylinder is 663 mm^3 and of the cuboid 587 mm^3.

The heat transfer coefficient, which follows from the experiments and the above equation, is an overall heat transfer coefficient. It consists both a free convection part and a radiative part.

By subtracting the radiative part from the overall heat transfer coefficient the convective part is calculated.

Both convective and radiative heat losses are weak functions of temperature over the measurement range.

3. Data reduction

The convective heat transfer coefficient can be calculated in a number of ways. Usually the temperature at the beginning of the cooling curve T_0 is used to calculate the overall heat transfer coefficient. This method can only be used if the heat transfer coefficient remains constant over the time interval used in the calculation.

To overcome this problem the initial temperature is regarded as the object temperature 10 seconds before the actual value T..

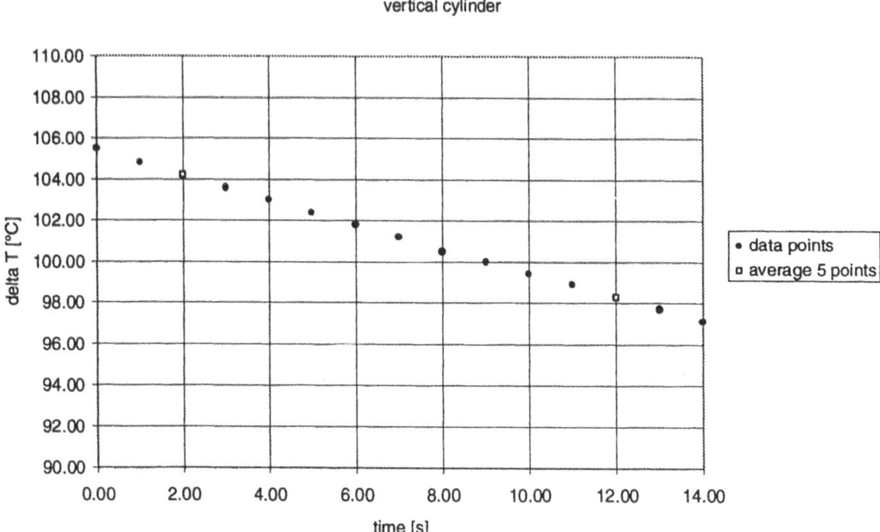

Figure 4. delta T versus time

The average of the first 5 data points is taken and the average of data points 11 to 15.

$$T_1 = \sum_{1}^{5} \frac{T_i}{5} \qquad T_2 = \sum_{11}^{15} \frac{T_i}{5} \tag{1}$$

The overall heat transfer coefficient is calculated with the use of these two average temperatures. The two average temperatures are taken 10 seconds apart, in order to minimise the error in calculating the overall heat transfer coefficient. This is shown graphically in figure 4 with the initial temperature at 2 seconds and the actual temperature at 12 seconds

The next overall heat transfer coefficient is calculated by using data points 6 to 10 and data points 16 to 21.

From these two average temperatures the overall heat transfer coefficient is calculated

$$h_{overall} = -\frac{\rho V C_p}{A(t_2 - t_1)} \ln\left[\frac{T_2 - T_{amb}}{T_1 - T_{amb}}\right] \tag{2}$$

The radiative part of the heat transfer coefficient at component temperature T_2 and ambient temperature T_{amb}.

$$h_{radiation} = \left[(T_2 + 273)^2 + (T_{amb} + 273)^2\right]\left(T_2 + T_{amb} + 546\right)\varepsilon\,\sigma \qquad (3)$$

The convective heat transfer coefficient can be calculated by subtracting the radiative part from the overall heat transfer coefficient.

$$h_{convection} = h_{overall} - h_{radiation} \qquad (4)$$

Results for the convective heat transfer coefficient for a vertical cylinder are shown in figure 5.

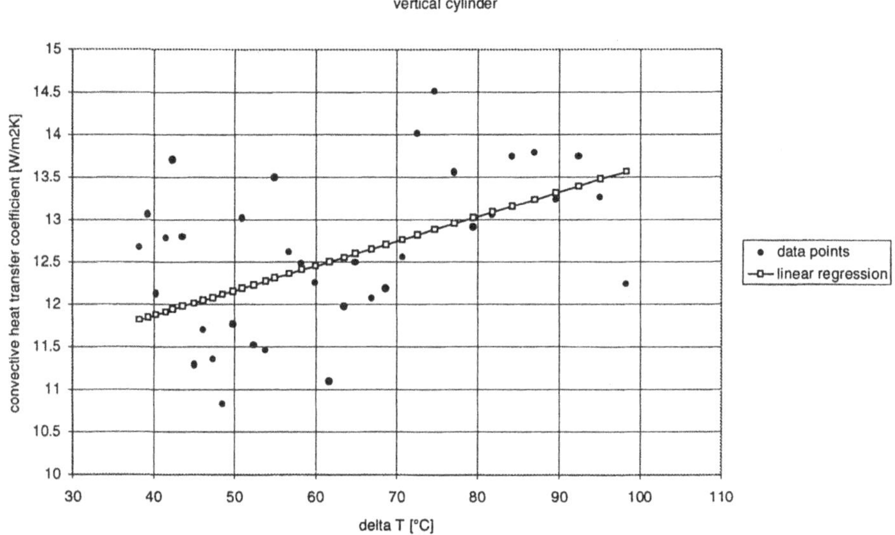

Figure 5. convective heat transfer coefficient of vertical cylinder

4. Statistical Analysis

As can be seen from figure 5 calculated overall heat transfer coefficients show large deviations from point to point. This is due to the nature of formula (2). Error analysis shows that small deviations in the temperatures T_1 and T_2 have large effects on the calculated heat transfer coefficients. Figure 6 shows the error in the overall heat transfer coefficient as a function of the error in T_1 and T_2 at a ΔT of approximately 100 °C. As can be seen an error of 0.2 °C in measured temperatures yields an error of 1 [W/m^{2}K] in the heat transfer coefficient.

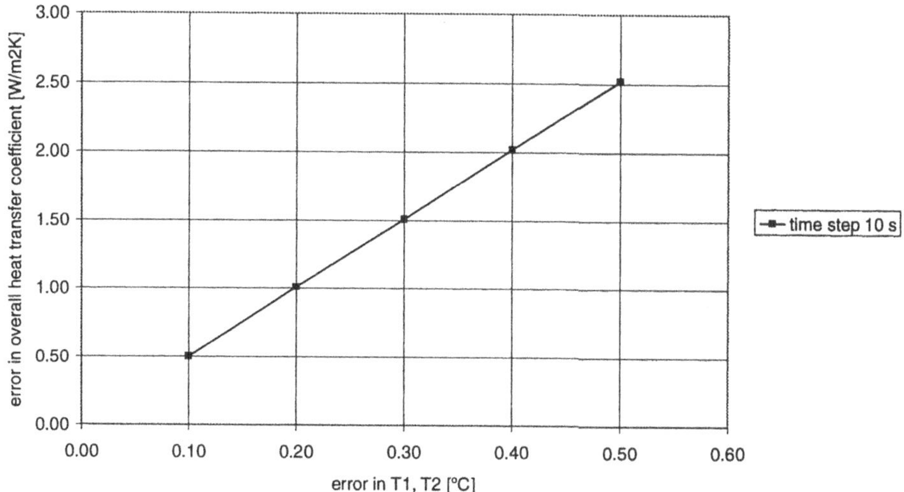

Figure 6: Error in overall heat transfer coefficient

In order to compare the results a regression model was used over a temperature difference range of 40 to 100 °C. The cylinder model is used as a reference model . Usually the correlation between the temperature difference and the convective heat transfer coefficient can be described by:

$$\overline{Nu} = C\,(Gr \cdot \Pr)^n\, K \tag{5}$$

The Nusselt number usually has a power n of 0.25 in the laminar flow region. Over the measured temperature range fitting the data with a power law does not yield a better fit than the use of linear regression. For the sake of simplicity the linear regression model was chosen. For this reason data obtained for our measurement range should not be extrapolated beyond the measured interval.

The linear models that were used are shown below.

cylinder $\Rightarrow$ $$\overline{h}_c = \beta_0 + \beta_1 T + \varepsilon_{full} \tag{6}$$

tilted cuboid $\Rightarrow$ $$\overline{h}_c = (\beta_0 + \beta_2) + (\beta_1 + \beta_4)T + \varepsilon_{full} \tag{7}$$

horiz. cuboid $\Rightarrow$ $$\bar{h}_c = (\beta_0 + \beta_3) + (\beta_1 + \beta_5)T + \varepsilon_{full} \quad (8)$$

The model for the cylinder is used as a reference model to test if the other models are significantly different from the cylinder.
If there are no differences between the cylinder, tilted cuboid and horizontal cuboid, the null hypothesis is valid, namely $\beta_2 = \beta_3 = \beta_4 = \beta_5 = 0$.

The so called F parameter can be used to test the null hypothesis.

$$F = \frac{(SSE_{red} - SSE_{full})/2}{SSE_{full}/(n-2)} \quad (9)$$

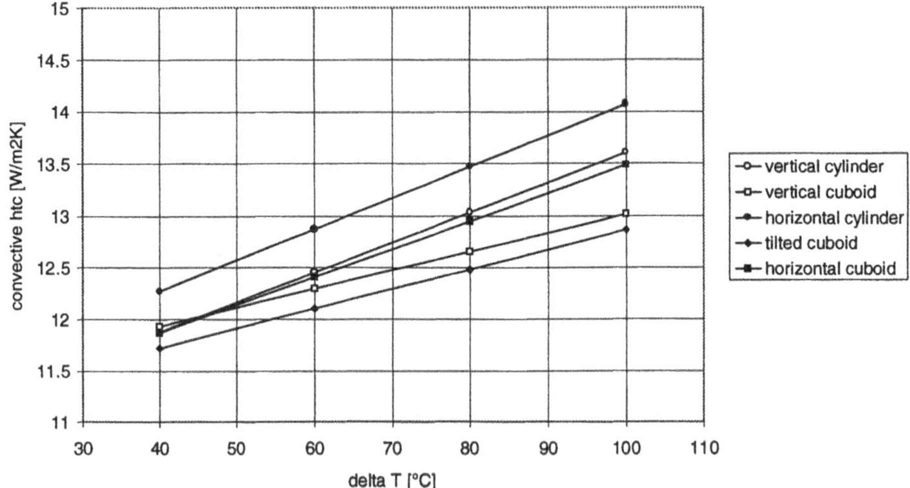

Figure 7. convective heat transfer coefficients for cuboids and cylinders

The F parameter test shows that all differences between cylinders and cuboids are significant. Figure 7 shows the linear fit of the convective heat transfer coefficient as a function of temperature difference.
Table I shows a comparison of convective heat transfer coefficients at three different object temperatures.

TABLE I. Convective heat transfer coefficient [W/m^{2}K] with 95% confidence interval

delta T [°C]	40	60	80	100
Horizontal Cylinder	12.3± 0.2	12.9 ± 0.1	13.5 ± 0.1	14.1 ± 0.3
Tilted Cuboid	11.7 ± 0.2	12.1 ± 0.1	12.5 ± 0.1	12.9 ± 0.3
Horizontal Cuboid	11.9 ± 0.2	12.4 ± 0.1	13.0 ± 0.2	13.5± 0.3
Vertical Cylinder	11.9 ± 0.2	12.5 ± 0.1	13.0 ± 0.2	13.6 ± 0.3
Vertical Cuboid	11.9 ± 0.2	12.3 ± 0.1	12.7 ± 0.2	13.0 ± 0.3

5. Conclusions

At a ΔT of 60 °C, a vertical cuboid shows 1% lower values than a vertical cylinder. A tilted (45 °angle) cuboid shows 6% lower values than a horizontal cylinder. A horizontal cuboid shows 4% lower values than a horizontal cylinder.
F test shows that differences between objects are small but significant in all cases. Differences between subsequent runs with the same object are not significant.
In summary, it can be concluded that significant differences in heat transfer coefficients exist between cylinders and cuboids with the same outer areas, at least for natural convection. However the differences are so small that in normal practice cylinders can be replaced by cuboids.

6. Acknowledgement

The partial support of the Commission of European Communities under ESPRIT contract DELPHI is acknowledged.

7. Notation

A	[m^2]	surface area
C_p	[J/kg°C]	heat capacitance
h	[W/m^2K]	heat transfer coefficient
n	[-]	number of data points
t	[s]	time
T	[°C]	temperature
V	[m^3]	volume
ϵ	[-]	emissivity
ρ	[kg/m^3]	density
σ	[W/m^2K^4]	Stefan-Boltzman constant

Subscripts

amb	ambient temperature
full	referring to full model
red(uced)	referring to reduced model
i	referring to measurement i

8. References

[1] J. Beck and K. Arnold: Parameter Estimation, *John Wiley & Sons*

[2] G. Box, W. Hunter, J. Hunter: Statistics for Experimenters, *John Wiley & Sons*

[3] R. Moffat: Using Uncertainty Analysis in the Plaaning of an Experiment, *Journal of Fluid Engineering, June 1985, Vol. 107.*

NATURAL CONVECTION HEAT TRANSFER OF METAL CUBOIDS FLUSH MOUNTED IN A HORIZONTAL PLATE

Ir. J.P.A. DRABBELS
University of Technology Eindhoven, the Netherlands

Abstract

The increased heat production per unit area in electronic equipment leads to a more critical heat balance. A more sophisticated and accurate way to calculate or measure the heat transfer from critical components to the local environment is needed. A lot of advanced computer packages exist for calculating the thermal behaviour of electronic equipment. If the heat transfer is strongly dependent on natural convection near horizontal surfaces, small changes in boundary or initial conditions could lead to significant differences in the calculated temperature- and velocity field (see Lasance [1]). The assumptions made in the calculation procedure or boundary conditions can lead to errors, that in the near future will not be within tolerable limits. Therefore an experimental apparatus is designed, that enables a better interpretation of numerical results corresponding to natural convection heat transfer near horizontal plates with discrete heat sources.
The experimental apparatus consists of a test cabinet, a specially designed test object and a temperature measurement and flow visualization technique. The experimental results are compared with numerical results, calculated with the finite volume package Flotherm from Flomerics. A parameter study is made in order to discuss the differences between the experimental and numerical results. After elimination of the important error sources, the overall agreement between experimental and numerical results is good.

Introduction

The heat generated in consumer electronics is dissipated to the environment by conduction, convection and radiation. Since it is not desirable to apply a fan, the convective heat transfer is strongly dependent on the temperature field and flow resistance, which are determined by the orientation of the printed circuit boards (PCBs) (horizontal, vertical), the relative positions of the PCBs and the distribution of the components on the PCBs. For spatial reasons, it is sometimes necessary to position the PCBs horizontally in the apparatus.
The reliability of electronic equipment decreases when the temperature increases. PCBs are designed so that even at high ambient temperature the junction temperature of all components remains below an upper temperature limit, generally 125 $^{\circ}$ C. A safety factor is often applied, because the heat transfer cannot be determined accurately.

E. Beyne et al. (eds.), Thermal Management of Electronic Systems II, 273-280.

However this has become more difficult, as technology advances the number of features increases and more functions are integrated into single components. Besides, the consumer asks for a more compact product. The number of components on a PCB and the size of the components is thus strongly increasing. The produced heat per unit area is strongly increasing. In order to guarantee a reliable product in the future, a more sophisticated and accurate way to calculate or measure the heat transfer from critical components to the local environment is needed. This article is directed to this issue. Section 1 describes an experimental apparatus that enables a better interpretation of numerical results corresponding to natural convection heat transfer near horizontal plates with discrete heat sources. Section 2 describes the results obtained with the apparatus. Section 3 describes the numerical calculation procedure and numerical results. The experimental and numerical results are compared in section 4. Section 5 gives some conclusions.

1. Test Apparatus

Figure 1 shows a schematic drawing of the test apparatus. It consists of a test cabinet, a test object, a temperature measurement technique and a flow visualization technique. The function of the above modules will now be discussed.

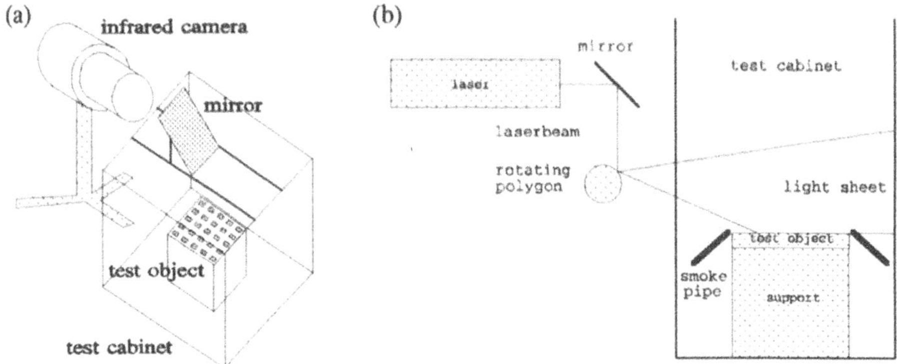

Figure 1 Schematic drawing of the test apparatus; (a) temperature measurement technique; (b) flow visualization technique.

1.1 TEST CABINET

The test object is placed inside an enclosure, called the test cabinet. The dimensions are equal to 0.6*0.6*1.6 m. This test cabinet does not have a significant influence on the temperature and velocity distribution near the test object. Two vertical walls are made of plywood, the other two vertical walls are partly made of polystyrene glass for visualization purposes.

1.2 TEST OBJECT

Figure 2 shows a picture of the test object. It consists of a polystyrene board (dimensions 200*200*20mm) and a matrix of 25 flush mounted aluminum components. Heat transfer by radiation between components does not have to be taken into account. The components are of equal size and shape. The dimensions (20*15*15mm) are representative for certain components in electronic equipment. Each component contains an electrical resistance of $\pm$ 100 Ω . By putting all resistances parallel to one another, all kinds of dissipation patterns can be created. The test object is placed on top of a polystyrene support (dimensions 200*200*200mm).

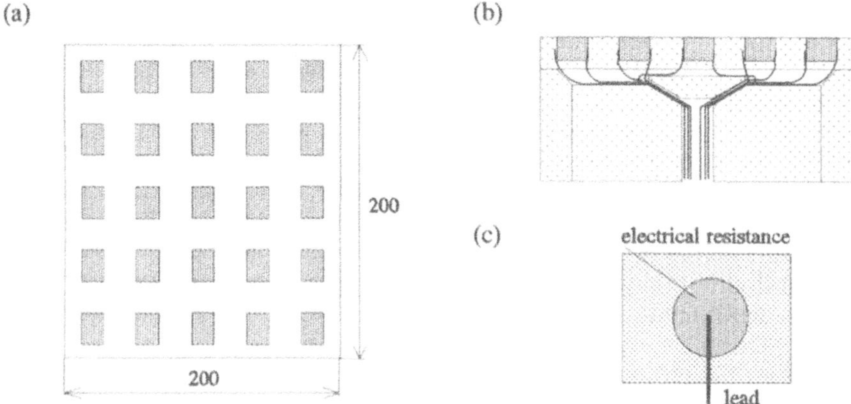

Figure 2 (a) Top view of test object; (b) cross section of test object and support; (c) cross section of one component.

1.3 TEMPERATURE MEASUREMENT TECHNIQUE

An infrared camera (AGA 680 thermovision) is chosen for measurement of the temperature distribution on the test object, because it is a contactless measurement method and it gives the complete temperature distribution on the test object. The absolute temperature is obtained by calibrating the camera with a temperature reference body. The uncertainty interval is approximately equal to $\pm$ 2 K.

1.4 FLOW VISUALIZATION TECHNIQUE

In order to explain the overall relation between the temperature- and velocity field, the flow pattern should be visualized. Several visualization methods are investigated. Injection of cigarette smoke was chosen for its simplicity, low cost and neutral buoyancy. A more detailed description of the method is shown in Moffat [2] and Drabbels [3]. Figure 1b shows a schematic drawing of the visualization technique. The cigarette smoke, produced by blowing air through a (filter) cigarette, is transported through hoses to both sides of the test object. A light sheet, produced by a laser and rotating polygon is positioned at the same row of the test object as the two hoses. The video camera is directed normal to the plane of the laser sheet. The light sheet (approximately 30 cm high) is very thin (less than 1 mm). It is now possible to obtain an impression of the local flow pattern.

2. Experimental results

2.1 MEASURED TEMPERATURE DISTRIBUTION

The temperature distribution on the test object is measured at three dissipation rates (0.4 W, 0.6W and 0.8 W per component) and several dissipation patterns. The results show that for the normal dissipation pattern (all components are dissipating heat) the temperature of the components in the center of the testobject is somewhat higher than the temperature of the components on the edge of the test object. The temperature gradients near the boundary of aluminum component and board material are rather high. The temperature distribution of the components is not completely symmetric. Deviations of approximately 3% of the absolute temperature difference do occur. This is mainly caused by small differences of dissipation rate per component (the consequence of small differences in electrical resistance of component and voltage drop over component). For a quantitative presentation of the component temperatures, the reader is referred to section 4.

2.2 VISUALIZED FLOW PATTERN

The flow pattern is visualized with the apparatus, shown in figure 1b. The air flow near the test object is mainly directed towards the center of the test object. The flow pattern is laminar in this region. Each powered component produces a plume, which rises towards the center of the test object. It thereby oscillates in a vertical plane with increasing amplitude further downstream, until a transition to a turbulent state takes place. The transition position depends on the dissipation rate and dissipation pattern. The position moves upstream when the dissipation rate gets higher. The dissipation pattern has a stronger influence on the transition position. If only the center component is powered, a single plume is produced. The plume is not disturbed by other plumes. Transition to a turbulent state takes place at approximately 1 m above the test object, sometimes even at 1.5 m above the test object. If more components produce plumes, the transition position moves upstream. If all components are powered, the transition position lies somewhere at 0.5 m above the test object.

3. Numerical Calculations

3.1 NUMERICAL CALCULATION PROCEDURE

The numerical calculations are performed with the finite volume package Flotherm from

Flomerics. The solution procedures in Flotherm are based on the techniques of Computational Fluid Dynamics (CFD). The flow visualization results show that the flow near the test object is laminar. It is assumed that the influence of the oscillating plumes and the transformation to a turbulent state at approximately 50 cm above the test object have a negligible influence on the temperature distribution of the test object. Therefore the calculations are performed steady state without a turbulence model. The problem will be treated as symmetric, so only one quarter of the cabinet is calculated. It is now possible to apply a finer mesh distribution.
The dimensions of the test object are equal to the dimensions in the experiment. The dimensions of the calculation domain (represents dimensions test cabinet) are chosen in such a way that a further enlargement of the dimensions does not have a significant effect on the temperature distribution of the components on the test object. The results show that the dimensions of the experimental test cabinet (0.6*0.6*1.6m) are probably large enough. At the cabinet walls the no-slip boundary condition is prescribed. The heat transfer coefficient at the outer walls is set to 20 W m^{-2} K^{-1}. At the bottom, the temperature is set to zero. The thermal conductivity of the components is set to 200 W m^{-1} K^{-1}. The thermal conductivity of the board material and the emission coefficient of the test object surface (heat loss by radiation) are varied. The mesh distribution is refined until the temperature of the components does not change substantially. The mesh is refined inside the test object at the component-board boundary and in the boundary layer just above the test object.

3.2 NUMERICAL RESULTS

Figure 3 shows the calculated temperature and velocity field near the test object in the horizontal and vertical plane at a dissipation rate of 0.8 W per component. The calculated component temperatures are written in each component. Large temperature gradients occur in the board near the boundary with the components. The velocities are directed towards the center of the test object. The velocity distribution is symmetric with respect to the diagonal (see figure 3a). On the symmetry line of the test object, each component produces a plume (see figure 3b). The flow between the components rises also. At larger distances from the symmetry lines, the vertical velocity component is smaller. The calculated and measured flow pattern is approximately the same near the test object. Further downstream the differences become significant, mainly caused by oscillations and transition to turbulence.

4. Comparison of Experimental and Numerical Results

A numerical parameter study is made to investigate the sensitivity of several parameters on the temperature distribution of the components on the test object. The important parameters are: the calculation procedure, initial and boundary conditions, the mesh distribution, the dimensions of the calculation domain, the material properties and emission coefficient. The measured temperature distribution is dependent on the accuracy of the infrared camera, small differences in electrical power per component and the heat loss through the leads of the electrical resistances. The influence of the most important parameters will now be discussed.

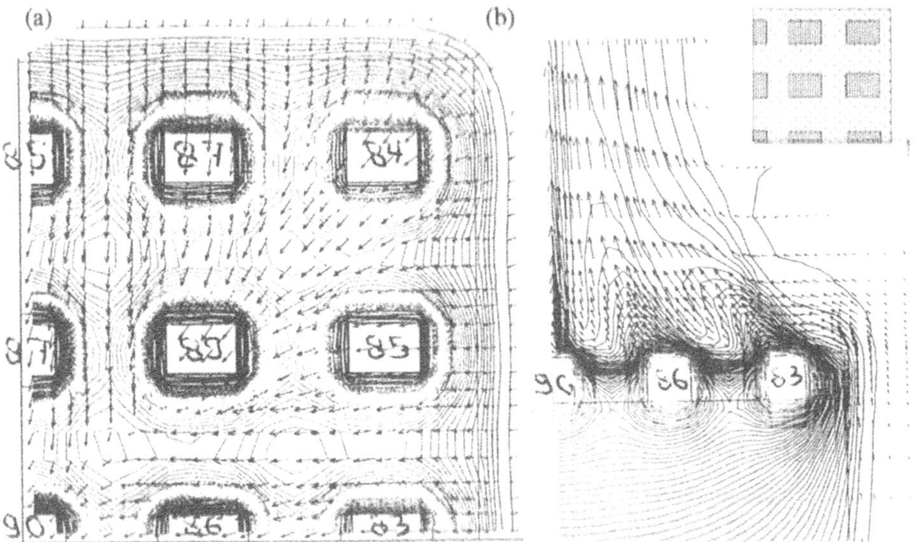

*Figure 3 Temperature- and velocity field near test object in the horizontal (a) and vertical plane (b) when all components are dissipating heat. $\Phi = 0.8$ W per component, $\lambda_{board} = 0.027 + 13\ 10^{-5} * \Delta T$ W m^{-1} K^{-1}, $\epsilon = 0.9$, mesh (39*39*41).*

4.1 THE EMISSION COEFFICIENT OF BLACK PAINT

According to VDI Wärmeatlas [4] and Philips [5], the emission coefficient of black oil paint varies somewhere between 0.9 and 0.96. The emission coefficient of polystyrene is approximately equal to 0.9. The influence of ϵ is calculated numerically with Flotherm. The calculated component temperature decreases approximately 1 K if $\epsilon = 0.95$ is applied instead of $\epsilon = 0.90$.

4.2 THE THERMAL CONDUCTIVITY OF THE BOARD MATERIAL

Polystyrene is chosen for the board and support for its low thermal conductivity. The value of the thermal conductivity of polystyrene, presented in literature, often differs. Values between 0.027 W m^{-1} K^{-1} and 0.1 W m^{-1} K^{-1} are found. One explanation for this large interval is the difference between extruded and expanded polystyrene. Since the thermal conductivity strongly influences the component temperature, its value is measured with with a Poensgen apparatus (see Drabbels [6]). The following relationship is found:

$$\lambda_{board} = 0.027 + 13 * 10^{-5} * (T_{board} - T_{\infty})\ \mathrm{W\ m^{-1}\ K^{-1}} \qquad (1)$$

with T_{board} the temperature of the board material and T_{∞} the ambient temperature. The corresponding accuracy is equal to $\pm$ 4%. Figure 4 shows the calculated temperature distribution of the components at three different values of λ_{board}. The results are presented in one quarter of the test object (see figure 3a). Flotherm calculates a large influence of the thermal conductivity of the board material. The uncertainty interval for the component temperature, corresponding to the uncertainty interval of formula 1 is approximately equal to ± 1 K. Since the thermal conductivity of the board material is approximately equal to the thermal conductivity of air, the influence of a contact resistance between components and board can be neglected.

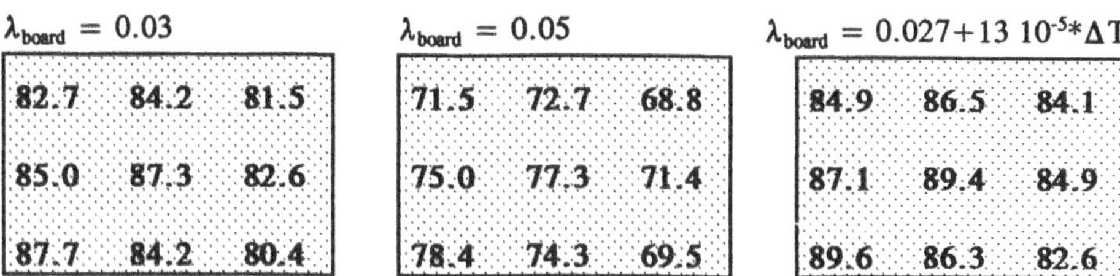

$\lambda_{board} = 0.03$

82.7	84.2	81.5
85.0	87.3	82.6
87.7	84.2	80.4

$\lambda_{board} = 0.05$

71.5	72.7	68.8
75.0	77.3	71.4
78.4	74.3	69.5

$\lambda_{board} = 0.027+13\ 10^{-5}*\Delta T$

84.9	86.5	84.1
87.1	89.4	84.9
89.6	86.3	82.6

Figure 4 *Calculated temperature distribution at $\lambda_{board} = 0.03$, 0.05 and $0.027+13\ 10^{-5}*\Delta T$ W m^{-1} K^{-1}. $\Phi = 0.8$ W per component, $\epsilon = 0.9$, mesh (39*39*41).*

4.3 HEAT LOSSES THROUGH THE LEADS OF THE ELECTRICAL RESISTANCES

The influence of heat losses through external surfaces is investigated in Drabbels [6]. The heat loss through the leads is a function of the thermal conductivity of the board material and leads, the diameter of the leads, the dimensions of the support and the distribution of the leads inside the support. The latter parameter results in a difference in heat loss through the leads between components in the center and components on the edge. Drabbels [6] calculated a heat loss between 13% (components in center of test object) and 16% of the total heat loss (components on the edge of the test object). The uncertainty interval for both heat losses should be smaller than ± 3%. The power applied at the numerical calculations should be compensated for these heat losses. Figure 5 shows the calculated temperature distribution at a power of 0.8 W per component, the calculated temperature distribution at a power distribution corrected for different heat loss through the leads and the measured temperature distribution. From this figure can be concluded that there is an excellent agreement between experimental and numerical temperature distribution.

calculated
0.80 W

84.9	86.5	84.1
87.1	89.4	84.9
89.6	86.3	82.6

calculated
0.70 W center
0.67 W edge

73.4	74.7	72.4
78.0	80.0	73.4
80.4	77.2	71.4

measured
0.80 W

74	72	68
78	76	73
77	76	74

Figure 5 *Calculated temperature distribution at $\Phi = 0.8$ W per component, calculated temperature distribution with Φ corrected for heat loss through leads. $\lambda_{board} = 0.027+13\ 10^{-5}*\Delta T$ W m^{-1} K^{-1}, $\epsilon = 0.9$, mesh (39*39*41) and measured temperature distribution at $\Phi = 0.8$ W per component.*

5. Conclusions

An experimental apparatus is designed for measurement of the temperature distribution on and for visualization of the flow pattern above a horizontal plate with discrete heat sources. The apparatus helps increasing the insight in the thermal behaviour of printed circuit boards at natural convection conditions. The temperature field is measured with an infrared camera. The complete temperature distribution can be obtained with an accuracy of $\pm$ 2 K. The flow pattern is visualized by injecting cigarette smoke in the test cabinet. This method is suited for a qualitative determination of the flow field.
The numerical calculations are performed with Flotherm. The calculated temperature distribution on the test object depends on several parameters. The investigated parameters are the dimensions of the calculation domain, the mesh distribution, the boundary conditions, the material properties (especially thermal conductivity of the board material) and the emission coefficient of the test object surface. The measured temperature distribution depends on the accuracy of the infrared camera and calibration procedure, small differences in the electrical power delivered to each component and heat losses through the leads of the electrical resistances. In order to compare the measured and calculated temperatures, the influence of the above parameters is determined as accurately as possible.
Although much attention is paid to the accuracy of the experimental apparatus, a high accuracy cannot be obtained. This is mainly caused by the uncertainty interval corresponding to heat losses through the leads and temperature measurement errors. Taken everything into account leads to the conclusion that the agreement between the calculated and visualized flow pattern near the test object and the agreement between measured and calculated temperature distribution is good.

References

1. Lasance C.J.M., Thermal management of air-cooled electronic systems: New challenges for research, *Thermal management of electronic systems, proceedings of Eurotherm Seminar 29 in Delft*, Kluwer Academic Publishers, pp 3-24, 1994.
2. Moffat R.J., Experimental methods for air cooling of electronic components, in Aung W. (editor), *Cooling techniques for computers*, Hemisphere publishing corporation, p.333, 1991.
3. Drabbels J.P.A., *Flow visualization*, Eindhoven University of Technology, Faculty of Mechanical Engineering, section WET, report 94.013, Eindhoven the Netherlands, 1993.
4. VDI Wärmeatlas, *Berechnungsblätter für den Wärmeübergang 5*, VDI Verlag, 1988.
5. Galenkamp H., Muyzenberg v.d.H., Nederlandse Philips bedrijven B.V., Center for Manufacturing Technology, report CTB591-94-6019, 1994.
6. Drabbels J.P.A., *Heat transfer from electronic components, mounted on a horizontal printed circuit board*, Eindhoven Institute for Continuing Education, Faculty of Mechanical Engineering, ISBN 90-5282-418-5, 1994.

5. MEASUREMENT TECHNIQUES

VISUALIZATION OF NATURAL CONVECTION IN INCLINED HEATED PARALLEL PLATES

ORONZIO MANCA, BIAGIO MORRONE AND SERGIO NARDINI
DETEC - Univeristá degli studi Federico II
P.le Tecchio 80, 80125 Napoli

1. Introduction

Natural convection has been thoroughly investigated due to its importance in many engineering applications. Furthermore, in the last years natural convection along isothermal and isoflux inclined plates has been analyzed because of its occurrence in several technical applications such as thermal control of electronic devices, solar collectors and so on. These situations of technical interest can involve thermal instabilities in plane air layers even at relatively low temperatures.

When a natural convection flow is induced by heating from below horizontal or slightly inclined plate with uniform wall temperature, the flow is potentially unstable due to the top-heavy situation occuring. The onset of the instability is caused when the buoyancy forces overcome the stabilizing effects of viscous and thermal diffusion. Many studies investigated the instability over inclined or slightly inclined plate heated with uniform temperature or uniform heat flux.

The horizontal and slightly inclined plate configurations have been widely investigated. One of the first studies on this topic has been carried out by Sparrow and Husar [1], who performed an experimental analysis by means of a visualization technique. They observed the occurrence of a secondary flow superimposed upon natural convection main flow on inclined plates. Similar investigations have been accomplished by Lloyd et al. [2], Haaland and Sparrow [3], Shaukatullah and Gebhart [4] and Pera and Gebhart [5]. De Graaf and van der Held [6] obtained experimentally that the convective heat transfer between parallel plates, heated from below, is characterized by longitudinal convection rolls whose axes are parallel to the primary natural convection flow. The occurrence of roll vortex instability in parallel-plate

E. Beyne et al. (eds.), Thermal Management of Electronic Systems II, 283-292.

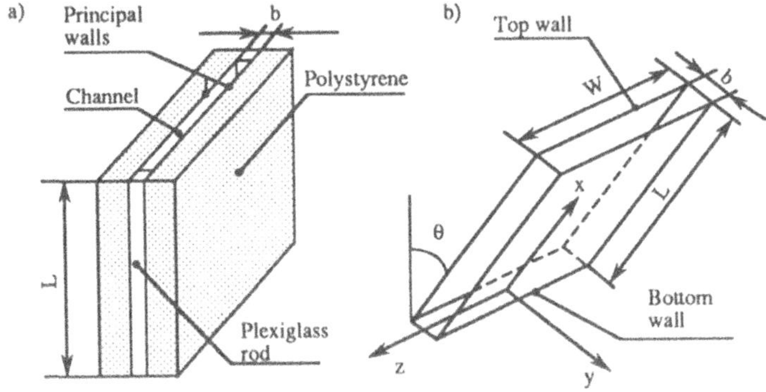

Figure 1. Sketch of the system: a) test section; b) geometry of the channel

horizontal and slightly inclined channels was investigated by Kurzweg [7]. Hart [8], Ruth et al. [9] and Clever and Busse [10] carried out both experimental and theoretical studies on the convective instability of fluid layers in inclined isothermal parallel plates heated from below. Kennedy and Khanel [12] accomplished an experimental investigation for inclined channel in natural convection flow, by means of a visualization technique. They observed a change in the thermofluidynamic regime for angles close to 70° respect to the gravity vector, that gave rise to three-dimensional longitudinal vortex rolls. Cheng and Kimura [13] presented flow visualization results in order to study the convective instability phenomena for natural convection flow in parallel-plates channels with an inclination angle from the vertical of 80° subjected to isothermal heating from below. They investigated the effects of air layer thickness on the flow patterns, and they observed three kinds of characteristic secondary flow.

In spite of the previous papers, to the knowledge of the authors it seems that there are no deep experimental analyses on the instability behavior of the fluid flow for natural convection in inclined channels with plates heated with uniform heat flux. In this paper an experimental analysis by means of a visualization technique of the air flow at different cross sections has been presented. The configurations have been analyzed with sub-horizontal and horizontal angles, 75°, 85° and 90° respect to the gravity vector.

2. Experimental apparatus

2.1. TEST SECTION

The test section is shown schematically in fig.1. The channel was made from two principal parallel walls with uniform heat flux, its side walls were unheated. Each principal wall consisted of two sandwiched phenolic fiber-bord plates, 400 mm high and 530 mm wide. The plate facing the channel was 3.2 mm thick and its surface adjacent the internal air was coated with a 35 μm thick nickel plated copper layer. The low emissivity of nickel (0.05) minimizes the radiation effect on heat transfer. The backside plate was 1.6 mm thick. Its back surface was coated with a 17.5 μm thick copperlayer, which was the heater. In order to reduce heat losses, a 150 mm polystyrene block was affixed to rear face of each plate. The side walls were made of plexiglass rectangular rods, machined within an accuracy of 0.03 mm. The spacing of the channel was mesured with an accuracy of $\pm 0.25\,mm$ by a dial-gauge equipped caliper, which could resolve 0.025 mm. The channel was 400 mm long and 475 wide and was open to the ambient along its top and bottom edges, fig.1b. It was secured to a tilting support frame. Discrete inclination were obtained by brackets and pinned joints, with an accuracy of $\pm 0.5°$. A finer adjustement ($\pm 0.05°$) was obtained in the range $85° - 90°$ from the vertical by micrometrical screw system. The entire apparatus was located within an enclosed room, sealed to eliminate extraneous air currents. The plates were heated by passing a direct electrical current through the heaters. The dissipated heat flux per board was evaluated with an accuracy of $\pm 2\%$ by measuring the voltage drop across the heaters and the electrical current passing through them. Fifteen thermocouples were affixed to the rear surface of the plates and embedded in the styrene to enable evaluation of conductive heat losses. Their maximum value was 28% of the Ohmic heat rate dissipated in each plate, in the investigated range of inclination angle $75° - 90°$. The ambient temperature was measured by shielded thermocouples placed near the leading edge of the channel.

A more detailed description of the experimental apparatus is reported in Manca et al. [14].

2.2. FLOW VISUALIZATION TECHNIQUE

Smoke for visualization was generated by burning incense sticks in a copper tube, connected to a compressor. The smoke was injected through a sort of glass heat exchanger to reduce the temperature of the smoke. The smoke was sent into a plenum and its temperature was controlled by means of a thermocouple. This value was close to that of the air ambient incoming into the channel. Then it was driven in the test section through a small slot situated along the leading edge of the channel. A sketch of the apparatus is reported in fig.2. The visualization was made possible by means of a laser sheet, generated by a He-Ne laser source.

The laser sheet was produced by placing a mirror near the end of the test section with an angle of 45° respect to the direction of the main flow, after which a cylindrical lens was placed to enlarge the beam as nedeed. A

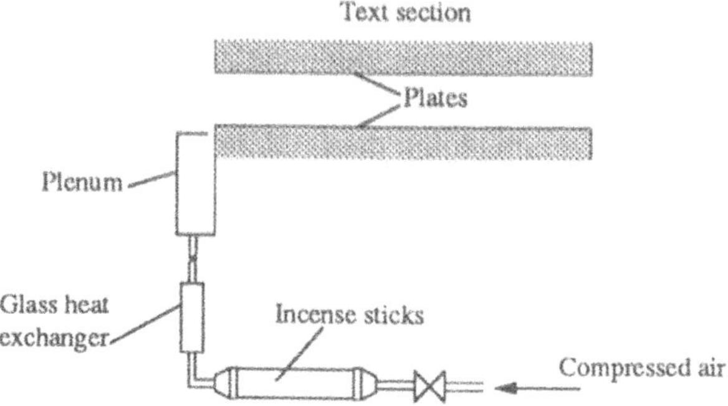

Figure 2. **Schematic view of the experimental apparatus**

fine regulation was allowed by means of a micrometer screw system, in order to get photos at different locations along the z axis. The same arrangement was used to obtain pictures of the secondary flow in the $y-z$ plane at several x locations. This procedure allowed to check whether or not the instability sets in.

3. Results and discussion

Experiments have been performed, with working fluid air, at three different angles of inclination respect to the gravity vector (90°, 85° and 75°), with one channel spacing, $b = 40\,mm$, the heat flux dissipated by the plates was set equal to 125 and 250 W/m^2, and three different heating modes were analyzed in this study. In the following these three modes will be indicated in Roman numbers as : I) both walls heated; II) top wall heated and the bottom one unheated; III) top wall unheated and the bottom one heated. These configurations produced Grashof numbers equal to $6.2 \cdot 10^4$ and $1.1 \cdot 10^5$, where the Grashof number is defined as:

$$Gr = \frac{g\,\beta\,\overline{q}_c\,b^5}{\nu^2\,k\,L}$$

and $\overline{q}_c$ is the average convective heat flux:

$$\overline{q}_c = \frac{1}{2L}\left\{\int_0^L q_{c,b}(x)\,dx + \int_0^L q_{c,t}(x)\,dx\right\}$$

and the properties were evaluated at the reference temperature:

$$T_r = \frac{\overline{T}_w + T_o}{2}$$

Figure 3. Longitudinal section, $\theta = 90°$, heating mode I

These investigations have been carried out to study qualitatively the instabilities that set in the channel, producing complex secondary flows that can modify the heat transfer coefficients. The first configuration analyzed is the symmetrically heated horizontal one (mode I, 90°). The value of the ohmic dissipated heat flux was equal to $125\,W/m^2$.

In figure 3 is shown the photograph depicting the longitudinal section, in the $x-y$ plane at $z=0$. It can be noticed that the motion of the fluid results symmetrical respect to the plane at $x = 200\,mm$, the lenght of the plate being $400\,mm$. The fluid gets into the channel in the lower region, with the leading edge on the bottom plate. At the inlet, the motion can be regarded as two-dimensional for x less or equal to $20\,mm$. Then, the flow begins to become unstable. After the appearence of the instability on the bottom plate, the main flow changes direction at $x \simeq 180\,mm$ and adheres to the top plate up to the exit of the channel. For this configuration, the main flow shows a *C-loop* behavior.

The photos of the longitudinal section, at $z = 0$, for the modes II and III, are shown in figure 4. It can be pointed out, fig.4a, that the configuration with mode II does not show any kind of instability, due to the favorable bottom-heavy fluid situation. In fact, it is evident that the layer of fluid entering the channel on the bottom plate, adheres to the top plate after the *C-loop* and gets out of the channel without generating any kind of secondary flow. In figure 4b, referring to the mode III, it is worth noting that the secondary motion is very close to the one of the mode I, figure 3, even if this configuration presents a greater penetration depth of the flow inside the channel, with a smaller inversion region and vortex zone. Further, it can be said that the onset of instability is at $x > 20\,mm$. A close behavior is obtained doubling the dissipated wall heat flux, for the mode III. The value of the axial coordinate at which the instability begins

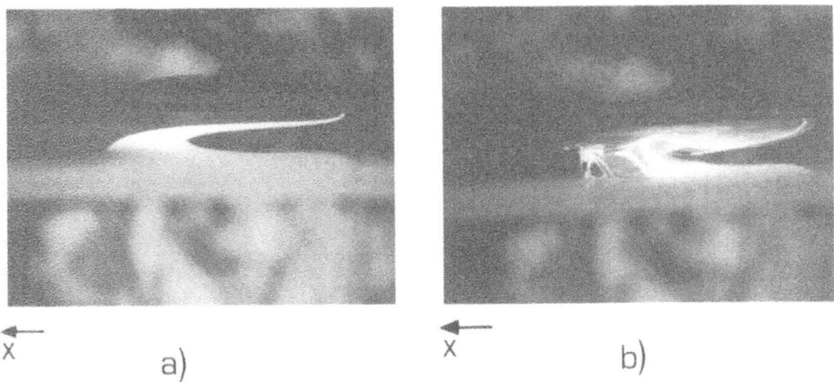

Figure 4. Longitudinal sections, $\theta = 90°$, $q_w = 125\,W/m^2$: a) mode II, b) mode III

to set in is about $30\,mm$. The base motion is characterized by the *C-loop* too, even if the vortex zone in the central region is bigger than the previous configuration.

In figure 5 are shown the cross-sectional views at $x = 20, 100$ and $180\,mm$ for the heating mode I. At $x = 20\,mm$, fig.5a, it is observed the onset of the instability due to a slight separation of the fluid from the bottom plate, while on the upper one there is a relatively chaotic motion. At $x = 100\,mm$, fig.5b the effects of instability grow and can be easily observed on the bottom plate periodical mushroom structures, while on the top one the chaotic region remains with a constant thickness. This effect is more remarkable at greater x, where there is a change in the secondary flow structures on both the plates, generating more chaotic structures, maybe due to the interaction of the mushrooms with the flow close to the top plate. The region close to the center of the channel seems to present a stable structure, $x = 180\,mm$ fig.5c, due to the inversion of the motion between the bottom and top plates.

The same cross-section views are reported in fig.6 for the mode III. In fig.6a it is noticed that at $x = 20\,mm$ there is no complete development of the secondary flow, and there is a beginning of the periodical motion along the z-direction, while on the top plate there is a developed completely chaotic secondary motion, as already noticed for the mode I. At $x = 100\,mm$ it can be detected the occurrence of periodical thermal mushroom-like structures on the entire gap of the channel. At greater x it seems to be a change from mushroom-like structures to chaotic and periodical vortices ($x = 180\,mm$, fig.6b).

In the following there will be presented the results for the mode III and for channels with an inclination equal to 85° and 75°. This is due to the fact

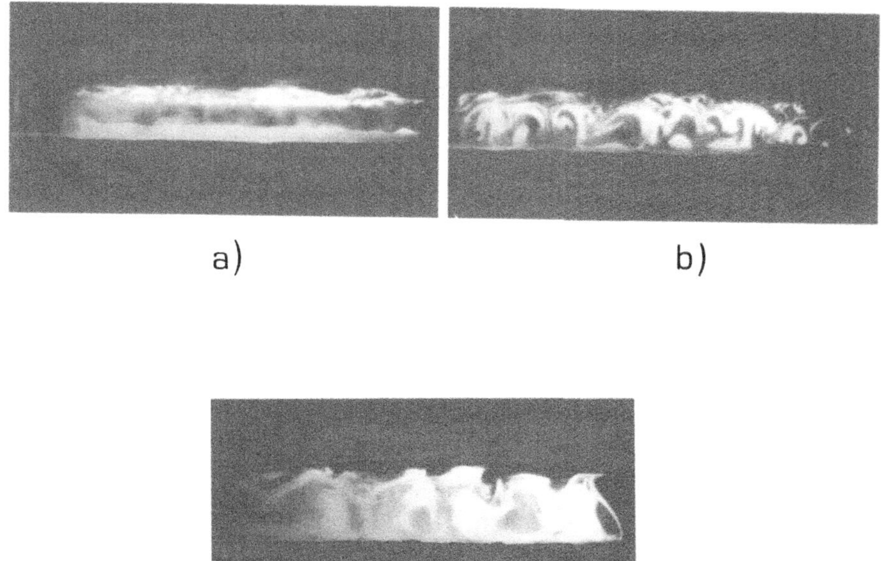

Figure 5. Cross-sections, $\theta = 90°$, $q_w = 125\,W/m^2$, mode I: a) $x = 20\,mm$, b) $x = 100\,mm$, c) $x = 180\,mm$

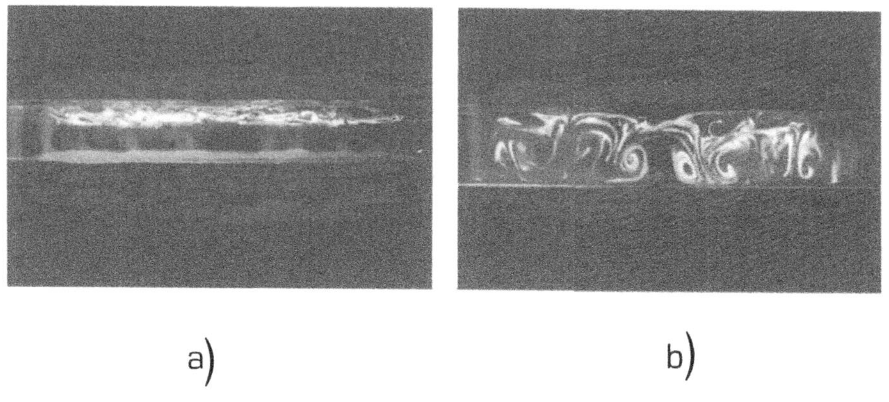

Figure 6. Cross-section, $\theta = 90°$, $q_w = 125\,W/m^2$, mode III: a) $x = 20\,mm$, b) $x = 180\,mm$

that these results are the ones qualitatively more significant. In figure 7, the views with an inclination angle of 85° are shown for the two investigated

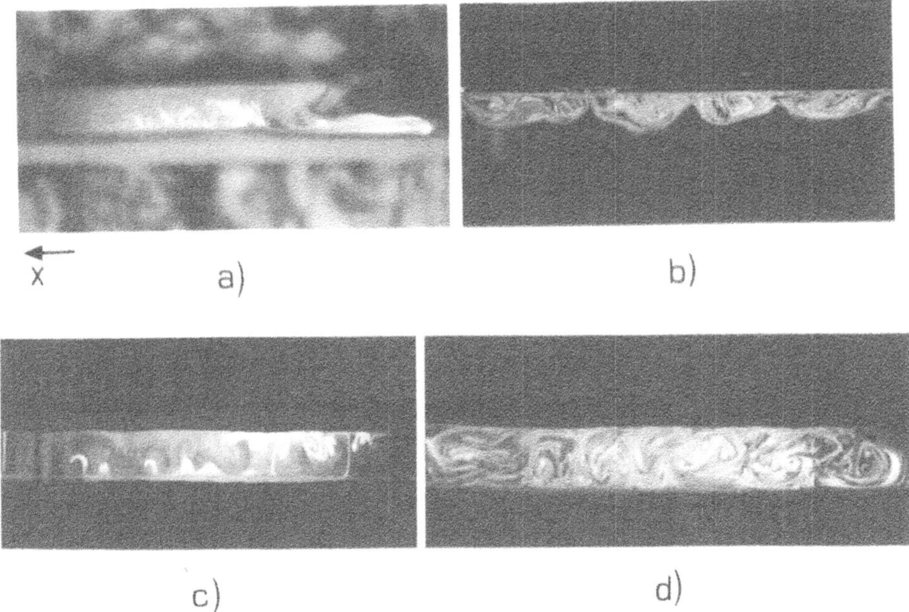

Figure 7. Inclination angle $\theta = 85°$, $q_w = 125\,W/m^2$: a) longitudinal section; cross-sections: b) $x = 398\,mm$, c) $x = 100\,mm$, d) $x = 300\,mm$

heat fluxes. It can be noticed in fig.7a that, in the plane $z = 0$, with a heat flux of $125\,W/m^2$ there is a complete filling of the channel section at $x \simeq 120\,mm$ due to the separation of the flow from the lower plate. The base flow is directed along the channel, but at the outlet it can be seen an inflow region on the lower plate, penetrating at $x \simeq 390\,mm$. This effect is due to the low value of the longitudinal buoyancy component. The rise of the plume at the channel exit produces a depression in proximity of the lower plate, that induces air from the ambient; this way a better cooling of the regions close to the exit is obtained, according to Manca et al. [14]. To this extent a cross-sectional view at $x = 398\,mm$ is reported in fig.7b. In figures 7c and 7d the cross-sections at $x = 100\,mm$ and $x = 300\,mm$ are reported. It is worth noting that there is transition from a mushroom-like structures to vortex roll ones as far as the secondary flow is concerned. The doubling of the wall heat flux produces very similiar structures to the previous configuration, but a vortex with the axis parallel to the z axis close to the separation region (between $x = 100\,mm$ and $x = 140\,mm$) was observed. In the figure 8 the longitudinal sections for the inclination angle equal to 75° are reported. In figure 8a it can be observed, for the lower heat flux, that there is detachment of the flow from the lower plate at $x \simeq 200\,mm$. The inflow is still present at the exit section. For the higher

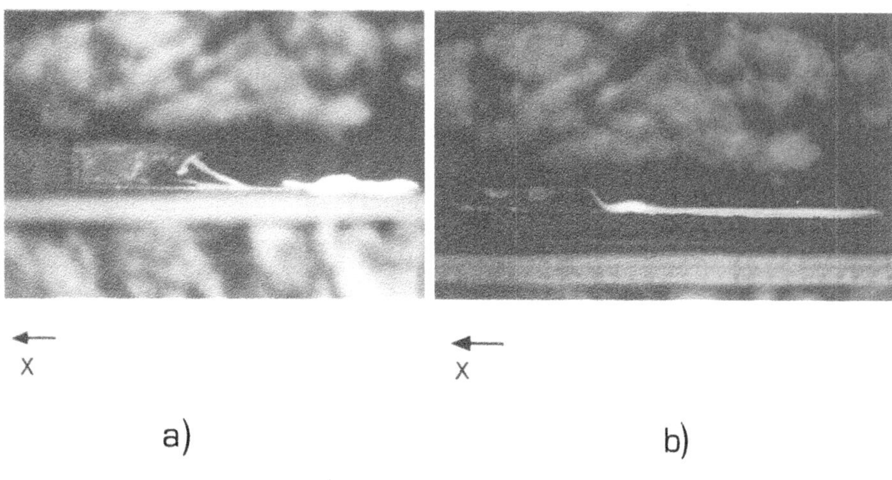

Figure 8. Longitudinal sections, $\theta = 75°$: a) $q_w = 125\,W/m^2$, b) $q_w = 250\,W/m^2$

value of the wall heat flux, fig.8b, the detachment takes place at about $x \simeq 250\,mm$; this is due to the fact that there is present a more efficient *chimney effect*, that tends to stabilize the motion. Here the inflow is still present with the same patterns of the previous configuration.

4. Conclusions

In this paper it has been investigated the phenomena of fluydinamic and thermal instabilities that can set in between two isoflux parallel plates, horizontal and slightly inclined respect to the gravity vector. The analysis focused on the occurrence of different types of secondary flows, such as mushroom-like and vortex roll structures. Further, there has been detected the presence of structures, or secondary motions, in the longitudinal planes never seen before, as the inflow occurring for the mode III with inclination angles of 85° and 75° respect to the vertical. More, for an inclination angle of 90° there is a *C-loop* base motion symmetrical respect to the plane at $x = 200\,mm$. The investigation that has been carried out qualitatively took into account the doubling of the dissipated heat flux; no substantial differences can be noticed for the two configurations.

5. Nomenclature

b	channel gap, $[m]$
g	gravity acceleration $[m/s^2]$
Gr	channel Grashof number
k	thermal conductivity $[W/mK]$
L	channel lenght $[m]$
q_c	convectiove heat flux $[W/m^2]$
T	temperature $[°C]$
x, y, z	cartesian coordinates $[m]$
Greek symbols	
β	volumetric coefficient of expansion $[1/K]$
θ	angle of inclination, $[°]$
ν	kinematic viscosity $[m^2$
Subscripts	
b	bottom wall
o	ambient value
r	reference value
t	top wall
w	wall

A bar over the symbols indicates a mean value

References

1. Sparrow, E. M. and Husar, R. B. (1969) Longitudinal vortices in natural convection flow on inclined plates, *Journal of Fluid Mechanics*, **Vol. 37**, pp. 251-255
2. Lloyd, J. R., Sparrow, E. M., and Eckert, E. R. G. (1972) Laminar, transition and turbulent natural convection adjacent to inclined and vertical surfaces, *International Journal of Heat and Mass Transfer*, **Vol. 15**, pp. 457-473
3. Haaland, S. E., and Sparrow, E. M. (1973) Vortex instability of natural convection flow on inclined surfaces, *International Journal of Heat and Mass Transfer*, **Vol. 16**, pp. 2355-2367
4. Shaukatullah, H., and Gebhart, B. (1978) An experimental investigation of natural convection flow on an inclined surface, *International Journal of Heat and Mass Transfer*, **Vol. 21**, pp. 1481-1490
5. Pera, L., and Gebhart, B. (1973) On the stability of natural convection boundary layer flow over horizontal and slightly inclined surfaces, *International Journal of Heat and Mass Transfer*, **Vol. 16**, pp. 975-984
6. De Graaf, J. G. A., and van der Held, E. F. M. (1953) The relation between the heat transfer and the convection phenomena in enclosed plane layers, *Applied Scientific Research*, **Vol. 3**, pp. 393-409
7. Kurzweg, U. H. (1970) Stability of natural convection within an inclined channel, *Journal of Heat Transfer*, **Vol. 92**, pp. 190-191
8. Hart, J. E. (1971) Transition to a wavy vortex regime in convective flow between inclined plates, *Journal of Fluid Mechanics*, **Vol. 48**, pp. 265-271
9. Ruth, D. W., Hollands, K. G. T., and Raithby G. D. (1980) On free convection experiments in inclined air layers heated from below, *Journal of Fluid Mechanics*, **Vol. 96**, pp. 461-479
10. Clever, R. M., and Busse, F. H. (1977) Instabilities of longitudinal convection rolls in an inclined layer, *Journal of Fluid Mechanics*, **Vol. 81**, pp. 107-127
11. Clever, R. M. (1973) Finite amplitude longitudinal convection rolls in an inclined layer, *Journal of Heat Transfer*, **Vol. 95**, pp. 407-408
12. Kennedy, K. J., and Khanel, J. (1983) Free convection in tilted enclosures, *ASME HTD* , **Vol. 28**, pp. 43-47
13. Cheng, K. C., and Kimura, T. (1991) Observation of convective instability phenomena in slightly inclined air layers heated form below: effect of air layer thickness, *ASME HTD*, **Vol. 178**, pp. 55-63
14. Manca, O., Nardini, S., and Naso, V. (1992) Experiments on natural convection in inclined channels, *ASME HTD*, **Vol. 212**, pp. 41-46

NEW JEDEC STANDARDS FOR THERMAL MEASUREMENTS.
Review and Examples

VALTER MOTTA
SGS-Thomson Microelectronics
Agrate B.[za] Italy

1. Introduction

Thermal design is becoming a key issue of many new electronic systems, due to the increase of operating temperature, speed, device density and miniaturisation.
For all these reasons, a good knowledge of the working temperature for a single integrated circuit or for a complex system is increasingly important.

Standardisation of thermal measurement methodology is needed in order to ensure data acquisition and exchange, in view of improved design rules at device, board and system level.
A JEDEC committee in these years developed some standards in order to define a common methodology for thermal measurement, using active devices or specific thermal designed chips with specific environmental conditions, in still air, and board design.

SGS-Thomson is one of the active members of the Jedec working group, named JC_15, which has developed and proposed new standard based on the experience of more than twenty Companies among the most important users and suppliers of microelectronics components.

One of the purposes of these activities is to produce some guidelines in the design of silicon thermal test patterns to optimise the measurements and minimise the differences, among users, due to non standard test chips or measurement methods.
Starting from the existing SEMI guidelines test method, a number of test chips have been designed and evaluated.
They all use the concept of having different chip sizes but the same structure for dissipating elements and temperature sensing elements on die.
This offers the opportunity of simple and repetitive measurement, with controlled and uniform power distribution and excellent resolution in response time, when thermal impedance is considered.

E. Beyne et al. (eds.), Thermal Management of Electronic Systems II, 293-300.

The goal of these rules is to guarantee a very high calibration of the temperature sensing diodes, in the order or ±2%, and the result is to obtain an accuracy in the thermal measurement better than ± 5%.
A method is also offered to calibrate the sensing diode and the measurement circuit.

2. A dedicated thermal test chip. Why?

Purpose of this activity is to offer a common methodology able to minimise the intrinsic imprecision connected to the use of a device which is not dedicated to thermal issues and to avoid huge differences in the results among Companies due to the use of non-standard dedicated devices or non standard methodologies.

The advantages connected to the use of thermal dedicated devices, in comparison to active devices or multipurpose test chips, are mainly the following:

- high levels of power and high resolution in the power control
- very simple and reproducible measurement control; in this way it is possible, and easy, to correlate different situations (function of different packages, materials, boundary conditions etc.)
- sensing element designed to guarantee a repeatable conversion factor between forward voltage and temperature
- heating and sensing elements completely separated from the electrical point of view
- uniform power distribution on the whole silicon area
- high control and resolution in the impedance measurements.

In order to maximise this advantages, it is mandatory to develop and design the devices following well-defined rules hereafter described.

3. Test pattern layout.

Following the recommendations already presented in some SEMI standards ([1] - [4]), in particular the SEMI guidelines test method #G32-86, the JEDEC committee is proposing a better definition of test chip design.

The thermal test chips are composed of one or more dissipating elements distributed uniformly on the silicon surface covering at least the 85% of the whole chip area.
This heating elements can be indifferently bipolar transistor or diffused resistors.
The choice between transistors or resistors depends on two main needs.
The first one is the necessity to guarantee a very high control in the dissipated power, mainly in the impedance measurement, and, in this case, the best choice is the bipolar transistor.

On the other hand, mostly for large devices, it is important to have a uniform power distribution on the whole active area, and, in this case, the diffuse resistors are the suggested choice.

An other important target is the power capability of the test chip: the minimum power level requested is 10 Watt. For a thermal device with large area is better to have higher capability for the power level; in this way it is possible to cover all the possible applications of the integrated circuit where sometimes a power up to 30 Watts is requested.

Besides the needs of a uniform and well-controlled dissipated capability, it is mandatory to be able to read the temperature increase due to the power applied.
For this reason, electrically insulated from the dissipating elements, one or more temperature sensors are placed in the more significant points.
When the power dissipation is uniform, the hottest point is in the chip centre, so this is the most important point where a sensor must be placed. Other interesting measurement points, for the reliability point of view, are corners and edges of the die.

An example of thermal test chip developed in SGS-Thomson is shown in the following figure: the heating elements are resistors (covering all the die surface with the exceptions of the edges where there are metal and interconnections), while the sensing diodes are four, placed in the centre, in the corner and in the middle of two edges, (named A, B, C, and D in the next figure). The total dimension of the die is 9x9 mm^2.

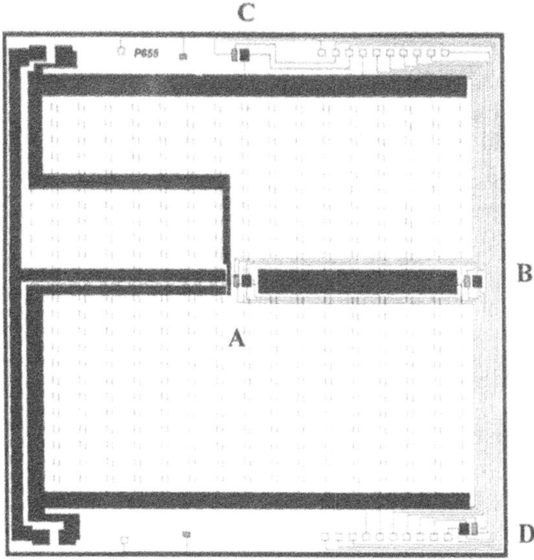

Figure 1 - Layout example of a test chip developed in SGS-Thomson

In figure 2 is represented the measurement bench for a test chip used in SGS-Thomson made by 2 bipolar transistor and a sensing diode, with 2x2 mm^2 of area.
The bench is very simple: only a pulse generator a DC and a current supplier and few voltmeters and amperometers are necessary.

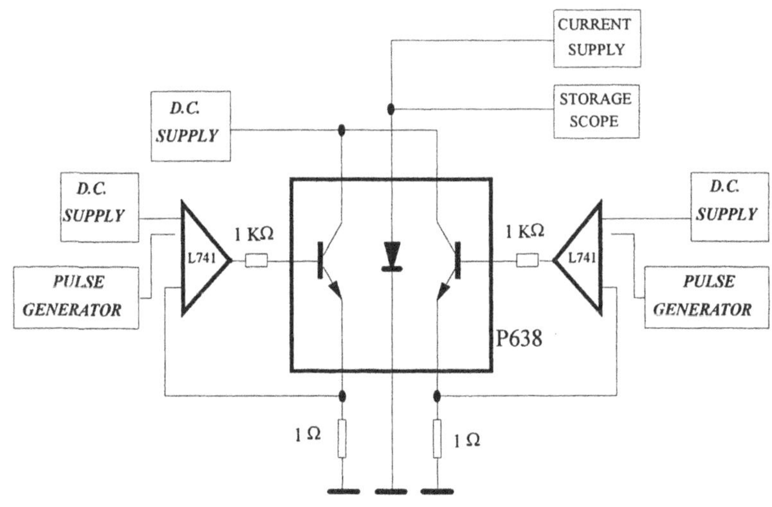

Figure 2 - Example of measurement bench

To cover all the applications, from the micropackages to the biggest PQFP's, only a single test pattern is not enough.
To have devices from few to hundreds square millimetres a family of standard devices, with well-defined dimensions, is suggested.
The family used in SGS-Thomson has been accepted in the JEDEC proposal as a good example of standard; the proposed dimensions for square and rectangular dice are reported in the following table.

TABLE 1. Test pattern family

Square sensors	Rectangular sensors
2 x 2 mm	2 x 4 mm
4 x 4 mm	4 x 6 mm
6 x 6 mm	6 x 12 mm
9 x 9 mm	9 x 18 mm
12 x 12 mm	12 x 18 mm
	12 x 24 mm
15 x 15 mm	
18 x 18 mm	18 x 24 mm
24 x 24 mm	

The choice of these dimensions is related to the possibility to fit a lot of die sizes using as much as possible an elementary cell; in fact most of the dimensions above reported can be obtained using only three elementary cells: 2x2, 6x6 e 9x9 mm^2.

4. Temperature sensitive parameter (TSP).

The temperature sensing elements should be able to monitor the temperature in all the possible applications, practically from 0 to 180 ^{0}C.

On the basis of the diffusion techniques, the sensor can be made in two different ways: with a diffused resistor or a P-N junction.
The resistor uses the resistivity dependence on the temperature therefore monitoring the variation in the resistivity it is possible to know the temperature.
Disadvantages of this method are the poor reproducibility of the resistor calibration and the difficulty to design a resistor as much as possible punctiform.

For all these reasons the sensor that offers better performances, and is becoming a standard, is a diode P-N.
This diode is forward biased with a current as low as possible, to avoid a self-heating of the sensor, but high enough in order to guarantee a reasonable voltage drop on the diode: the suggested value is between 100 μA and 5 mA.

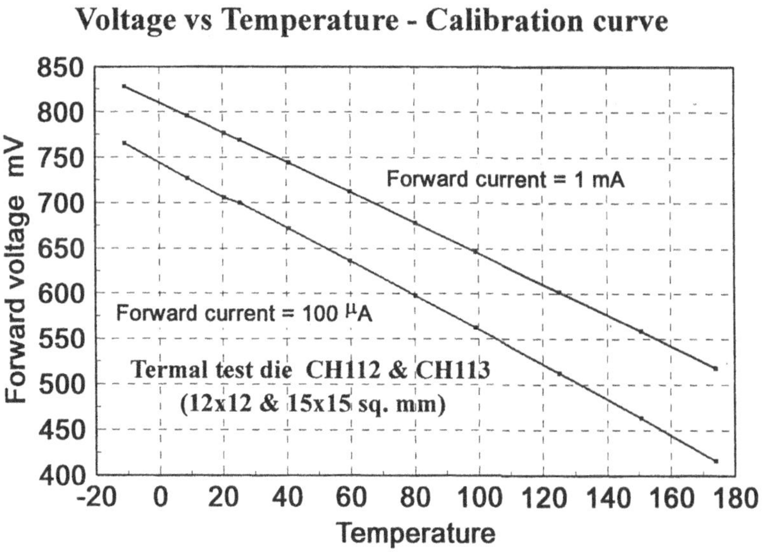

Figure 3 - Voltage/Temperature relationship on a sensing diode

The forward voltage drop is strictly related to the mobility of the electrons in the silicon and therefore to the temperature of the diode; for this reason is easy to correlate the diode temperature and its voltage drop.
In Figure 3 is plotted the calibration curve for a thermal test chip, developed in SGS-Thomson, with two different forward currents: 100 μA and 1 mA.

The relationship voltage-temperature, as shown, in first approximation, is linear.
The slope of the curve gives the K-factor of the sensing diode, and it can be obtained from the following formula:

$$\mathbf{K = -(T_2 - T_1) / (V_{F2} - V_{F1})} \qquad (1)$$

Where: T_1 and T_2 = equilibrium temperature at first and second calibration point
V_{F1} and V_{F2} = forward voltage at first and second calibration point

To minimise the measurement errors Jedec suggests to use a temperature range as large as possible and in any case at least equal to 50 °C.
On the basis of typical dopant levels used in the diffusion technology, the K-factor value is about 0.5 °C/mV.
If the calibration factor, in the working range, does not change for an amount higher of the 5% of its value, the K-factor can be considered as a constant parameter.

Besides, if the variation for the K-factor is higher than 5%, then also the calibration factor has to be treated as a function of the temperature. This means that the K-factor also has to be described as a mathematical function of the temperature.
The best representation for this function is a polynomial expression.
To calculate the K-factor at a fixed temperature (2) must be used, while the junction temperature can be obtained from (3).

$$\mathbf{K_i = - (T_{i2} - T_{i1}) / (V_{f2} - V_{f1})} \qquad (2)$$

$$\mathbf{T_j = [\, K_i \bullet \Delta V_f + T_x \,] \quad ^\circ C} \qquad (3)$$

where:

$\mathbf{K_i}$ = K-factor calibration in the i-th interval
$\mathbf{T_{i1}}$ $\mathbf{T_{i2}}$ = equilibrium temperature at the beginning and the end of the i-th interval
$\mathbf{V_{f1}}$ $\mathbf{V_{f2}}$= voltage measured at the beginning and the end of the i-th temperature interval
$\mathbf{T_j}$ = junction temperature
$\mathbf{\Delta V_f}$ = voltage drop on the sensing diode
$\mathbf{T_x}$ = reference temperature

In the following figure is reported an example of the relationship between the K-factor and temperature.

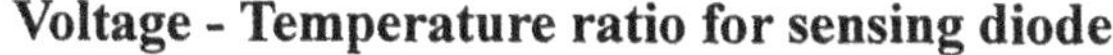

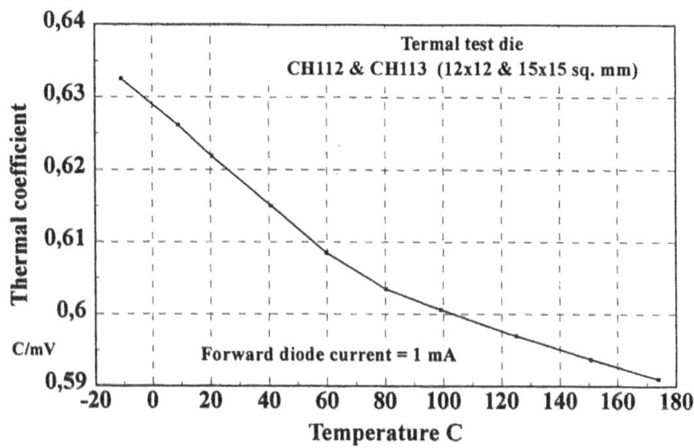

Figure 4 - K-factor curve versus the temperature

For this test pattern the polynomial interpolation of the K-factor versus the temperature, obtained from the previous curve, is:

$$K(T) = 0.629 - 3\text{-}38 \cdot 10^{-4} \cdot T - 1.12 \cdot 10^{-6} \cdot T^2 + 3.35 \cdot 10^{-8} \cdot T^3 - 7.55 \cdot 10^{-11} \cdot T^4 \qquad (4)$$

TQFP (10x10x1.4) Rth(j-a) on boards

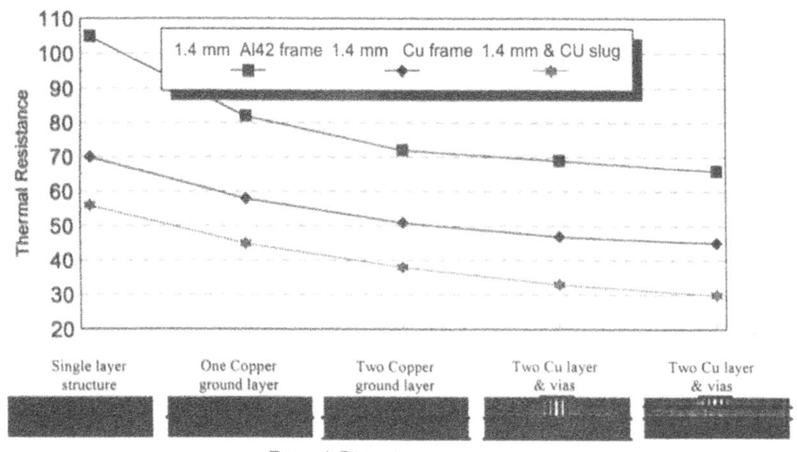

Figure 5

5. Example of thermal test pattern application

In the figure 5 is shown an example of information obtainable using a thermal test die: in this case the thermal behaviour of a package depends on its structure and on the board structure. This information can help a lot to evaluate develop and design both packages and PC boards.

In the lower part of the figure is reported the section of the PCB's to better understand the board structure, the number and the position of the ground copper planes.

6. Acknowledgements

Thanks to all the members of JC_15 committee for the effective co-operation end collaboration that allowed to define a large number of standards in the thermal management.

A special thanks to Bob Bond and Tiao Zhou that co-ordinated the SGS-Thomson activities in the Jedec committee.

7. References

[1] SEMI test method #G32-86, SEMI guideline for unencapsulated thermal test chip.

[2] SEMI test method #G43-87, Test method, junction to case thermal resistance measurement of moulded plastic packages.

[3] SEMI test method #G38-87, Still and forced air thermal resistance measurement for integrated circuit packages.

[4] SEMI test method #G42-87, Specification, thermal test board standardisation for measuring junction to ambient thermal resistance of semiconductor packages.

[5] Methodology for the thermal measurement of component packages (single semiconductor device), JC15.1 subcommittee test proposal, JEDEC, 1994.

[6] Integrated circuit thermal measurement - Electrical test method (single semiconductor device), JC15.1 subcommittee test proposal, JEDEC, 1994.

[7] Environmental conditions - natural convection (still air), JC15.1 subcommittee test proposal, JEDEC, 1994.

[8] JEDEC JC15.1 Thermal test board proposal for low conductivity PCB#3, 1994.

EXPERIMENTAL STUDY ON LOCAL CONVECTIVE HEAT TRANSFER FROM A PROTRUDING CUBICAL COMPONENT

E.R. MEINDERS, T.H. VD MEER AND C.J. HOOGENDOORN
Delft University of Technology, Applied Physics, Delft, the Netherlands

AND

C.J.M. LASANCE
Philips Research, Eindhoven, the Netherlands

Abstract

An experimental method with uncertainty analysis is presented to examine the local convective heat transfer from cubical protruding obstacles in a vertical channel flow. Results for a single component are discussed to validate the experimental method and measurement techniques. Infrared thermography is used to measure the surface temperature distributions. Flow visualisations are carried out for interpretation of the heat transfer results.

1. Introduction

The development of powerful integrated circuits together with the growing minimisation forces an accurate and efficient thermal management of electronic circuitry, since overheating has been experienced as an important failure of operation. Knowledge on convective heat transfer is therefore of growing relevance. Printed circuit boards (PCB's) contain several different components sizes which makes the usage of dimensionless heat transfer correlations with one characteristic length-scale improper. Besides this, a reference temperature, as required for defining heat transfer coefficients (htc), is hard to determine. Correlations between local htc's and local flow properties (velocity fluctuations, turbulence intensities and length scales), are more appropriate to apply in PCB-design. The adiabatic surface temperature is a suitable reference temperature in the definition of htc's [1].

E. Beyne et al. (eds.), Thermal Management of Electronic Systems II, 301-308.

Local convective heat transfer from cubical bodies is a quite new subject of investigation. Some research effort has been undertaken by Natarajan *et al.*, [2]. They investigated the local mass transfer of a cube in a channel flow using the naphthalene sublimation technique. Almost the same configuration was investigated, however for higher Reynolds numbers which restricts comparison with present study. Two-dimensional local heat transfer from a square prism to an airstream was investigated by Igarashi [3], [4]. In those particular studies free stream velocities ranged between 6 and 28 m/s. The surface of the prism was covered with a thin stainless steel sheet for creating a constant heat flux boundary condition.

This paper is essentially focussed on an experimental method to determine local convective heat transfer coefficients for cubical components. First results on convective heat transfer from a single component to turbulent channel flow are discussed. These results are used to validate the method and measuring techniques. Furthermore, a qualitative interpretation of heat transfer coefficients is made with usage of flow patterns obtained with visualisation techniques.

Nomenclature

D	cube dimension (m)	T_{amb}	ambient temperature (K)
d	epoxy layer thickness (m)	U_∞	free stream velocity (m/s)
d_h	hydraulic diameter (m)	ϵ	emissivity
F	view-factor	λ	thermal conductivity (W/mK)
h_{ad}	adiabatic htc (W/m^{2}K)	λ_{air}	thermal cond. of air (W/mK)
h	htc (W/m^{2}K)	ν	kinematic viscosity (m^2/s)
Pr	Prandtl number	σ	Stefan-Boltzmann const. (W/m^{2}K^4)
Re_L	Reynolds number ($=U_\infty L/\nu$)	ϕ''_{cond}	conductive heat flux (W/m^2)
T_{co}	copper temperature (K)	ϕ''_{rad}	radiative heat flux (W/m^2)
T_{sur}	surface temperature (K)	ϕ''_{conv}	convective heat flux (W/m^2)
T_{ad}	adiabatic temperature (K)		

2. Test facility and experimental method

The test configuration consists of a vertical windtunnel section with cross section 500 × 50 mm. The cubical component is positioned on one of the vertical walls. The aspect ratio width/cube-dimension was 50/15=3.3. Free-stream velocities range between 1-5 m/s. The cubical component consists of an internal copper cubical core covered by a thin epoxy layer. A schematic drawing of a composed component is shown in figure 1. Heat is generated in the centre of the component by a dissipating resistance wire. An advanced controller is designed for powering the component. The stability of the adjusted internal temperature is good (within 0.1 ^oC) because of the low time-response of the component. The conductivity of copper is about

1000 times larger than the conductivity of epoxy (390:0.24 W/mK). The thickness-ratio epoxy layer-copper is almost 1-5. Hence, the thermal resistance of the epoxy layer is much larger compared to that of the copper core. The thermal resistance for heat transfer to the ambient is even

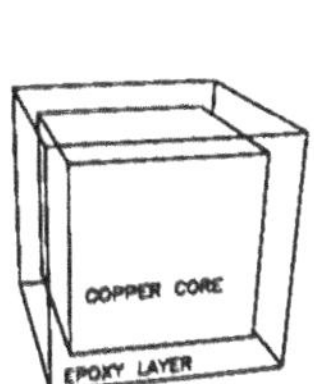

Figure 1. schematic picture of the composed component

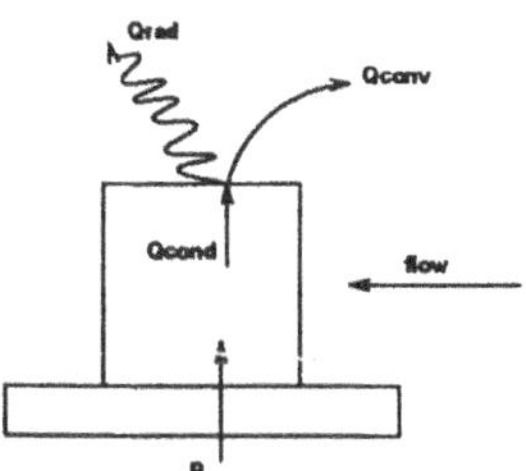

Figure 2. balance of heat fluxes at the surface of a component

larger. Therefore, the temperature decay across the copper core is negligible compared to the epoxy-layer which implies an almost uniform copper temperature, even for relative high fluxes. This internal copper temperature was measured with a thermocouple. The outer surface temperature distribution of the epoxy layer was determined with Infrared thermography. An 1-D approximation of the conductive flux is given by the temperature difference between the inner and outer surface of the epoxy layer. However, the three-dimensional (3D) geometry forces a 3D approach since lateral fluxes appear. These lateral fluxes cause deviations of more than 30 percent of the total convective heat transfer. The temperature distribution in the 3D epoxy layer was therefore determined numerically by solving the Laplace equation with the measured surface temperatures as boundary conditions. With this solution, local conductive fluxes through the wall are known. At the surface of a protruding component, local equilibrium of heat fluxes determines a local convective heat flux ϕ''_{conv}:

$$\phi''_{conv} = \phi''_{cond}|_s - \phi''_{rad} \tag{1}$$

with $\phi''_{cond}|_s$ the conductive flux and ϕ''_{rad} the radiative flux. A schematic drawing is depicted in figure 2. The local adiabatic htc reads [5]:

$$h_{ad} = \frac{\phi''_{cond}|_s - \phi''_{rad}}{T_{sur} - T_{ad}} = \frac{\lambda \frac{\partial T}{\partial n}|_s - \epsilon F \sigma (T^4_{sur} - T^4_{amb})}{T_{sur} - T_{ad}} \tag{2}$$

with T_{ad} the adiabatic surface temperature as reference temperature. The adiabatic temperature is the surface temperature a component obtains when it is abstained from power supply during operation of all other components. Since the adiabatic surface temperature can be examined locally, the adiabatic heat transfer coefficient is an appropriate parameter for describing local heat transfer. Radiative heat transfer is presented in a simplified expression for illustration purposes. In reality, influences from neighbouring components, shrouding wall and base plate have to be taken into account. Radiative heat fluxes can be calculated when the surface temperature distribution, emissivity, view factors, etc are known. With knowledge of the layer dimensions, thermal conductivity and the temperature distribution in the epoxy layer, local conductive heat fluxes can be calculated, finally resulting in local htc's. Heat losses through the base plate and lead-wires are not of interest in this approach and can be disregarded. However, second-order effects are to be expected when heat losses by conduction via lead-wires and baseplate preheat the upstream airflow.

3. Uncertainty in experimental results

An evaluation of experimentally obtained data is possible with an accurate error discussion. The standard single sample uncertainty analysis recommended by Moffat [6] is used for calculating the uncertainty in heat transfer coefficient. An 1-Dimensional approximation of the temperature gradient at the surface can be considered to estimate its contribution to the total uncertainty in h_{ad}.

$$\frac{\partial T}{\partial n}\Big|_s = \frac{(T_{co} - T_{sur})}{d} \tag{3}$$

with T_{co} the copper temperature , T_{sur} the surface temperature and d the epoxy layer thickness. The resulting uncertainty in h_{ad} becomes:

$$\begin{aligned}(\delta h_{ad})^2 = &\left(\frac{\frac{\partial T}{\partial n}|_s}{(T_{sur} - T_{ad})}\right)^2 (\delta\lambda)^2 + \left(\frac{\lambda}{d(T_{sur} - T_{ad})}\right)^2 (\delta T_{co})^2 + \\ &+ \left(\frac{4\epsilon\sigma F T_{amb}^3}{T_{sur} - T_{ad}}\right)^2 (\delta T_{amb})^2 + \left(\frac{h_{ad}}{T_{sur} - T_{ad}}\right)^2 (\delta T_{ad})^2 \\ &+ \left(\frac{\lambda(T_{co} - T_{sur})}{d^2(T_{sur} - T_{ad})}\right)^2 (\delta d)^2 \\ &+ \left(\frac{\sigma F(T_{sur}^4 - T_{amb}^4)}{T_{sur} - T_{ad}}\right)^2 (\delta\epsilon)^2 + \left(\frac{4\epsilon F\sigma T_{sur}^3 + \lambda/d + h_{ad}}{T_{sur} - T_{ad}}\right)^2 (\delta T_{sur})^2\end{aligned} \tag{4}$$

The material constants λ and ϵ were determined experimentally. Locally at the surface of a component, the emissivity can deviate due to differences in layer thickness of the used black paint. Also the thermal conductivity of epoxy can deviate due to anisotropic structures and/or internal stresses. Sensitivity studies proved that these uncertainties do not attribute significantly to the total error in the heat transfer coefficient and are estimated to be less than 2%. As discussed in previous section, the three dimensional geometry causes lateral effects which effect the temperature distribution in the epoxy layer and indirectly the temperature gradient at the surface. Grid refinement was applied to examine a numerical uncertainty in $\frac{\partial T}{\partial n}|_s$. The epoxy layer thickness was carefully created with an imprecision of less than 1% (deviations in thickness approximately 0.01 mm). The most important source of uncertainties in this approach is caused by the surface temperature of the component. Two additional methods (black bodies and reference objects) are applied to calibrate the IR system. The achieved uncertainty in the surface temperature reading was between 0.5 and 1.0 ^oC. The internal copper temperature was measured within 0.1 ^oC. The uncertainty in the ambient temperature is not importantly affecting the total error. Surface temperatures varied between 50 and 65 ^oC. The copper temperature was about 75 ^oC. Both the ambient and adiabatic temperature were around 20 ^oC. With epoxy layer thickness d=1.5 mm, thermal conductivity λ=0.24 W/mK and emissivity ϵ=0.93, the uncertainty in the adiabatic heat transfer coefficient ranged between 5 and 10%.

4. Results

Figure 3. Smoke visualisation of the flow pattern around a cube at Re=10^4.

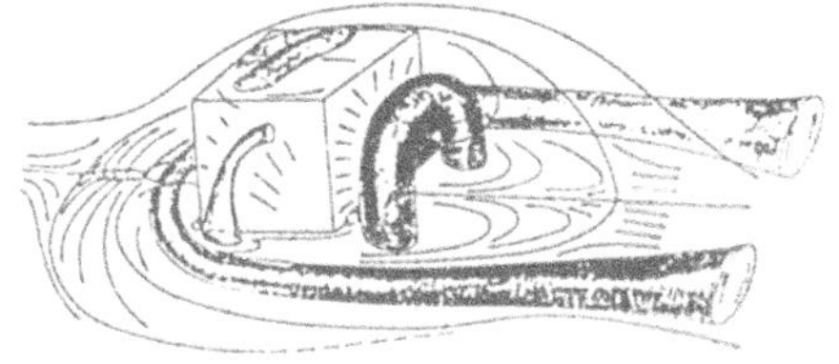

Figure 4. Sketch of flow patterns around cube from Larousse *et al.* with Re=10^5

For two different free-stream velocities, U_∞=4.0 and U_∞=5.8 m/s, local convective heat transfer from a single component in channel flow was examined. The cubical element was mounted at 0.5 m down stream in vertical position. The hydraulic diameter d_h of the channel, ratio of 4 times the cross-section area and the perimeter [7], was 9.1 cm. The Re-numbers based on this hydraulic diameter for the two situations were Re_{d_h}=2.4 10^4 and Re_{d_h}=3.5 10^4. Transition from laminar to turbulent channel flow occurs at $Re_{d_h} \geq 2300$ [7]. For both free stream velocities, the channel flow was turbulent. A flow visualisation of the flow patterns around a single cubical component is shown in figure 3. The flow structures in present study, obtained with visualisation techniques, are similar to those described by Larousse *et al.* and Hunt *et al.* [8], [9]. A sketch of the flow around a cube, abstracted from Larousse *et al.* [8], is depicted in figure 4. The slightly larger Reynolds number is not of relevance since this schematic picture is only used for illustration. The typical horseshoe vortex appears quite obvious from both figures. Local adiabatic htc's are discussed for two different cross-sections, exhibited in respectively figures 5 and 6. The htc's calculated on the edge of the cubical component are not presented because of too large uncertainties. The front face is exposed to stagnation flow. To

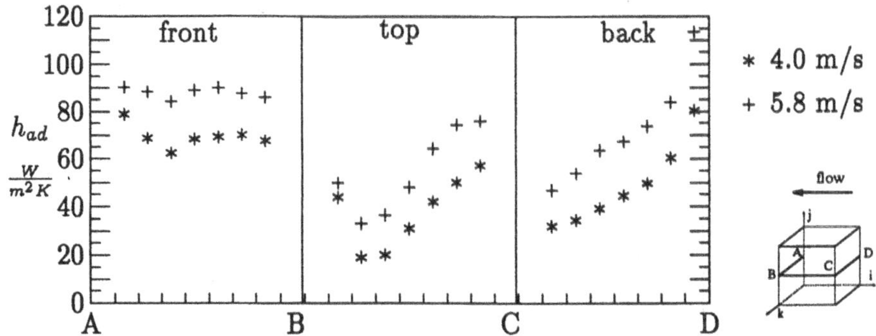

Figure 5. Local htc's along a cross-section perpendicular to the base plate.

validate the data, a comparison is made with impinging flow heat transfer. The heat transfer in the stagnation point of a body of revolution is given by [10], [11]:

$$h = 0.763\lambda_{air}\left(\frac{\beta}{\nu}\right)^{\frac{1}{2}} Pr^{0.4} \tag{5}$$

with β the velocity gradient at the interface between boundary layer and free stream parallel to the plate. For this gradient, we can substitute the expression for a flat disk in a uniform stagnation flow [11]: $\beta = U_\infty/D$, with D the diameter of the disk. In present study, the disk diameter is replaced by the cube dimension D. The heat transfer coefficient in the stagnation point is independent of position which can be observed in figure 5. For free stream velocities U_∞=4.0 and U_∞=5.8 m/s, these local htc's become respectively h=71.9 W/m^2K and h=86.5 W/m^2K. The measured htc's for the single component in the stagnation point are respectively 70.2 W/m^2K and h=90.1 W/m^2K. Within the claimed uncertainty, the agreement is excellent. The impinging flow at the front face, results in an increased heat transfer near the stagnation point. The horseshoe vortex cause a recirculation region close to the base plate resulting in a minimal htc. At the top

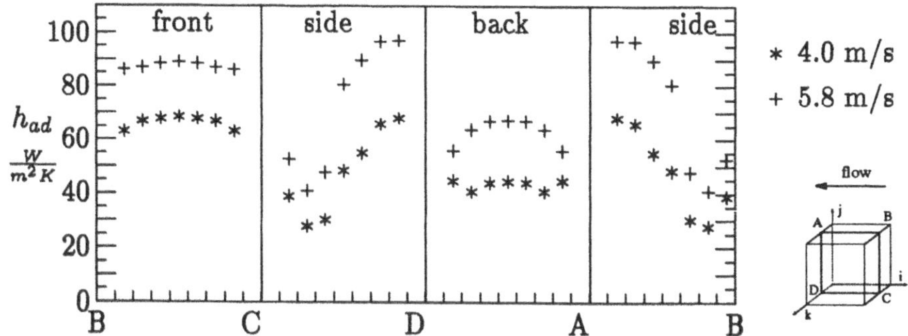

Figure 6. Local htc's along a cross-section parallel to the base plate.

face, the htc decays to a minimal value at approximately 1/3 from the leading edge. Again, this is caused by a recirculation region. In down-stream direction, the htc increases to a maximal value which is explained by a reattachment point of the flow. The back face exhibits an increasing htc which possibly results from again the recirculation behind the cube. Closer to the base plate, htc's are increasing. Natarajan [2] observed similar phenomena however, his experiments were performed for higher Reynolds numbers (two decades). In figure 6 htc's are depicted for a cross-section parallel to the base plate. At both side faces, again two extrema can be observed. A minimal htc indicates a recirculation point and the maximal htc again a reattachment region. The near face exhibits in the centre an almost uniform profile which decays to both edges. The htc's from the back face can not be explained easily. Possibly the recirculating flow decreases the htc. Large differences in local htc can be observed from above two plots. (local htc's range from 30 - 80). Especially, faces with strong recirculating flow conditions (side faces and top face) exhibit remarkable differences. For four faces, surface averaged heat transfer coefficients are presented in next table. The last column shows the surface averaged htc for the total component. Both side faces exhibit equal htc's since symmetry is applied. The adiabatic htc, averaged

TABLE 1. surface averaged h_{ad} $[W/m^2K]$

v [m/s]	front	top	side	back	total
4.0	68.9	38.3	51.7	48.1	51.7
5.8	90.1	55.3	78.4	69.3	74.3

over all five surfaces and indicated with total, is in agreement with results from literature [12], although slightly different situations were considered. For a free stream velocity v=4.0 m/s this averaged h was approximately 50 W/m^2K and approximately 80 W/m^2K for free stream velocity v=5.8 m/s.

5. Conclusions

- The experimental method appears to be an accurate and satisfying method to determine local heat transfer coefficients.
- First results on surface averaged adiabatic htc's are in agreement with values from literature. Local htc's were evaluated with impinging flow heat transfer. Pronounced differences in local htc's can be observed, especially for faces with strong recirculations.
- The used experimental techniques are suitable for accurate temperature measurements. Infrared temperature readings appeared to be the largest source of uncertainty and, momentary, infrared image restoration techniques are explored to minimise this error to 0.3 ^oC.
- Global flow patterns are obtained with visualisations techniques. These results satisfy for a first interpretation of local heat transfer coefficients.

References

1. Moffat, R.J., and Anderson, A.M. (1990) Applying heat transfer coefficient data to electronics cooling, *Journal of Heat Transfer*, **Vol. 112**, pp. 882-890
2. Natarajan, V., and Chyu, M.K. (1994) Effect of flow angle-of attack on the local heat/mass transfer from a wall-mounted cube, *Journal of Heat Transfer*, **Vol. 116**, pp. 552-560
3. Igarashi, T. (1985) Heat transfer from a square prism to an air stream, *Int.J.Heat Mass Transfer*, **Vol. 28 no. 1**, pp. 175-181
4. Igarashi, T. (1986) Local heat transfer from a square prism to an air stream, *Int.J.Heat Mass Transfer*, **Vol. 29 no. 5**, pp. 777-784
5. Incropera, F.P. and de Witt, D.P. (1981) *Fundamentals of heat and mass transfer*, 2nd. ed., John Wiley and Sons
6. Moffat, R.J. (1988) Describing the uncertainties in experimental results, *Experimental Thermal and Fluid Science* **Vol. 1 no. 1**, pp. 3-17
7. Burmeister, L.C. (1983) *Convective heat transfer*, John Wiley and sons
8. Larousse, A., Martinuzzi, R. and Tropea, C. (1991) Flow around surface-mounted, three-dimensional obstacles, *8th Symp. on Turbulent Shear Flows*, TU-Munich/Germany, 1, pp. 14-4-1/14-4-6
9. Hunt, J.C.R., Abell, C.J., Peterka, J.A. and Woo. H. (1978) Kinematical studies of the flows around free or surface-mounted obstacles; applying topology to flow visualisations, *J. Fluid Mech.*, **Vol. 86, part 1**, pp. 179-200
10. Sibulkin, M. (1952) Heat transfer near the forward stagnation point of a body of revolution, *J. Aeron. Sci.*, **Vol. 19**, pp. 570-571
11. Meer, Th.H. van der (1987) *Heat transfer from impinging flame jets*, PhD-thesis
12. Anderson, A.M. and Moffat, R.J. (1992) The adiabatic heat transfer coefficient and the superposition kernel function: part 1 - data for arrays of flatpacks for different flow conditions, *Journal of Electronic Packaging*, **Vol. 114**, pp. 14-21

MODELLING OF AXIAL FANS FOR ELECTRONIC EQUIPMENT

J. HENNISSEN[1], W. TEMMERMAN[2], J. BERGHMANS[1], K. ALLAERT[2]
[1] *K.U.Leuven, Dept werktuigkunde, Celestijnenlaan 300 A, B-3001 Heverlee (Belgium)*
[2] *Alcatel Bell, F. Wellesplein 1, B-2018 Antwerpen*

Abstract

The work presented concerns the modelling of a fan used for forced convection cooling of a subrack of printed board assemblies (PBAs) as a separate component for accurate CFD simulations.
Detailed measurements of the fan outlet velocity profile (axial, tangential and radial components) have been made for one fan operating point in a measurement set-up apart from the PBAs. CFD simulations have been made for a simplified version of a subrack, comparing a few elementary fan models.
Results show that incorporation of the hub into a model of the axial fan is necessary. The tangential velocity at the outlet is not negligible. The value of the radial velocity is small throughout the fan outlet plane and may be ignored in a CFD model.

1. Introduction

Detailed measurements of the fan induced velocity pattern which occurs in a printed board assembly (PBA) showed that this pattern is highly non - uniform, resulting in inadequate cooling of parts of the structure [1]. Due to the compactness of the assembly the lack of uniformity of the velocity profile at the fan outlet has important repercussions on the flow distribution over the different card 'channels'. CFD simulation of the flow can only be successful if it takes this non-uniformity into account.

A simulation which used detailed measurements of the velocity pattern upstream of the cards as input conditions, showed important progress in predicting the velocity patterns in the PBA assembly [2]. It is felt however that this approach can't be generalised, as every change of the configuration would require detailed

E. Beyne et al. (eds.), Thermal Management of Electronic Systems II, 309-318.

measurements of a velocity profile, in order to make accurate CFD simulations possible.

The work presented here studies the feasibility of measuring the outlet velocity profile of the fan in a measurement set-up, apart from the PBA assembly, and using this velocity profile as input boundary condition for the CFD code, independent from the downstream equipment configuration.

2. Experimental set-up

The fan studied is an axial fan of the firm PAPST, type 4148XP, which has an outer diameter of about 120 mm.

The experimental set-up is based on existing set-ups for measuring fan characteristics. Some adaptation allows for the new tasks. Fans designed for use without duct are to be tested without ductwork [3]. Usually, chamber test methods are used for these kinds of tests : to combine the requirements 'free outlet' and adjustable pressure drop, a chamber is connected to the outlet of the fan. The pressure drop is controlled by use of a variable nozzle at the outlet of the chamber. Environmental disturbances of the flow at the inlet are avoided by providing an inlet chamber. An overview of the design is shown in figure 1.

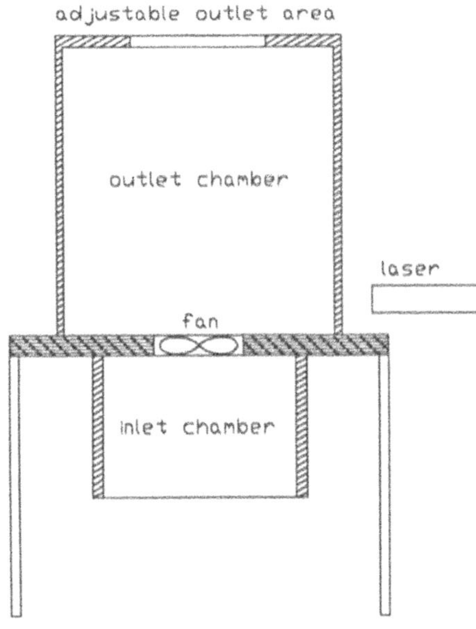

fig 1 : measurement set-up

To avoid influencing the flow profile with the 'outlet chamber', the chamber has to be sufficiently large. ANSI standard 210-85 specifies that the outlet chamber should have a cross-sectional area of at least sixteen times the area of the fan outlet for axial fans [3, 7.3.1]. For inlet chambers, it specifies that the cross-sectional area has to be at least five times the area of the fan inlet [3, 7.3.2]. An upper limit to the applicable flow chamber radius is set by the maximal focus distance of the LDV system used, which equals 600 mm. In accordance with these limits and with other practical limitations, the outlet chamber diameter is 800 mm, and the inlet chamber 600 mm.

3. Some features of the LDV measurement technique

Velocity measurements are performed with an LDV (Laser Doppler Velocity) measurement system. It consists of a SPECTRAPHYSICS 5 W all lines argon laser and a DANTEC 2D beam splitter and Flow Velocity Analyser. This system measures the velocity in a small volume formed by the intersection of two laser beams originating from the same source. The distance between the parallel beams is set at 60 mm and focusing optics are used with a focal distance of 600 mm.

The LDV measurement technique is based on the Doppler frequency shift of the light beam reflected by moving particles. It therefore requires the presence of tracer particles in the flow field. These particles must be light enough to ensure that they follow the air flow. On the other hand they have to generate sufficient reflection of the laser beam. Incense smoke particles prove to be a good solution for these two requirements.

A measurement consists in taking 500 accepted samples (measurement signals which satisfy a required signal to noise ratio), and calculating the mean value and the root mean square value of the velocity in two directions, perpendicular to the laser beam. A 3 D-traverse system allows to focus on different points in the flow field.

4. Results

Measurements of the fan outlet profile were made for one fan operation point. The chosen point is the one corresponding to the free (unobstructed) air flow condition. A comparison of velocity profiles measured with and without a test chamber surrounding the fan outlet showed only small differences, which validates the outlet chamber.

To obtain a fan velocity profile which is usable even for compact PBAs, the fan outlet profile has to be determined as close as possible to the fan outlet plane. The measurements in a plane with a downstream distance from the fan outlet plane of 7 mm satisfy this requirement. Measurements in two extra planes at downstream distances 27 and 47 mm were made. The (2D)-velocity profile was determined for two traverse directions (laser movement directions), perpendicular to each other. The traverse directions were chosen such that the first direction leads to values of the axial and radial velocities, and the second to values of axial and tangential velocities. This is sufficient to characterise the flow pattern in the whole plane, if axial symmetry holds.

In figure 2, a comparison is made of the measured velocities in the planes at distance z=7mm and z=47 mm. The difference between the velocity profiles is striking.
For the axial velocities (figure 2a and 2c), the peak values of the plane at position z = 7 mm downstream of the fan outlet are clearly higher than those at the plane z=47 mm. As could be expected, the velocity profile tends to flatten as the distance from the fan outlet increases. Yet the flattening process is not strong enough to cause a uniform flow at z = 47 mm. The effect of the hub is still clearly present here.

Axisymmetry does not hold : the velocity profile at the left side (front side) of the centre is different from the one at the right side (back side).

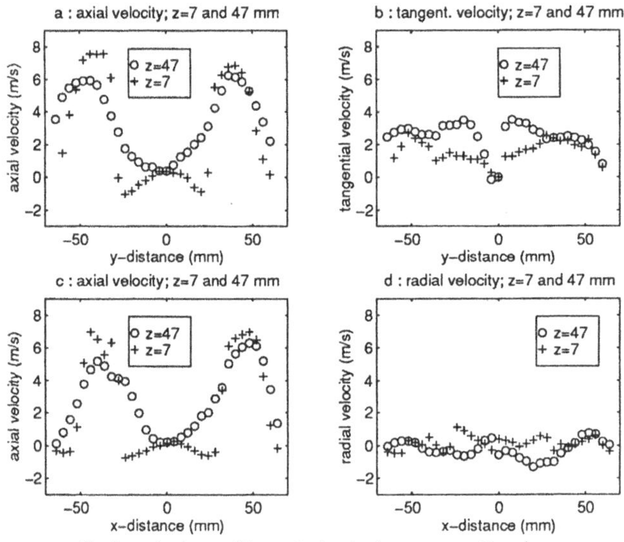

fig 2 : velocity profile evolution in downstream direction

For the radial velocities (fig. 2d), a profile is observable at z = 47 mm downstream of the fan, which is less at z = 7 mm downstream. Probably, recirculation flow causes these patterns. Nevertheless the absolute values of these velocities are low.

The tangential component of the velocities measured in the plane z = 7 mm downstream of the fan is smaller than the one at z = 47 mm (fig. 2b), although no swirl augmenting components are present downstream of the outlet. The explanation for this phenomenon will be given below.

Comparison of the results at z=27 mm with those at z=7 mm leads to similar conclusions.

The main geometrical deviation from axisymmetry is due to the presence of three studs (fan supports) at the outlet plane. Measurements in the plane z = 7 mm downstream at two traverse directions rotated over 120° with the fan centre as axis of rotation led to a significantly better agreement between the velocity profiles. This shows that the studs are the main cause of the asymmetry in the velocity profiles.

To obtain an overall view of the flow profile at the outlet of the fan, the velocity vector was measured in several locations in the plane 7 mm downstream of the fan outlet plane. These points form an orthogonal grid with point to point distances of 5 mm. To get a 3D-view of the velocity vector with a 2D measurement system, every location of the grid was tested two times (with the laser beam rotated over 90 °). This implies that the vertical component of the velocity was measured twice in every point. The grid points too far from the free outlet area (e.g. over the hub) were skipped, because, due to seeding problems in areas with zero velocity, it would take too long to obtain accurate values.

In figure 3, the measured axial velocities are shown as a function of the radial position only. The extended range of values for a given radius illustrates the fact that axisymmetry does not hold. A combination of radius and angle is necessary to describe the flow pattern at the fan outlet.

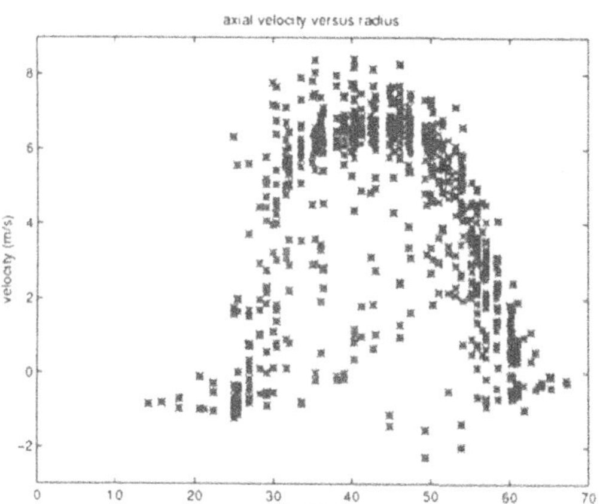

fig 3 : axial velocity as a function of radial position

To obtain a representative set of data for a given angle, the data obtained have been reprocessed with the aid of MATLAB. For a grid of points with radius 5,10, 15, ..., 65 mm and angle -180, -170, -160, ..., 170 an approximated value of the velocity in three directions (axial, tangential and radial) is calculated, based on the cartesian velocities of the measured grid, via an inverse distance interpolation method.

Figures 4a and 4d show the variation of the axial velocity as a function of the angle for two radii. Two lines per diagram are present : one per perpendicular traverse direction of the laser beam. As no data filtering or curve fitting has been done, the curves follow an erratic path, due to the error in individual measurements. Nevertheless some clear trends are observable. The axial velocity is nearly constant over all angles. The position of the studs (situated at angles -135, -15 and 105 °) leads to a sudden fall of the axial velocity. At both sides of the studs the axial velocity is locally higher than the mean axial velocity, and this phenomenon is more accentuated at the side corresponding to the positive fan rotation direction (which is the negative angle direction in the chosen coordinate system.). Apparently, the influence of the studs on the upstream flow pattern is small. The flow is deflected by the studs, resulting in a contraction of streamlines at both sides of the studs and a local velocity rise.

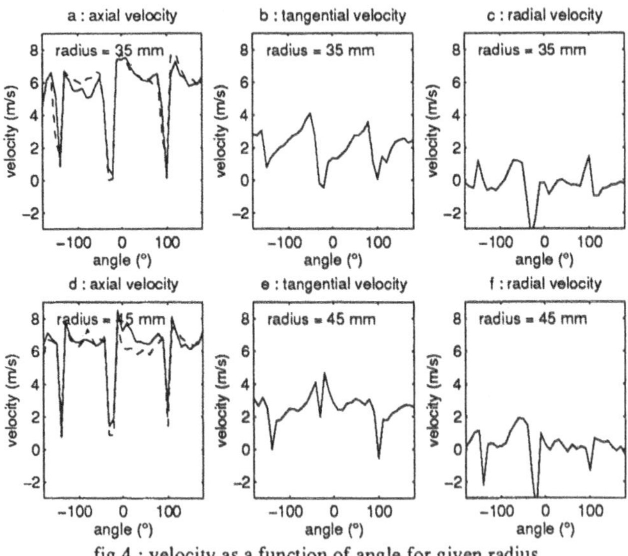

fig 4 : velocity as a function of angle for given radius

Figure 4b and 4e show the variation of the tangential velocity as a function of the angle for two radial positions. A saw teeth pattern is observable. At the studs, the tangential velocity falls to zero. Downstream of the studs (positive fan rotation direction, negative angle) the velocity increases rapidly, until a maximum of about 4 m/s is reached. This corresponds to the highest values of the tangential velocity measured in the measurements for the plane z=47 mm downstream of the fan (figure 3). At lower angles it decreases again, but less steeply.. .The formerly mentioned phenomenon that the tangential velocity seemed to increase with increasing distance from the outlet plane may now be explained : former measurements were made in a plane close to the studs. The studs produced a kind of shadow effect which cut down the tangential velocities at z=7 mm downstream of the outlet plane. This shadow effect didn't occur at z=47 mm downstream of the outlet plane.

Figure 4c and 4f show the variation of the radial velocity as a function of the angle for given radial positions. Apart from the neighbourhood of the stud angle, the value of the radial velocity is small. It is uncertain whether the contraction of streamlines or measurement inaccuracies due to beam reflection effects are responsible for the high local values near the studs.

5. CFD calculations using fan models with varying level of detail

In [1] LDV-experiments are reported on a PBA assembly. The assembly consisted of 2 x 2 fan units and 2 subracks with 9 PCBs each, mounted conforming to ETSI Standards. Two different PBA structures were tested, one using metal core PCBs and a second using standard PCBs with heat sinks mounted on the IC's. A front and side view of the test assembly is shown in figure 5.

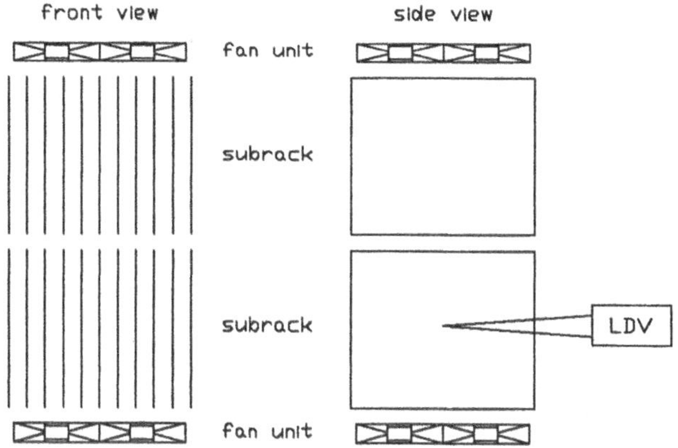

fig 5 : PBA assembly of [1]

Detailed measurements of the vertical velocity pattern are available for one card channel in 3 horizontal planes. One plane is located 21 mm above the bottom fan-unit, the second plane is located at the lower edge of the lower subrack, the third at the top edge of the lower subrack (285 mm above fan unit).

Flotherm simulations have been performed on a structure similar to this assembly. The system modelled for CFD calculations considers only the lower subrack. The modelled structure is restricted to 5 card channels (6 PBAs) and one fan unit (two fans fixed in the bottom plate, which cover the bottom plate partially). Symmetry considerations allow the results of these simulations to be extrapolated to the real assembly. Distances between the different cards and position of the cards relative to the fan unit are chosen in accordance with the system considered in [1]. Individual heat sinks or turbulence generators are not considered. The turbulence effects caused by these components are accounted for by specifying increased surface roughnesses and an average height of components, based on the heat sink height. The cards contain also a DC/DC converter. The dimensions of this converter have been respected in the simulated structure, modelling this component as a flow blocking element.

To be able to calculate velocity distributions, and given the 3D nature of the geometry simulated, 4 variables are to be calculated essentially, being the pressure and the three cartesian velocity components. This asks for at least 4 differential equations to be solved. Turbulence modelling has been incorporated using the standard k-ε-model. In the k-ε-model, the (local) value of turbulent viscosity is calculated from the values of k (turbulence energy) and ε (rate of k-destruction, kinetic energy dissipation rate), which in their turn are calculated by means of two extra transport equations.

Temperature will be an essential variable to incorporate when heat transfer effects are to be considered. It is thought however that temperature effects will affect velocity distributions only in a limited way, given the forced flow character brought about by the fans. Therefore, the effect of temperature as an independent variable has been neglected. Properties of the fluid (laminar viscosity, density) are taken to be constant and are evaluated at absolute pressure 1 bar and temperature 20 °C.

Two fan models were compared.
In a first model, the fan is represented as a rectangular element on the bottom plane of the structure. A uniform velocity profile is established, perpendicular to the plane in which the fan is situated. The dimensions of the fan were set so as to give an area equivalent to that of the actual circular fan. The mass flow entering the structure is fixed to 0.05 kg/s. The value of k (turbulence energy) for the entering mass is fixed to 1.7 m/s. This value showed to be an acceptable averaged value according to the measurements.
For ε (kinetic energy dissipation rate), no measurement results are available. Developers of the Flotherm software code advise following relation, which showed to be applicable to developed flows :

$$\varepsilon_{in} = 0.1643 \,.\, k^{3/2}{}_{in} \,/\, L_{in} \,,$$

where k_{in} is the formerly specified turbulence energy at the inlet, and L_{in} one tenth of the diameter of the fan diameter. This leads to a value for ε_{in} of 32.9 m^2/s^3.
In a second model, an element was added to the fan of the first model, which blocks the flow over part of the plane, representing the hub. The total flow is now applied over the unblocked part of the surface. Swirl has also been added, by specifying a (constant) tangential velocity over the unblocked part of the fan surface, being 3 m/s.

Calculations were performed using a calculation grid with cell to cell mean distances of about 5 mm. Some deviation from this refinement option was necessary, due to the geometrical properties of the structure (structural boundaries have to match the underlying grid). Careful choice of the grid refinement between two key-points (anchor points of the grid, matching the structural boundary locations) resulted in smooth cell size variations, which are important to obtain accurate results. The resulting grid contains 55 x 35 x 57 cells.

Figure 6 compares the simulation results with the measurements obtained in [1].

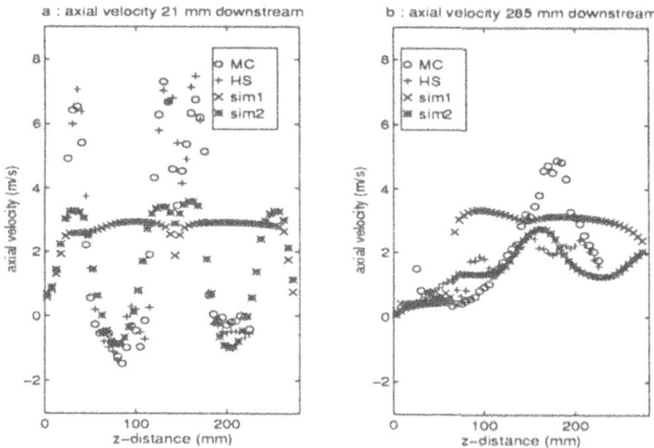

fig 6 : comparison of measurements and simulation results

Figure 6a gives a view of the axial velocity profiles measured or calculated in horizontal plane 1, which is located 21 mm above the bottom fan-unit and 12 mm below the card boundaries.

The velocity profiles measured in plane 1 for two card configurations, one with heat sinks (symbol '+') and one with metal core (symbol 'o') are about the same, an illustration of the fact that the influence of downstream conditions on the velocity profile near the fan outlet is of minor importance.

The velocities calculated in simulation 1 (symbol 'x') lead to a flat profile, which clearly does not correspond to reality. The axial velocity profile calculated in simulation 2 (symbol '*') gives better agreement, but the peaked character of the real system is not predicted, due to the lack of agreement between the real fan geometry (circular) and the modelled one (rectangular). This may cause inaccurate predictions in the mass flow distribution over different card channels. As far as a correct flow distribution is important, which normally will be the case, a more accurate description of the fan geometry is necessary.

Figure 6b gives a view of the axial velocity profiles measured or calculated in horizontal plane 2, which is located near the top edge of the lower subrack (285 mm above the fan unit). Agreement between flow simulation 2 and measurement results for the heat sink geometry is improved, due to the damping effect of flow resistances on the velocity profile non-uniformities : wall friction and turbulent mixing cause non-uniformities at the fan outlet to disappear at rather small distances downstream. Velocity profile details at the fan outlet are of minor importance to obtain an accurate velocity profile at large distance from the fan, as far as the mass flow distribution in the different card channels is accurate.

6. Conclusions

For the axial velocity at the outlet of a typical axial fan used for the cooling of PBA assemblies, an important deviation from a uniform velocity profile at the outlet is present, mainly due to the presence of a hub in the axial fan. Axisymmetry isn't fullfilled, the velocity variations for a given radius are large. The three studs (fan supports) at the outlet plane are found to be the main cause of this asymmetry. Due to the regular positioning of the studs, an angular symmetry approximation, in which the velocity pattern is repeated over 120°, is in agreement with measurements.
To describe the axial and especially the tangential velocity profile of a fan accurately, a model which takes radius AND angle into account is necessary. The studs function as a local flow blockage. For the axial velocities, this results in a flow profile which is approximately flat, apart from a small region near the studs, where a velocity peak balances the region of no flow above the studs. For the tangential velocity a saw tooth profile occurs, with a zero value and a peak value in the neighbourhood of the studs in the fan rotation direction. For the radial velocities the deviations from zero velocity are small.

To obtain an accurate velocity profile at large distance from the fan, the velocity profile details at the fan outlet are of minor importance. An accurate description of the velocity profile at the fan outlet nevertheless will be necessary to obtain a correct mass flow distribution over the different card channels. Results show that incorporation of the hub into a model of the axial fan is necessary. Taking the tangential velocity at the outlet into account is also important to be able to predict heat transfer coefficients. A fan model using constant velocities at the free outlet section of the axial fan is acceptable but the geometrical representation of the fan in the simulated CFD structure should be as accurate as possible, given the coarseness of the grid.

7. References

[1] H. BRUNEEL, B. BEERNAERT, G. MORTIER, J. DECLERCQ, B. BOESMANS, W. TEMMERMAN, W. NELEMANS, E. LAUWERS, "Supporting experiments for CFD based thermal design of telecommunication equipment", Eurotherm Seminar no. 29, Delft, june 1993

[2] W. TEMMERMAN, W. NELEMANS, T. GOOSSENS, "Air flow modelling requirements nearby fans.", Flotherm user's conference, Boston, may 1993

[3] ANSI/AMCA 210-85,ANSI/ASHRAE 51-1985, "American National Standard : Laboratory methods of testing fans for rating.", 1985

6. THERMOMECHANICAL MODELLING

THERMAL DEGRADATION OF POWER MODULES

L TIELEMANS[*], G GREGORIS[+], L DE SCHEPPER[x], M D'OLIESLAEGHERS[x]
Destin N.V.[*]
Wetenschapspark 1, B-3590 Diepenbeek
Alcatel Espace[+]
Av. JF-Champollion 26, F-Toulouse
L.U.C. (IMO)[x]
Wetenschapspark 1, B-3590 Diepenbeek

Abstract

One of the main problems in power devices is the increase of the junction temperature due to internal heat dissipation. Different packaging technologies have different thermal management. This management changes during the life time of the device due to thermal mechanical stresses. The degradation of the thermal resistance is one of the main limitations in life time of power devices. In this study different substrate technologies have been compared on their degradation during reliability testing. Thermal shock (liq. to liq.), thermal cycling (air to air) and power cycling test were applied. For power cycling test, the degradation of the devices is measured as a change in thermal resistance (Rth). For the DCB substrate technology, this change is measured at different power dissipation levels to study the influence of temperature difference between junction and case on the degradation behaviour.

Reliability tests

One of the goals of this study was to compare thermal shock tests (passive stress) with the power cycling life test (active stress). The thermal shock (liq. to liq) and thermal cycling (air to air) were applied according to Mil Std 883 method 1011 and 1010 both under test condition C. For power cycling a test system was specially developed to measure thermal resistance during power cycling on the same test bench. The thermal resistance of the power devices was tested according to MIL Std 883 method. To be able to compare the power cycling test with the thermal cycling and shock test (cycles -65°C to 150°C) the stress conditions for power cycling were chosen as close as possible to these stress conditions. The system was developed to cool the sink temperature up to -50°C. By choosing the power so that the junction temperature is about 125°C (to avoid degradation of the transistor parameters, which implies a change in calibration curve) a maximal initial temperature difference between junction and case of 175°C was applied in this study.

E. Beyne et al. (eds.), Thermal Management of Electronic Systems II, 321-328.

Devices

As package technologies, DCB (Direct bounded Copper), IMS (Insulated metal substrate) and thick film on ceramic were studied in a L4 design of experiment (DOE) (table 1), where as other variables solder, solder thickness and die size were taken into account.

Table 1

no	substrate	die size [mm x mm]	solder	thickness [μm]
1	DCB	13,4x13,4	Pb Sn	80
2	DCB	12x12	Ag Sn	100
3	DCB	5,6x4,5	Pb Ag Sn	120
4	Thick film	13,4x13,4	Pb Ag Sn	100
5	Thick film	12x12	Pb Sn	120
6	Thick film	5,6x4,5	Ag Sn	80
7	IMS	13,4x13,4	AgSn	120
8	IMS	12x12	Pb Ag Sn	80
9	IMS	5,6x4,5	Pb Sn	100

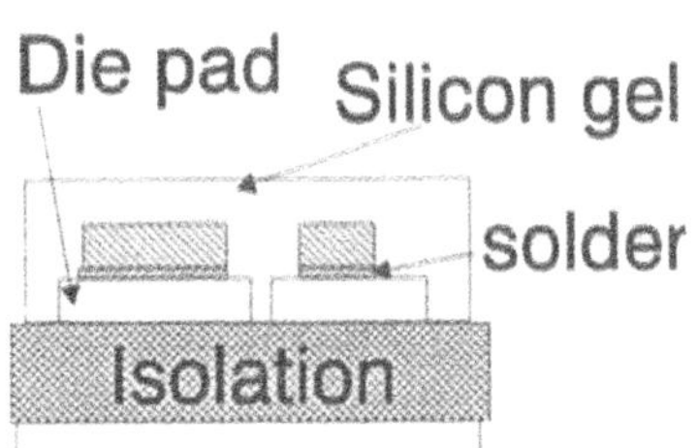

Fig 1: Schematic view of

Unfortunately, 2 combinations (no. 7 and 8) could not be measured due to technology or production problems. Only devices with no or very small initial voids in the solder die bond are used in this study.

Results

Table 2 and 3 show the cumulative failures for thermal cycling and thermal shock tests (cycles -65°C to 150°C). Three failure modes were found during these tests: (A) cracks at interface between die and solder on the diode (fig. 2); (B) cracks at interface between die and solder on the transistor (fig. 3); (C) delamination between metallic layer and ceramic (fig. 4)

For the determination of the thermal resistance Rth during power cycling the case temperature, V_{ge}, I_c were continuously measured during the power cycling. The power cycles were realised by constant current I_c.pulses. To compare the degradation during power cycling, all samples have to be submitted to the same stress conditions. The case temperature was cooled to -50°C and the power dissipation was chosen so that the junction temperature reaches about 125°C at the beginning of the test. Different degradation behaviours could be observed for the different substrate technologies (fig. 5). All tests were stopped due to lift off of the bond wires. No failures of mode A,B,C were seen. Instead some solders show void formation (fig. 6) table 4. This failure mechanism could not be observed in the thermal cycling/shock tests.

table 2: thermal shock **thermal cycling**

no.	10 cycles	20 cycles	50 cycles	mode	100 cycles	200 cycles	300 cycles	mode
1	8/14	11/14		A	16/16			A,B,C
2	7/14	9/14		A	16/16			A,C
3	2/5	2/5	4/5	A	2/5	2/5	4/5	A,C
4	1/9	1/9	3/9	A,B	0/10	0/10	6/10	A
5	4/9	8/9		A	8/9			A
6	4/13	7/13		A	2/11	7/11	11/11	A
7	9/11			A,C	11/15			A,B,C
8								
9	2/14	4/14		A	2/16	2/16	5/16	A

defects / sample size

Fig. 2 : Cracks in solder die bond of IGBT

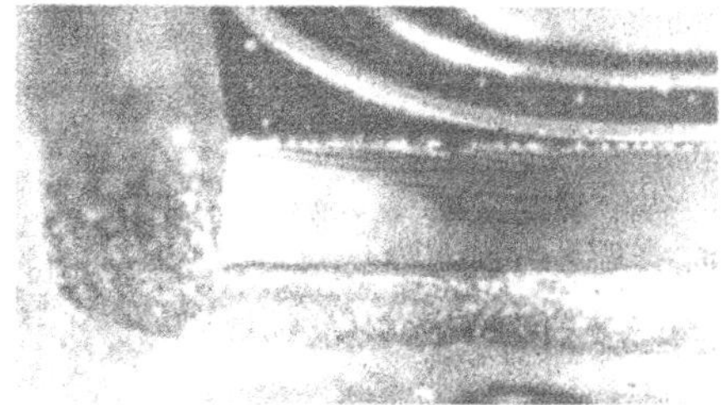

Fig. 3 : Cracks in solder die bond of diode

Fig. 4: Delamination between ceramic and Cu layer of DCB substrates

table 4: power cycling

no.	solder	mode
1	PbSn	holes
2	AgSn	small holes
3	PbAgSn	no holes
4	PbAgSn	very small holes
5	PbSn	holes
6	AgSn	holes
9	PbSn	holes

DCB
T_c= -50C
Rth [K/W]
10A
10A
50A
50A
75A
75A
power cycles

IMS
T_c= -50C
Rth [K/W]
10A
10A
power cycles

Thick film
T_c= -50C
Rth [K/W]
10A
10A
50A
75A
50A
75A
power cycles

Fig.5: Degradation of thermal resistance of DCB (a), thick film (b) and IMS (c) type of devices.

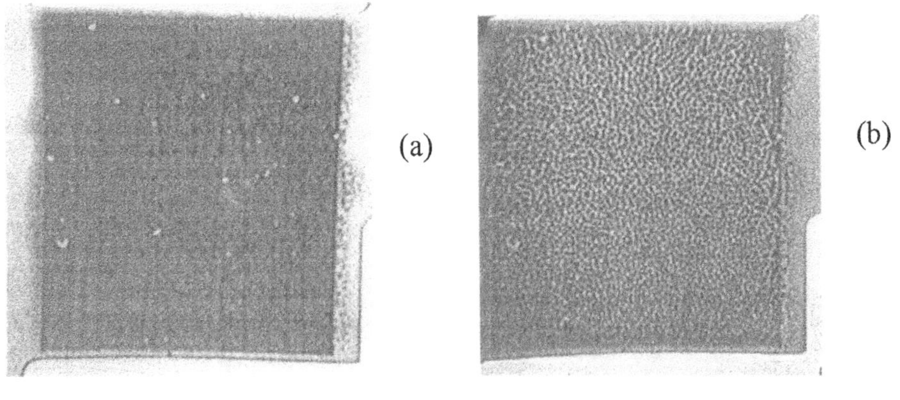

(c)

Fig.6: Microfocus X-ray of the solder die bond (sample 1). (a) initial, (b) after test, (c) after test: cross section

Correlation between thermal schock/cycling and power cycling

To compare the MTF of the thermal schock/cycling tests with the degradation of the thermal resistance during power cycling, a measure for this last degradation has to be defined. As degradation measure for power cycling, the inverse of the degradation velocity was taken (failure criterion / number of power cycles to reach the failure criterion).

For linear degradation: MTF = failure criterion / degradation velocity.

The MTF of the thermal cycling and shock test show a good correlation with the degradation of the thermal resistance observed during power cycling (fig. 7). This correlation between power cycling and thermal shock/cycling is a mathematical correlation. The failure modes are different and therefore also the degradation mechanisms. The information obtained out of thermal shock/cycling (passive stress) concerning reliability can not be translated to power cycling (active stress).

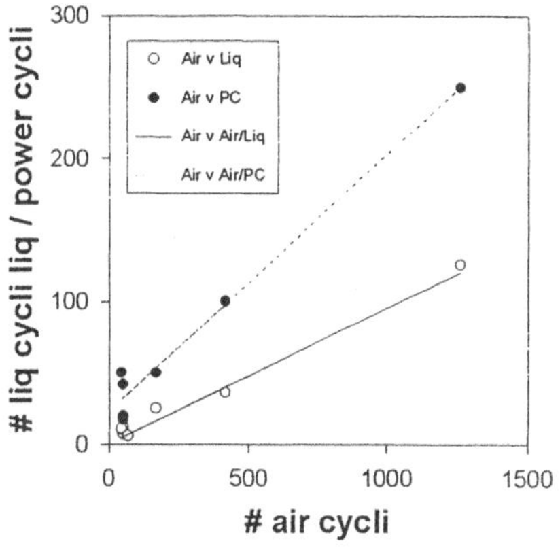

Fig. 7: Correlation between power cycling and thermal schock/cycling tests

DOE-results of power cycling

To analyse the results of the DOE experiment the initial thermal resistance and the change in Rth after # of power pulses was taken (table 5). For the sample no 7 and 8 a fictive result was taken, taking the following observations into account: For the 50A and 75 A dies the change is in all other cases almost the same and 2 times as large as for the 10A die.

The lowest initial Rth is obtained using DCB, the largest die. The influence of the solder type is neglectable. An increase in solder thicness seems to have a positive influence on the decrease of the Rth. The lowest degradation is obtained using IMS, PbSnAg-solder and the smallest die.

The substrate technology with the best performace has the worst life time behaviour. The Rth of DCB degrades up to 40% during power cycling, meaning that the junction temperature will increase with 40% during life time.

table 5: DOE-experiment

no	Rth [K/W]	Δ Rth [%] after 6450 power pulses
1	0,40	0,58
2	0,50	0,50
3	1,27	0,24
4	0,69	0,04
5	0,75	0,20
6	2,90	0,20
7	**1,00**	**0,10**
8	**1,00**	**0,10**
9	2,27	0,10

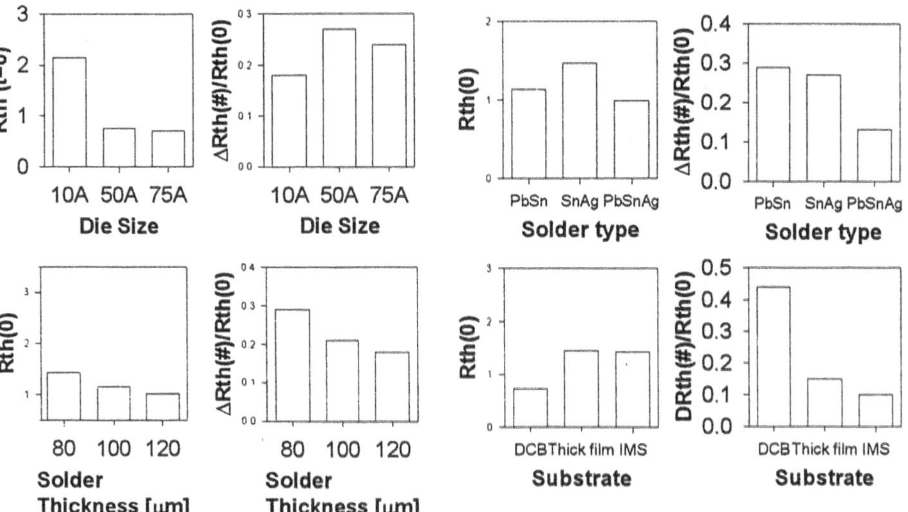

Fig. 8: Analysis of DOE experiment

Acceleration of test

To extrapolate these tests to real life conditions, the acceleration factor for power cycling has to be determined. This can be derived from tests at different power dissipation levels. While in most applications the devices are used by maximal allowed power, the junction temperature during power pulses will be always at it's maximum. Therefore the acceleration factor was studied with tests performed with an initial junction temperature at 125°C and different case temperatures (fig. 9, table 6). Out of these measurements it can be concluded that the test can be accelerated to using higher ΔT (fig. 9).

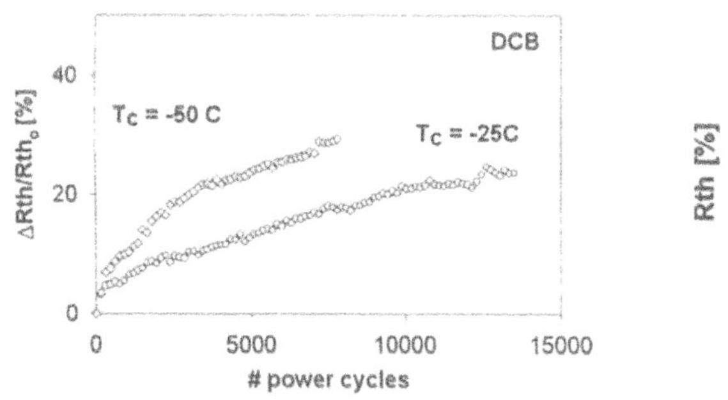

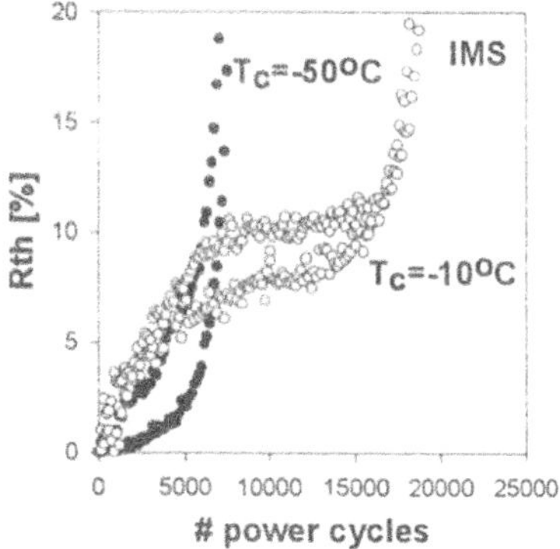

Fig.9 : Rth degradation of DCB and IMS type at different ambient temperatures.

Thermal Impedance

The test system developed measures the Vge change during the end of the pulse and records also the first 100 msec of the cooling curve. Therefore not only the degradation of the (static) thermal resistance can be studied, but also the change in dynamical thermal conductivity (thermal impedance) during power cycling (fig. 10).

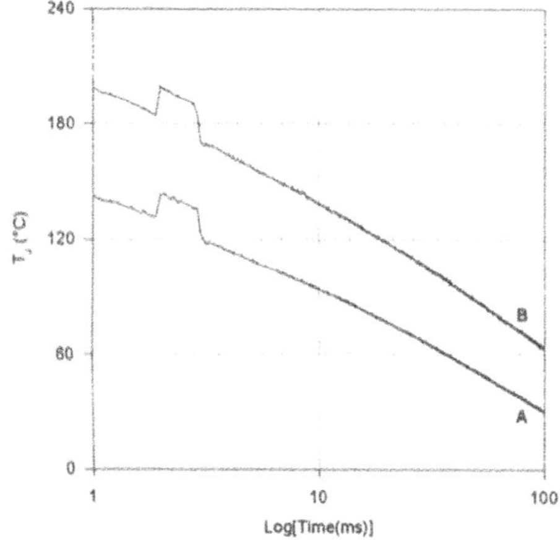

Fig.10: Cooling curves of sample 3 before (A) and after (7800 cycles) (B).

Wire bond lift off

All power cycling tests were stopped while bond wire lift off occured. During power cycling the power dissipation increased (fig. 11). This increase in V_{ge} is probably due to the increase in contact resistance.

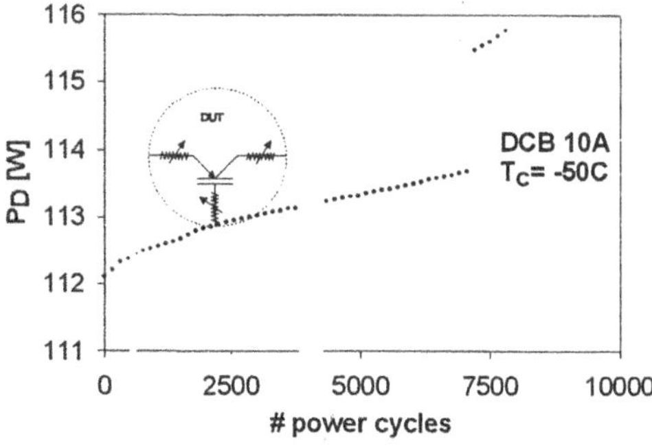

Fig.11: Change in power dissipation during power cycling test

In the first approximation, the TTF for the bond lift offs seems to be independent on the sample type (fig.12). The acceleration for bond lift off can be described with an Arrhenius type of degradation mechanism, where ΔT is the driving force: $TTF = A * \exp(\frac{B}{T_j(0)-T_a}$ or with the Coffin Manson model: TTF~1/ΔT^2 . Within the tested range, no difference could be made between the 2 models, although the prediction obtained by extrapolation to real life conditions (ΔT=80°C) give quite different results.

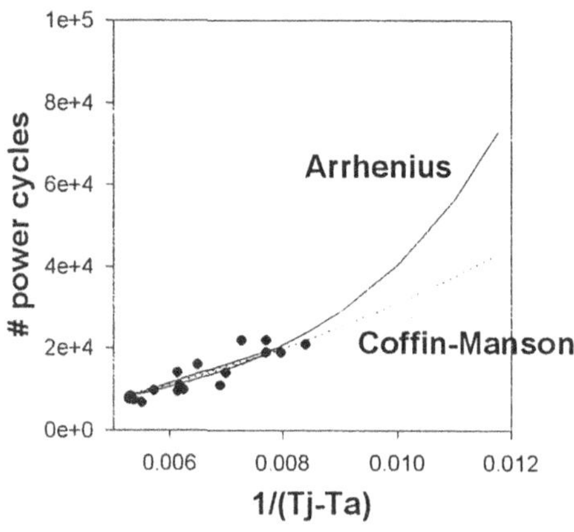

Fig.12: TTF for bond lift off observed at different stress levels $\Delta T= T_j-T_a$.

(T_j(t=0) was always about 125°C.)

Acknowledgement

This research was performed in the frame of the BRITE EURAM "Shortest"-project. The authors have to thank Mr P. Siliprandi and F Baio of IBM SEMEA for performing the temperature cycling and shock test.

THERMAL AND THERMOMECHANICAL MODELLING OF BALL GRID ARRAY PACKAGES

Thierry Fromont
BULL S.A.
BP 68 78340 Les CLayes sous Bois FRANCE

ABSTRACT

Ball grid array is a promising packaging technology. To exploit its benefits, a good knowleldge of its thermal and thermomechanical behaviour is mandatory.

Ball grid array allow to a higher silicon to printed circuit board (PCB) area ratio, sometimes increasing dramatically the power density at the package level. Elevated die temperature can degrade reliability and moreover decrease electrical performance (e.g CMOS applications).

Thermal analysis of package's structure and printed circuit board stackup are performed for a better understanding of the importance of the constitutive elements of the package allowing to an enhanced thermal coupling of both parts for cost effective thermal management.

The reliability of the solder balls is a key concern. Fatigue failure, induced by the thermal expansion mismatch between the package and the FR4 board, is one aspect of perticular importance. The plastic strains within the solder alloy computed in the finite elements analysis are used as inputs in the Engelmaier equation to estimate the fatigue life of the assembly.

1 THERMAL ANALYSIS

The thermal analysis was performed on cavity up packages, since it is asssumed that cavity down packages are well suited for heat sink attachment if needed. Two body sizes for both plastic (PBGA) and high temperature cofired ceramic (CBGAH) structures were investigated .

-a 20 mm body package, 2.5mm thick, with a 9 mm square chip and 320 IO's in 4 peripheral crowns of balls, including 56 power IO's,

-a 42 mm body package, 2.5mm thick, with a 15 mm square chip and 624 IO's in 6 peripheral crowns of balls, including 144 power IO's.

E. Beyne et al. (eds.), Thermal Management of Electronic Systems II, 329-338.

An infinite array of packages soldered on an FR4 board with two grid pitch per package size was assumed:

-32mm (respectively 60mm) as representative of high density of VLSI packages

-45mm (respectively 90mm) to simulate a lower VLSI density on the PCB

A reference temperature of 20°C was chosen and a uniform density of power was applied at the top of the chip (1 W per chip). The heat transfert coefficient was assumed to be 20°C/W m² (this is an average value for a 1m/s air speed) and applied to free areas of the PCB and of the package. Air temperature rise was neglected.

1.1 PACKAGES THERMAL ENHANCEMENTS

To enhance thermal performances of PBGA's, several improvements are achievable:

-The first is to add copper layers (used as power planes). Therfore these internal layers are connected to the equivalent layers of the substrate. By the way there is a small thermal coupling between the package and the PCB.

-The second is to add some balls (thermal balls), arranged in a 1.27mm grid matrix below the cavity and connecting them to the ground plane of the substrate.

-The third consists in connecting these thermal balls to the backside of the chip using thermal vias. Therefore, the thermal coupling between the chip and the PCB is increased, so that we can consider the PCB as an Heat Sink.

CBGAH thermal performances are enhanced by adding balls under the chip area and to connecting them to the PCB ground plane.

Table 1 below, summerises the configurations modelled .

TABLE 1 : Thermal configurations

Type	Internal copper planes	Thermal balls under cavity	Thermal vias under cavity
PBGA			
1	No	No	No
2	Yes	Yes	No
3	Yes	Yes	Yes
CBGA			
4	N/A	No	No
5	N/A	Yes	No

1.2 FINITE ELEMENT MODEL

The thermal steady state conduction analysis with imposed convective boundary conditions was performed using the general purpose finite element program ANSYS[1] to simulate the cooling conditions of the packages in a user environment.

Due to symmetries in the structure the model can be reduced to one eighth of a package and associated PCB. A typical finite element representation is shown in Figure 1. Elements used in the model are 3D isoparametric solid (ANSYS Stiff 70) for all parts excepted for vias, solder connections and air gap under the cavity which are modelled using 3D conducting bars (ANSYS Stiff 33) .

The thermal analysis considers conductive heat transfert in each part of the system, wich include FR4, internal power planes, via stucture, solder balls, internal layers and vias, die attach and chip. Previously, a separate micro finite element analysis was performed to get the equivalent thermal conductivity of the copper planes to take into account the effects of the holes needed by the non connected vias.

Material properties used in the model are defined in table 2.

TABLE 2 : PROPERTIES OF MATERIALS

Material	Thermal conductivity(W/mk)
Epoxy	0.4
Silicon	110
Adhesive	1
Copper planes	210
Tin Lead	30
Ceramic	17

FIGURE 1 : TYPICAL MESHING

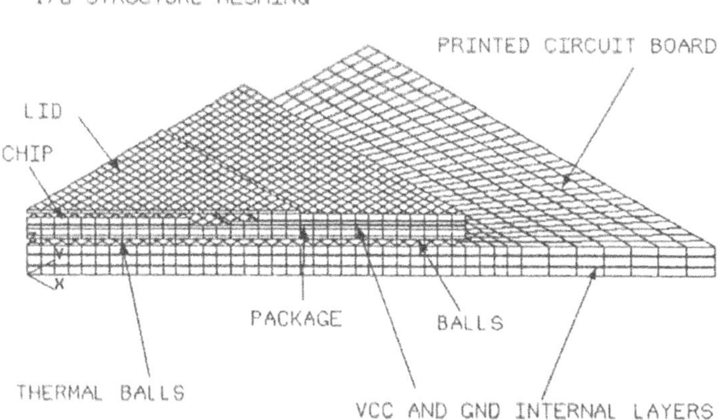

1.3 RESULTS

An example of the 3D temperature profile is given in figure 2.

The thermal resistance junction to air defined as:

(maximum silicon temperature-air temperature)/Power dissipated

Results are reported in table 3.

TABLE 3 : THERMAL RESISTANCES JUNCTION TO AIR (°C/W)

Package Type	Wide Body Area per Package		Small Body Area per Package	
	90x90	60x60	45x45	32x32
1	80	83	136	140
2	19	21.5	37.5	44.5
3	11	13.5	16.8	23.5
4	12	14	24.5	31
5	8.5	10.5	20	26.5

FIGURE 2 : TYPICAL TEMPERATURE PROFILE

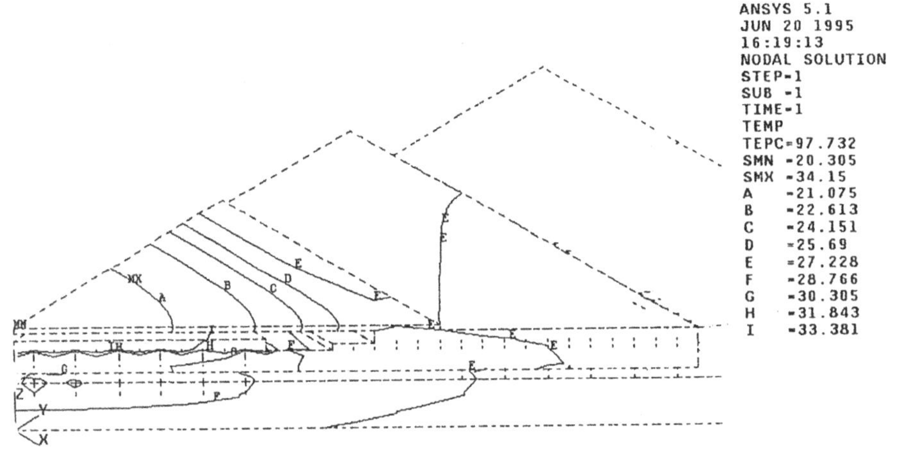

1.4 EXPLOITATION OF RESULTS

It clearly appears that the initial version of the PBGA offers a very poor thermal coupling between package and PCB. This is the reason why increasing the board size does not decrease significantly the global thermal resistance.

The first improvement (consisting in copper layers within the package). increases the thermal coupling but the junction to board still remains high. Adding thermal balls to the previous structure improves the junction to board thermal resistance: but the best results are achieved by using thermal vias. providing a direct path from chip to PCB to the heat generated within the silicon.

Due to the thermal conductivity of the package. ceramic ball grid arrays offer a better thermal coupling between chip and peripheral balls. This explains why thermal performances are better than with plastic packages and why improvements are less efficient.

Typical power dissipation capabilities of the packages. in machine environment are reported in table 4.The environment is assumed to be a 1m/s air flow at -60°C air with a maximum allowed junction temperature of 105°C.

TABLE 4 : Maximum Power Dissipation(W)

Package Type	Wide Body Area per Package		Small Body Area per Package	
	90X90	60X60	45X45	32X32
1	0.8	0.8	0.5	0.5
2	3.4	3	1.7	1.5
3	5.9	4.8	3.9	2.8
4	5.4	4.6	2.7	2.1
5	7.6	6.2	3.2	2.4

2 THERMOMECHANICAL ANALYSIS

The thermomechanical analysis was performed on cavity up PBGA's and CBGA's.

The Ceramic Ball Grid Array is a 1.4 mm thick ceramic base populated with a full 26X26 balls. Balls are made of .9mm diameter high temperature melting non eutectic Tin Lead alloy (90%Pb 10%Sn). and soldered on the printed circuit board with eutectic Tin Lead alloy. Solder alloy volume on the printed cicuit board has been optimized for fatigue failure (experimentally and correlated with previous modelling results). High temperature cofired ceramic (CBGAH) and low temperature cofired ceramic (CBGAL) are investigated and compared to a 3mm thick ceramic base (CBGAH Thick) for which we have experimental results.

The Plastic Ball Grid array is a 42 mm body package with 6 peripheral crowns of functionnal balls and a 12X12 matrix of thermal balls under the cavity. Balls are made of eutectic solder alloy. Soft solder alloy ball are very sensitive to the package weight. so that the gap between the PBGA and the PCB is function of the solder ball volume and the package weight. Therefore two gap value (0.37 mm and 0.49 mm) are used for modelling (respectively PBGA S Gap and PBGA L Gap).

The 0-100°C thermal cycles were chosen as the stress to which packages mounted on PCB are subjected.

2.1 APPROACH

Ideally. the model would be a three dimensional single model (3D Model) including appropriate portion of the BGA and of the PCB. with sufficient details in the solder balls region of interest to accurately determine the plastic strain distribution. Unfortunately. such a model would request an enormous computer time.
A second approach would consists in the development af a coarse 3D model to represent the structural coupling between component and PCB and to determine the deformations at the ball boundaries. These deformations are then used as input boundary conditions to a more detailed model of the solder balls. However. computer time remains to much important.
Therefore. we reused the approach defined in a previous analysis. which has been sucessfully correlated with experimental results. This approach consists in a two dimensionnal model of the structure. the crossing plane being the diagonal plane of the package.

A reference temperature of 125°C (assuming stress free structure) was chosen and symmetry displacements conditions were applied at the boundaries of the model. A serie of temperature steps of 1°C was chosen to simulate the thermal cycles. but internal steps were generated to ensure numerical convergence.
Post processing at 100°C and 0°C allow to the computation of the plastic strain variation within the solder joint.

2.2 MODEL

In both cases the PCB model includes both dielectric and copper power and ground layers. The PBGA model includes also a 15mmX15mm chip bonded with 100 microns of adhesive in the cavity, whereas it has been neglected in the ceramic package since it is assumed that the structural influence of the chip on the solder balls can be neglected due to the stiffness of the ceramic.

For time computing savings, the meshing of the joints is splitted in two :

-coarse meshing group for structural effects

-fine meshing group for accurate plastic strain computation.

The elements used in the model are 2D isoparametric solids with non linear capabilities and plane strain option (ANSYS STIFF 42)

A typical 2D model meshing is shown in figure 3 (PBGA).

The physical properties used in the model for purely elastic materials, are shown in table 5. Non linear properties are used for solder alloys

TABLE 5 : PROPERTIES OF ELASTIC MATERIALS USED IN MODEL

Material	Young Modulus $1X\ 10^{10}$ Pa	Poisson Ratio	TCE $1X10^{-6}$
Ceramic High Temperature	25.5	0.27	6.5
Ceramic Low Temperature	11.7	0.27	6.7
Copper	11.7	0.35	16.8
FR4	1.5	0.33	17
BT	3.3	0.33	15

FIGURE 3: TYPICAL MESHING

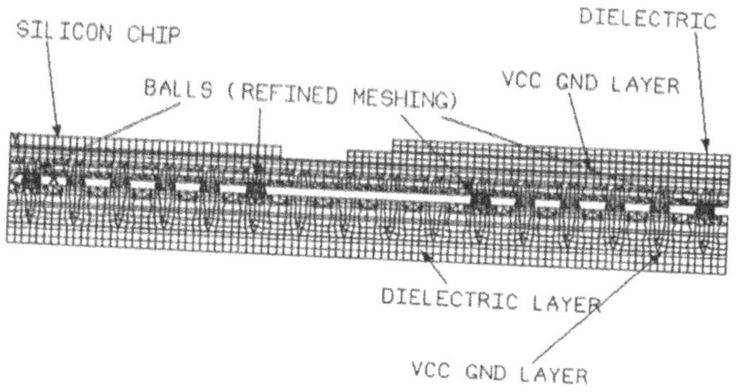

2.3 RESULTS

Maximum plastic strain variations during 100°C to 0°C thermal cycle within the eutectic solder joint are reported in table 6.

Plastic Ball Grid Array

A typical distribution of the variation of the plastic strain during 100°C to 0°C thermal cycle within the eutectic solder joint of a corner joint is shown in figure 4. For the large gap condition (PBGA LG), inside a given solder joint the maximum value of this variation (therefore preferential failure sites) are located along the interface between the component and the solder, generally on the side oriented away from the package center, excepted for the ball located at the periphery of the die cavity.

The solder joints exhibiting the greatest plastic strain variation are located below the cavity. This is probably due to the fact that the local coefficient of thermal expansion is modified by the chip. For peripheral joints, the strain increases when going away from the center of the package.

Small gap condition assemblies (PBGA SG) have the same behaviour, excepted that in some cases preferential failure sites are equally located on interface with package or printed circuit board. However plastic strain is not very sensitive to the gap.

Ceramic Ball Grid Array

Generally , inside a given solder joint the maximum value of this variation (therefore preferential failure sites) are located along the interface between the component and the solder, generally on the side oriented away from the package center. This variation increases when going away from the center of the package.

TABLE 6 : MAXIMUM PLASTIC STRAIN (%) IN CBGA AND PBGA SOLDER JOINTS

Distance To axes in mm	0.635	6.9	10.8	13.3	15.9	19.7
PBGA L.G.	0.25	0.24		0.19		0.21
PBGA S.G.	0.23	0.21		0.18		0.20
CBGAL	0.24		0.25		0.39	
CBGAH	0.25		0.27		0.67	
CBGAH THICK	0.26		0.4		1.2	

FIGURE 4 : TYPICAL PLASTIC STRAIN PROFILE IN PBGA SOLDER JOINT

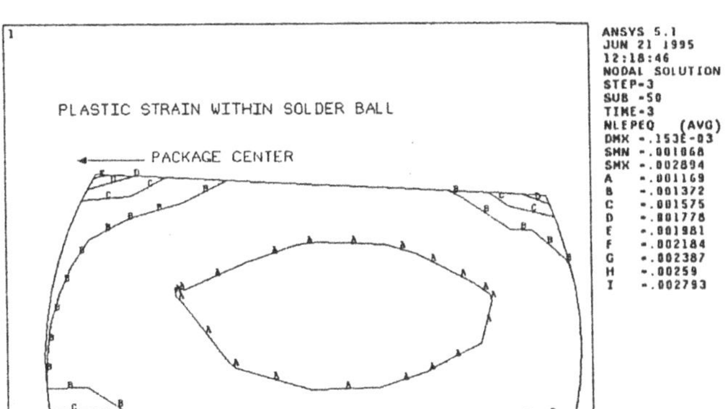

2.4 RELIABILITY MODEL

The mean number of cycles to failure can be estimated from Engelmaier solder fatigue life model and the maximum plastic strain variation computed by F.E.A. For the package structure and solder ball shapes assumed. the computed values of the mean solder life expectancy, using the Engelmaier correlation are reported in table 7.

PBGA mounted on PCB is better than CBGA mounted on PCB.

Gap between PBGA and PCB is not a very sensitive parameter. First failures should occur in the balls located under the chip (6000 cycles under the chip and 8000 cycles for corner joints).

Thickness of the ceramic package has a great influence on the solder joint life. The CBHAH Thick results are well correlated with experimental results.
The nature of the ceramic is also an important parameter. Due to its lower stiffness. life expectancy is doubled when using Low Temperature Ceramic

TABLE 7 : ESTIMATED MEAN NUMBER OF CYCLES BEFORE FAILURE

Distance to axes in mm	0.635	6.9	10.8	13.3	15.9	19.7
PBGA L.G.	6100	6500		9700		8200
PBGA S.G.	7000	8200		10000		8900
CBGAL	6500		6100		2900	
CBGAH	6100		5300		1150	
CBGA H THICK	5700		2700	500		

3 CONCLUSION

The thermal analysis of BGA packages reveals the importance of their constitutive elements. Efficient improvements of thermal performances of PBGA are found to be achievable by the use of internal copper layers and thermal vias and balls. By the way it is possible to have PBGA 's with a power dissipation range close from CBGA's.

Thermomechanical behaviour of PBGA is very good. Improvements are possible for CBGA through thickness optimization and material choice. Influence of heigh of solder connections not investigated in this work is also a way of improvement.

Figures of merit reported allow to optimize package structure and material selection for cost effective thermal managment and reliability.

ACKNOWLEDGMENT

The work presented here has been funded in part by Research Projects from the European Community (Esprit Project 8248 CHIPPAC). The author also thank the Bull Packaging Department staff for their fruitfull discussions.

REFERENCES

[1] Swanson Analysis Systems,Inc. ANSYS User's Manual for Revision 5.1

THERMAL AND THERMO-MECHANICAL EVALUATION OF A 'CHIP IN MOULDED INTERCONNECT DEVICE'

F. CHRISTIAENS, E. BEYNE, J. ROGGEN
IMEC
Kapeldreef 75, B-3001 Heverlee, Belgium

J. VAN PUYMBROECK, M. HEERMAN
Siemens LPT
Siemenslaan 4, B-8020 Oostkamp, Belgium

Abstract
In this paper, a thermal and thermo-mechanical evaluation study of a new packaging approach, called CIMID (Chip in Moulded Interconnect Device), will be presented. The thermal performance of a prototype CIMID structure has been investigated by means of finite element analysis and experimental thermal characterisation techniques. Thermally induced stresses were calculated for a uniform cooling condition, which simulated the cooling down process after the encapsulation and curing process. The thermo-mechanical behaviour of the CIMID assembly was compared to those of a chip on PCB assembly.

1. Introduction

A '**C**hip **I**n **M**oulded **I**nterconnect **D**evice' (CIMID) technology has been developed using a new interconnection and packaging approach that combines elements of 'Moulded Interconnect Device' (MID), 'Chip-On-Board' (COB) and thin film MCM-D technologies [1]. A schematic view of a typical CIMID package cross-section is shown in figure 1.

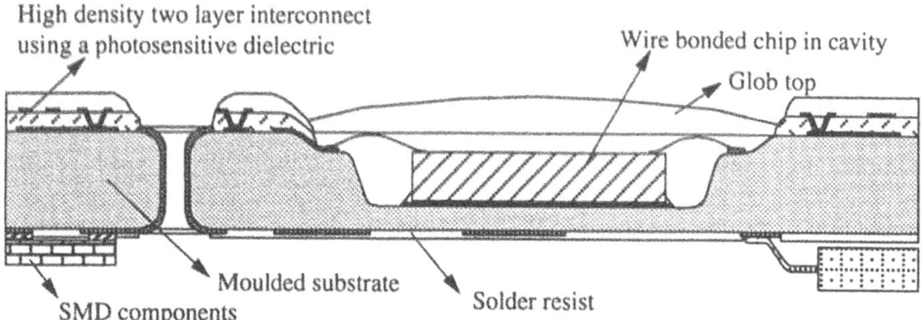

Figure 1. Cross section of a CIMID assembly.

E. Beyne et al. (eds.), Thermal Management of Electronic Systems II, 339-348.

The package body is fabricated by injection moulding from high temperature thermoplastics. The moulding process allows substrate features such as cavities for attaching bare die, via holes, and plastic studs for connection to a PCB. Metallisation of the thermoplastic substrate is performed using electroless and electrolytic copper deposition. A high speed laser defines the conductor patterning on the metallised surfaces.

In this paper a thermal and thermo-mechanical evaluation of the CIMID technology is presented. Since the thermal conductivity of thermoplastics is extremely low [2] special attention is required during the package thermal design phase in order to cope with increasing chip heat fluxes. Heat conduction through the CIMID package has been studied by means of finite element analysis. A detailed 3-D finite element model was constructed using the commercial finite element code SYSTUS™.

Thermal simulations were performed for two different types of package thermal boundary conditions:

1) Isothermal bottom surface and insulated top surface (= experimental cold plate condition). The effect of the metallisation and thermal vias on the junction-to-case thermal resistance was numerically investigated. It will be shown that the presence of a quasi continuous metallisation improves heat spreading through the substrate.

2) Uniform convective heat transfer coefficient (HTC) to all boundaries. For these boundary conditions the metallisation really plays a vital role as heat spreader. Without metal layer, convective heat transfer is limited to a small surface area due to the low thermal conductivity of the substrate. The thermal resistance reduces significantly by increasing the metal thickness to 20 μm.

The thermal resistance of a CIMID test structure has also been experimentally characterised. Four cooling environments were investigated: temperature controlled heat-sink, liquid jet impingement with FC-70, still air, and still fluid FC-70. The measured thermal resistance values were in accordance with the simulations.

Finally, the geometry of the thermal model was used to analyse the thermally induced stresses in the CIMID assembly under uniform cooling conditions, simulating the thermal load after curing of the glob top material. A similar model was constructed for a COB on FR4 structure with the same dimensions.

2. Finite Element Modelling of Heat Conduction through the Package

2.1. GEOMETRY

A complex 3-D finite element model of a typical CIMID-package was constructed in order to investigate the heat dissipation capabilities of this new packaging approach. A cross-section of one half of the modelled geometry is shown in figure 2. The structure consists of a 24 x 24 mm^2 substrate with central cavity in which a 5 x 5 mm^2 silicon die is glued using an epoxy adhesive (thickness = 30 μm). The thermoplastic substrate is manufactured by injection moulding of Ultem 2200 material, and has a thickness of 400 μm under the central cavity. The overall thickness of the substrate is 900 μm. The finite element geometry was limited to one quarter of the actual structure. Nevertheless, adiabatic boundary conditions at the symmetry planes mimic the real geometry.

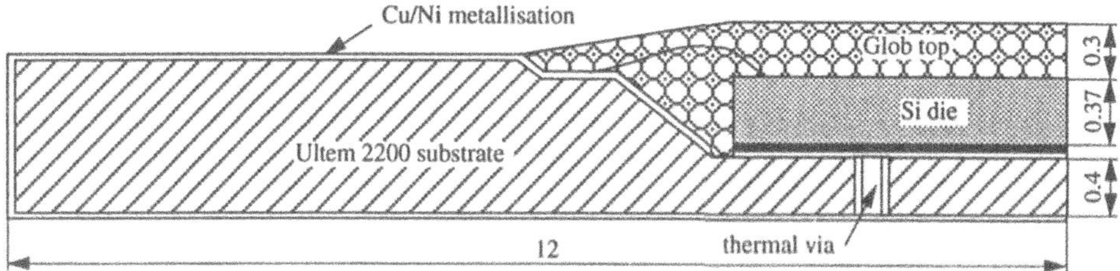

Figure 2. Cross-section of one half of the modelled CIMID structure.

Thermal vias were not geometrically modelled. Though, the presence of metallised vias was taken into account by locally increasing the substrate thermal conductivity under the die. Calculation of the local thermal conductivity is based on 1 dimensional heat flow through a thermal network consisting of a few metallised vias in parallel with the remaining Ultem material.

An epoxy encapsulation (glob top) protects the die against the environment. A technological advantage of CIMID is the intrinsic glob top barrier formed by the edges of the cavity. Substrate, die, die attach, and encapsulant are modelled using linear 3-D elements. The substrate is covered with a metallisation consisting of 15 μm copper and 5 μm nickel. This metallisation is modelled as a continuous layer using 2-D shell elements. The complete finite element model consists of 12615 elements and 10869 nodes.

2.2. MATERIAL PROPERTIES

Table 1 lists the set of material properties used for the thermal and thermo-mechanical simulations. Note the poor thermal conductivity of the thermoplastic material Ultem 2200.

TABLE 1. Thermal and thermo-mechanical properties.

Material	k (W/mK)	E (GPa)	ν	CTE (ppm/°C)
Ultem 2200	0.21	5.6	0.40	25
FR 4	-	17.4	0.3	15
Silicon	150	170	0.26	2.6
Encapsulant	0.62	10	0.25	15
Epoxy adhesive	1	6.08 if $T < T_g$	0.34	50 if $T < T_g$
$T_g = 85$ °C		3.71 if $T > T_g$		180 if $T > T_g$
Cu/Ni	308	-	-	-

2.3. BOUNDARY CONDITIONS AND LOADS

Two types of thermal boundary conditions were considered for the thermal analysis part:
1) isothermal bottom surface + adiabatic top surface.
2) uniform heat transfer coefficient applied to all surfaces.
These boundary conditions represent the experimental conditions of the cold plate, fluid bath (h = 135 W/m^{2}K), and still air (h = 13.5 W/m^{2}K) thermal characterisation methods. The heat source in the active layer of the silicon die is represented by a uniform heat flux.

3. Discussion of the Thermal Simulation Results

3.1. ISOTHERMAL BOTTOM SURFACE

The nodes in the bottom surface of the substrate were kept at a temperature of 0 °C. The edges and the top side of the substrate and glob top were assumed adiabatic. This type of analysis corresponds to the experimental cold plate method and provides qualitative insight in the geometrical and material parameters that affect the internal heat conduction from the active chip side to the package external surfaces.
The relevant thermal resistance R_{jc} is defined as:

$$R_{jc} = \frac{T_j - T_{bottom}}{Power}$$

First, we considered a thermoplastic substrate with 4 thermal vias (diameter = 600 µm and wall thickness = 20 µm). Therefore, the thermal conductivity under the cavity was locally increased from 0.21 to 2.01 W/mK. In this case, the dissipated heat mainly flows through the vias, while the rest of the substrate is quasi isothermal (no heat spreading).

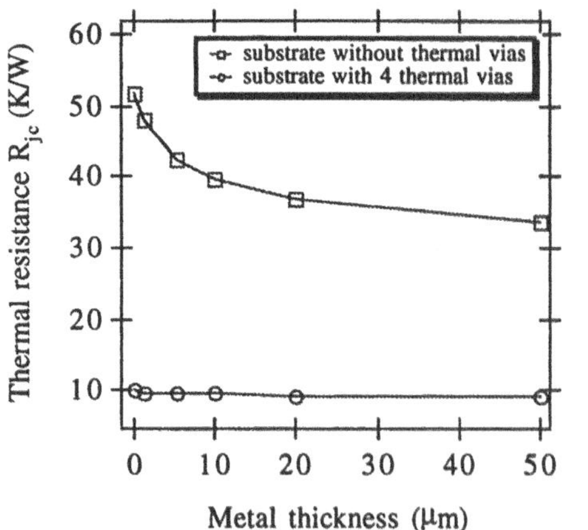

Figure 3. Influence of the metal thickness on the thermal resistance R_{jc}.

Figure 4. Temperature distribution in the CIMID-package without thermal vias.

As depicted in figure 3, the metal layer does not enhance heat conduction when metallised vias are present under the cavity. Though, without thermal vias, a significant amount of heat is flowing through the metallisation from the top surface along the sides into the isothermal bottom. Figure 4 illustrates the heat spreading effect in the low conductive substrate due to the high thermal conductivity of the metallisation. In the absence of thermal vias, a 20 µm thick Cu/Ni layer reduces R_{jc} with 25 %. This value is still 3.5 times higher than R_{jc} for the same package with only 4 thermal vias.

3.2. UNIFORM HEAT TRANSFER COEFFICIENT

In order to analyse the thermal performance of the CIMID-package in a more practical cooling environment, heat conduction was modelled under convective boundary conditions A first set of simulations deals with natural convection in air. Therefore, a uniform heat transfer coefficient of 13.5 W/m^{2}K was applied to all surface elements of substrate and encapsulation. The ambient temperature was set equal to 0 °C.

The relevant thermal resistance R_{ja} is defined as:

$$R_{ja} = \frac{T_j - T_{ambient}}{Power}$$

Figure 5 depicts the calculated R_{ja} as function of metal thickness. Without metallisation, the heat has to be transferred into the air across a very small surface area and R_{ja} is governed mainly by a huge external convective thermal resistance. In this case, the effect of thermal vias is insignificant. Metallisation of the substrate with a thin layer of Cu/Ni reduces R_{ja} drastically. The metal clearly acts as a heat-sink, making the total surface area available for convective heat transfer to the surrounding air. Thermal vias further reduce R_{ja}, due to a less resistive heat flow path between chip and bottom metallisation.

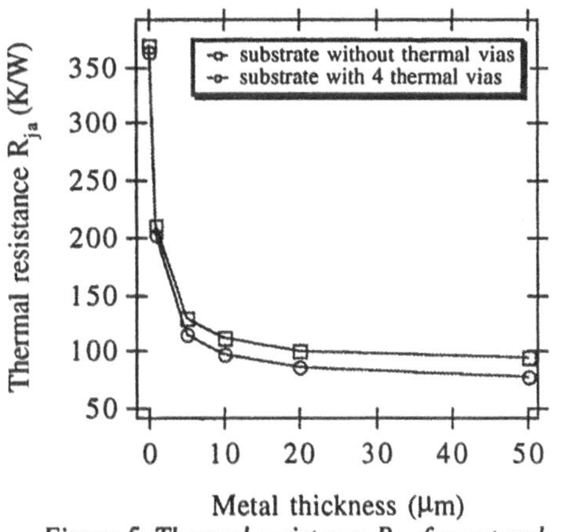

Figure 5 .Thermal resistance R_{ja} for natural convection cooling in air, h = 13.5 W/m^{2}K.

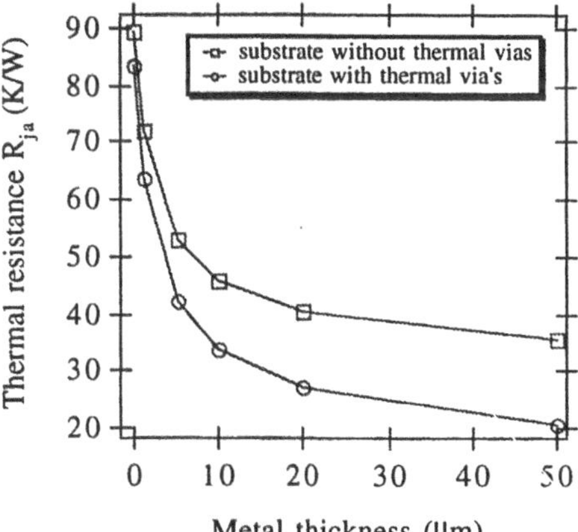

Figure 6. Thermal resistance R_{ja} for natural convection cooling in a fluorinert liquid, h = 135 W/m^{2}K.

A similar set of simulations was performed for a heat transfer coefficient of 135 W/m^2K, which represents natural convection cooling in fluorinert liquid FC-70.

4. Thermal Resistance Measurements

In order to validate the finite element results, practical thermal resistance measurements were performed on a CIMID test structure. The test structure consists of a metallised Ultem 2200 substrate with dimensions 45 x 45 mm^2 and several cavities as depicted in figure 7. A 5 x 5 mm^2 thermal test die was mounted in the corresponding 6 x 6 mm^2 cavity with 4 thermal vias, diameter 600 μm, wall thickness = 20 μm. The test chip contains two diffused resistors for heating and three diodes that are used as thermometers for indirect sensing of the junction temperature. Prior to the actual thermal resistance measurement, calibration of the diode voltages versus temperature was done in a thermostatic bath filled with the fluorinert liquid FC-70, within a temperature range 20 - 90 °C.

The experimental thermal resistance evaluation was performed under the following environmental conditions: cold plate with insulated top surface, liquid jet impingement scheme with FC-70, natural convection FC-70, and natural convection in air. A description of the liquid jet impingement configuration is given in [3]. Figure 8 depicts the measured chip temperature rise as function of power dissipation. Thermal resistance values and measurement uncertainties for 95 % confidence interval are summarised in table 2.

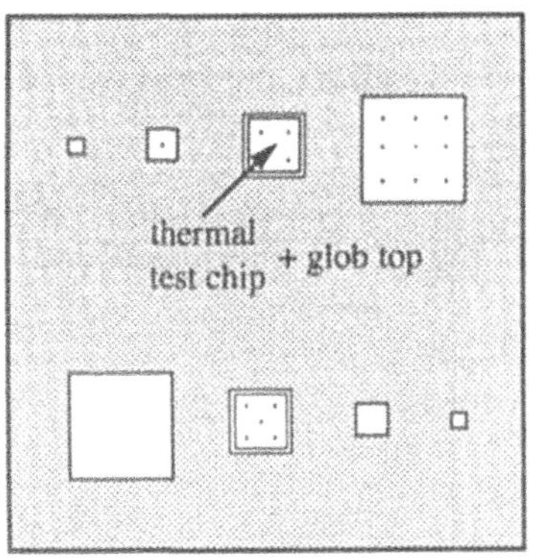

Figure 7. Lay-out of the CIMID test substrate.

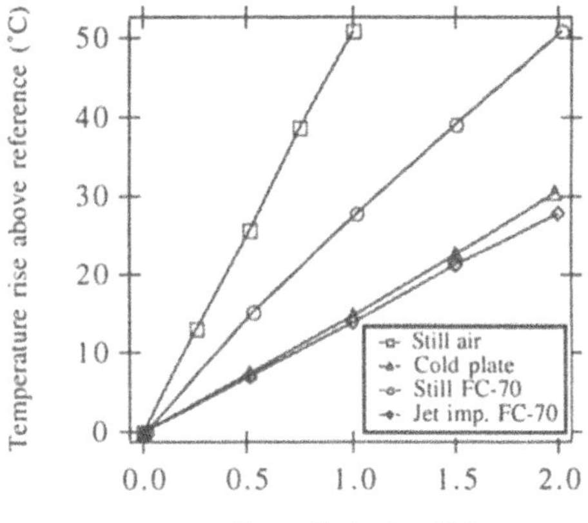

Figure 8. Thermal resistance measurement for CIMID test structures.

For the cold plate measurements, R_{jc} was 50 % larger than the initial simulations predicted. After careful examination of the test structure, we could conclude that the discrepancy is mainly due to geometrical differences between the modelled package and the thermal test structure. The thickness of the test substrate was 600 μm and the thickness of the die adhesive seemed to be 2 or 3 times larger than modelled. Taking into account these real parameters, the finite element model predicted an internal R_{jc} of

14.1 K/W which is much closer to the measured value. Keep in mind that R_{jc} is very sensitive to uncertainties in the diameter and wall thickness of the thermal vias.

TABLE 2. Overview of the measured thermal resistances (K/W) and measurement uncertainties (95 %).

	still air		still FC-70		jet imping. FC-70		cold plate	
Power (W)	R_{ja}	ΔR_{ja}	R_{ja}	ΔR_{ja}	R_{ja}	ΔR_{ja}	R_{jc}	ΔR_{jc}
0.25	51.50	1.30						
0.5	51.06	0.85	29.23	0.52	14.07	0.36	14.82	0.38
0.75	51.16	0.70						
1	50.90	0.61	27.22	0.32	14.07	0.20	14.85	0.20
1.5			26.17	0.25	14.02	0.16	15.14	0.15
2			25.29	0.22	13.99	0.13	15.27	0.12

In contrary to the cold plate arrangement, thermal resistance measurements under natural convection in air and FC-70 were lower than expected. As the surface area of the test substrate was about 3.5 times larger than modelled, the external convective thermal resistance is much lower than expected. This phenomenon confirms the important heat-sink effect of the metallisation. The higher internal resistance of the substrate and die attach can not be observed in this mounting arrangement, because R_{ja} is mainly governed by convective heat transfer. Interpretation of the resistance values for natural convection in FC-70 reveals that the convective heat transfer coefficient heavily depends on temperature.

5. Thermo-Mechanical Simulations

As most package types, a Chip In Moulded Interconnect Device is a multimaterial structure. A good thermo-mechanical compatibility between the materials is thus a primary requirement in order to guarantee the mechanical and electronic functionality of the package. Due to the mismatch in coefficient of thermal expansion, complex 3-D stress and strain distributions will be introduced when the structure is subjected to temperature changes. During the assembly of the CIMID part, the materials will be repeatedly heated and cooled, e.g. curing of dielectric layers, curing of the die adhesive and the encapsulant, soldering, ... Residual stresses will always be present before the CIMID is operational. The present study is devoted to a qualitative comparison of the residual stress distribution in a CIMID package and a similar Chip on Board (COB) assembly on a standard FR4 (epoxy/glass laminate) substrate. The considered load is a uniform temperature drop from the encapsulant curing temperature to room temperature. Stresses were studied by means of linear elastic finite element analysis.

5.1. MODEL GEOMETRY

As far as the CIMID assembly is concerned, we refer to the geometry of the thermal modelling study. A similar model was constructed for a COB assembly. The same chip, die attach, and substrate dimensions were used as for the CIMID model (see figure 9). The thickness of the FR4 substrate is 1 mm. Thermal vias and metallisation layers were neglected in this thermo-mechanical modelling study. Because of symmetry, only one quarter of the complete geometry was modelled.

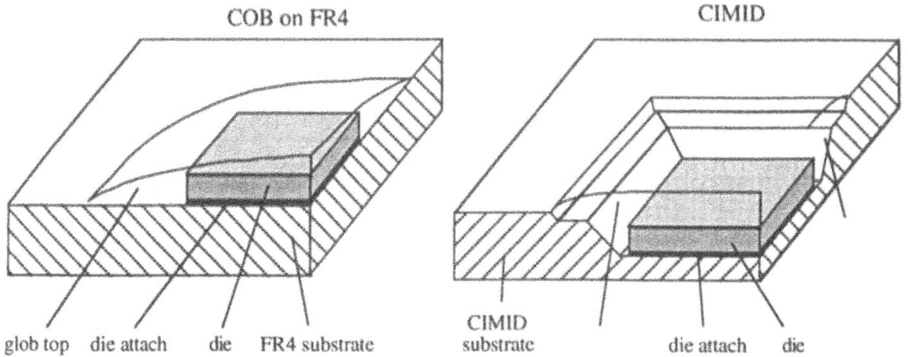

Figure 9. Quarter of the modelled COB and CIMID packages.

5.2. MATERIAL PROPERTIES

The mechanical properties of the CIMID materials are listed in table 1, together with the thermal properties. A comparison of the mechanical properties of Ultem 2200 and FR4 is given in table 3.

TABLE 3. Mechanical properties of Ultem 2200 and FR4 [2].

Property	Ultem 2200	FR4
Young's modulus (GPa)	6	17.4
Poisson ratio	0.4	0.3
Tensile strength (MPa)	140	333
Flexural strength (MPa)	210	503
Coefficient of thermal expansion (ppm/°C) x, y	25	15
z	47	60

5.3. BOUNDARY CONDITIONS AND LOADS

Symmetry in the geometry and load was taken into account by reducing the actual geometry to one quarter, as shown in figure 9, and applying appropriate displacement constraints at the nodes of the symmetry planes. In addition, the centre of the substrate was fixed in order to prevent a rigid body translation.

The structures were assumed to be stress-free at the curing temperature of the encapsulation (glob top) material. The thermo-mechanical load consists of the following three parts:

1) Linear shrinkage (0.3 %) of the encapsulation at the curing temperature (= 150 °C).
2) Cooling from 150 °C to the glass transition temperature of the die attach (= 85 °C).
3) Cooling from 85 °C to room temperature (= 20 °C).

Loads 1 and 2 were simulated simultaneously, while the third load was treated separately, with modified properties for the epoxy adhesive. The final stress distribution was obtained by linear superposition.

5.4. SIMULATION RESULTS

The results of the thermo-mechanical simulations for CIMID and COB are summarised in table 4. Comparison of the relevant stress components in the substrate, die, and die attach shows that:

1) In spite of the high TCE of the Ultem 2200 material, the stresses induced in the CIMID substrate are acceptable and comparable to those of FR4. This results from the lower Young modulus of the Ultem, compared to FR4. Tensile stresses occur under the die, with a maximum in the centre of the substrate. Shear stresses are highest under the chip corners. Peel stress is positive under the edges of the chip and negative under the encapsulant.
2) In the die, the maximum principal tensile stress occurs at the centre and is a peel stress. Principal stresses, which are mainly compressive, and shear stresses in the die corners are considerably lower for CIMID compared to COB.
3) The maximum shear stress in the die attach of the COB geometry is twice as high as for the CIMID geometry.

TABLE 4. Comparison of simulated stresses in the substrate, die, and die attach for CIMID and COB on FR4 geometries.

Location	Stress component (stress in MPa)	CIMID	COB
Substrate	Von Mises stress	32.8	49
	Normal in-plane stress	20.9	24.9
	Shear stress	11	18.9
	Peel stress	6.7	8.6
Die	Principal stress (tensile)	11.6	11.6
	Principal stress (compressive)	-234.0	-305.5
	Shear stress	74.3	90.6
	Peel stress	11.6	11.6
Die attach	Von Mises stress	95.8	105
	Normal in-plane stress	107	108
	Peel stress	45.5	47.7
	Shear stress	17.9	34.7

6. Conclusions

In spite of the low thermal conductivity of today's thermoplastic materials, the internal (conductive) thermal resistance of a CIMID-package can be reduced significantly by adding thermal vias in the substrate under the cavity. Finite element simulations of the heat flow through the package revealed that only a few thermal vias are sufficient to decrease the thermal resistance between junction and isothermal bottom with a factor 5. The inherent presence of a continuous metal layer is advantageous from a thermal point of view. The metallisation acts as heat spreader and thereby decreases the internal and external thermal resistance under convective cooling conditions. The influence of the metal thickness was numerically investigated.

The calculated Von Mises stresses in the thermoplastic Ultem 2200 were lower than those in the similar FR4-substrate. The principal stresses in the die were mainly compressive for both assemblies. The shear stresses at the corners of the die were substantially lower for the CIMID assembly. This thermo-mechanical analysis has proven that, although thermoplastic materials do not offer the mechanical properties of FR4 epoxy/glass laminates, their stiffness and strength are more than adequate for use as circuit board substrates.

Acknowledgements

The authors gratefully acknowledge M. Van de Peer for wire bonding of the test structures. The work described in this paper was supported by the Flemish Ministry of Economic Affairs.

References

[1] Beyne, E., Vanhoof, R., Christiaens, F., Roggen, J., Van Puymbroeck, J., Heerman, M., Gouwy, G., Bulcke, P.: "CIMID, a Technology for High Density Integration of Electronic Systems", Proceedings of the 27th International Symposium on Microelectronics, ISHM 1994.

[2] Funer, R.E., James, D.B.: "Advances in Thermoplastics for Electronic Applications", in Polymers for Electronic and Photonic Applications, edited by Wong, C.P., Academic Press, Inc., San Diego, pp. 333-387, 1993.

[3] Christiaens, F., Beyne, E., Temmernan, W., Allaert, K., Nelemans, W.: "Experimental Thermal Characterisation of Electronic Packages in a Fluid Bath Environment", Proceedings EUROTHERM Seminar No 45, 1995.

THERMAL AND MECHANICAL CHARACTERIZATION OF ELECTRONIC PACKAGES IN EXTREMELY HIGH FREQUENCY APPLICATIONS BY MEANS OF FINITE ELEMENT ANALYSIS

J.-P. Sommer [1], R. Dudek [1], B. Michel [1], M. Boheim [2], W. Hager [2]
[1] *Fraunhofer-Institut für Zuverlässigkeit und Mikrointegration (IZM) Berlin*
Gustav-Meyer-Allee 25
D-13355 Berlin
[2] *Daimler-Benz Aerospace AG, Sensorsysteme Ulm*
Wörthstr. 85
D-89077 Ulm

1. Introduction

The determination of possible failure mechanisms and advance in reliability are essential aims in modern micro electronics. Thermal and mechanical strength cannot be avoided in technological processes and under use. In general, this leads to three dimensional stress and strain fields with concentrations at special locations, e.g. at interfaces, edges, or material imperfections. One reason is the thermal mismatch between several materials that are combined in a package.
It is well known that there is a correlation between stress and strain maxima and the reliability of electronic packages. But it is very difficult to define these relationships precisely because of the diversity of influences. Therefore, special cyclic test conditions have been defined, and most of the electronic devices have to pass these tests for a defined number of cycles without failure.
Advanced finite element modelling is a well suited tool to study thermal and mechanical field distributions and to determine parts with large gradients or strength concentrations. Finite element modelling can be combined with special high precision measurement techniques for determination of displacements, deflections, or deformations in the micro region, basing on optical methods as on scanning electron microscopy (SEM) as well. Numerical and experimental results can be compared and combined. In general, this leads to a more precise knowledge of the field quantities and allows to improve both the input data for the numerical evaluations and the accuracy in the measurement technique.
The authors will give a short survey about their approach in this field. For example, some problems arising in millimeter wave technology are pointed out. Especially, the thermomechanical behaviour of millimeter wave modules based on GaAs technology was investigated.

E. Beyne et al. (eds.), Thermal Management of Electronic Systems II, 349-359.

2. Special demands on electronic packages in millimeter wave technology

One way to end up in a high productive, low cost fabrication of millimeter wave modules is to replace metallic packages and cases by plastic ones. There are some special requirements to such a synthetic material and the corresponding housings:

- electrical functionality integrated, therefore high mechanical precision necessary,
- very small changes in dimensions and shape during the whole life time,
- reproducable shrinking and very small warping,
- thermal stability,
- hermeticity,
- metallisable in high surface quality,
- low coefficient of thermal expansion (CTE),
- good heat transfer properties,
- low dielectric losses.

That's why the plastic compounds must be chosen very carefully to mimimize the influences due to global and local thermal mismatch [1]. On the other side, the package should be realized in such a way, that the unavoidable local stress concentrations will be diminished and crack initiating effects are reduced [2] - [4]. Therefore, solder and adhesive joints of any parts of the package are of special interest in the field of millimeter wave packaging as well as special components like feedthroughs, edges of chips, and interconnections between bond pads and wires in glob top packages.

3. Mechanical stresses in millimeter wave packages

Stresses due to several kinds of loading have to be taken into account. In addition, many problems in defective packages in microelectronics can be related to stresses due to the various stages of processing. They will be more acute as packages become more complex in geometry and in the number of materials that are combined. To decrease risk of failure caused by residual stress, it is imperative to understand the nature and the origins of this kind of stress. The stated problems can be classified into five groups [5], [6]:

- Thermally induced stresses
  This kind of stresses is caused by the thermal expansion mismatch between two or more material components. This is a problem that is ever present in microelectronics because of the heating up and cooling down processes. Temperature gradients may be of additional importance.

- Film (layer) stresses
  Residual stresses in thin films in most cases cannot be avoided. Various films are used for different functions like masking, passivation, dielectric insulation, and electric conduction. In general, thermal and the so called "intrinsic" stresses are overlaying. Intrinsic stresses are caused by film growth processes and their reasons are often not completely understood.

- Stresses due to lattice mismatch
  They are caused by lattice mismatch between a localized area and its surrounding matrix or between an epitaxial film and its substrate.
- Stresses due to machining
  They are related to the occurrence of local plastic deformation and cracking in surface regions of the components. Typical machining processes that may lead to residual stresses are finishing treatments of silicon wafers or ceramic substrates by grinding or polishing.
- Stresses due to embedded structural elements
  They often differ widely in the local surrounding of an embedded structure, e.g. grains or depositions. The origin of these stresses is a superposition of intrinsic (embedded body) and thermally induced stresses.

4. Finite element analysis and deformation measurement

A FE methodology in general consists of the following steps [7]:

- Physical modelling
  includes extraction of the essential details of the complex structure and loading conditions.
- Meshing (pre-processing) and completion of the FE model
  will be supported in a wide manner by commercial CAD-tools like I-DEAS or PATRAN.
- Numerical solution
  using powerful commercial (ABAQUS, ANSYS) or special research software tools, e.g. ASTOR (established for internal use only).
- Extraction and visualisation of the essential results
  A very large amount of result data (temperatures and fluxes in heat transfer analysis, displacements, strains and stresses in mechanical evaluations) at each node and corresponding to each time step must be reduced. Modern post processors as well as user specific or task adapted graphic modules enable the graphic representation of the field quantities to be analysed.
- Assessment of the results and modification of the FE model
  regarding to the pre-set development goals, e.g. optimization of geometric details like rounding edges, varying thicknesses of layers or changing loading conditions (thermal control).

In addition, the authors use the measurement techniques shown in Figure 1 to determine the response of the tested microstructure due to changes in the loading conditions. Comparing FE modelling procedure with these high precision measurement techniques, some essential preferences and disadvantages given in Figure 2 can be noted.

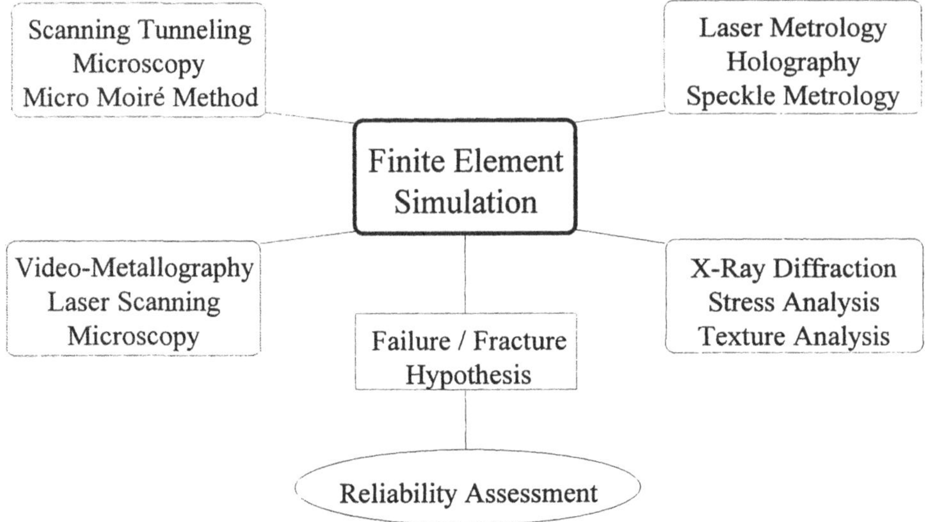

Figure 1. Interaction between FE modelling and measurement techniques.

Finite Element Modelling

Advantages

- model size free choosable
  2d - 3d
  kind of material law
- easy to modify the basic model
  sensitivity analysis
- results available not only at the surface
- no test structures needed

Disadvantages

- results not better than the model
  geometric representation
  material properties
  description of initial and boundary conditions

Laser Deformation Metrology

Advantages

- at real part
- no misrepresentation
  by incorrect model assumptions

Disadvantages

- many metrological tasks need
  special equipment and adjustment
- trouble due to surrounding influences cannot always be eliminated

Figure 2. Features of FE modelling and measurement techniques.

To solve a given problem numerical and experimental approaches should be combined wherever this will be possible. The confidence in the results and in the predicted thermo-mechanical reliability and life time can be essentially increased by combining the numerical and experimental techniques. The visualization of the calculated and the measured deformations can be performed on the screen at the FE workstation, [8]-[10].

5. Examples

5.1. THERMO-MECHANICAL BEHAVIOUR AT A FEEDTHROUGH

Millimeter wave technology based on metallic cases is widely mastered. To open the civil market for radar applications it is necessary to decrease cost essentially. Therefore, plastics instead of metals will be of special interest in combination with very efficient technologies like injection moulding.

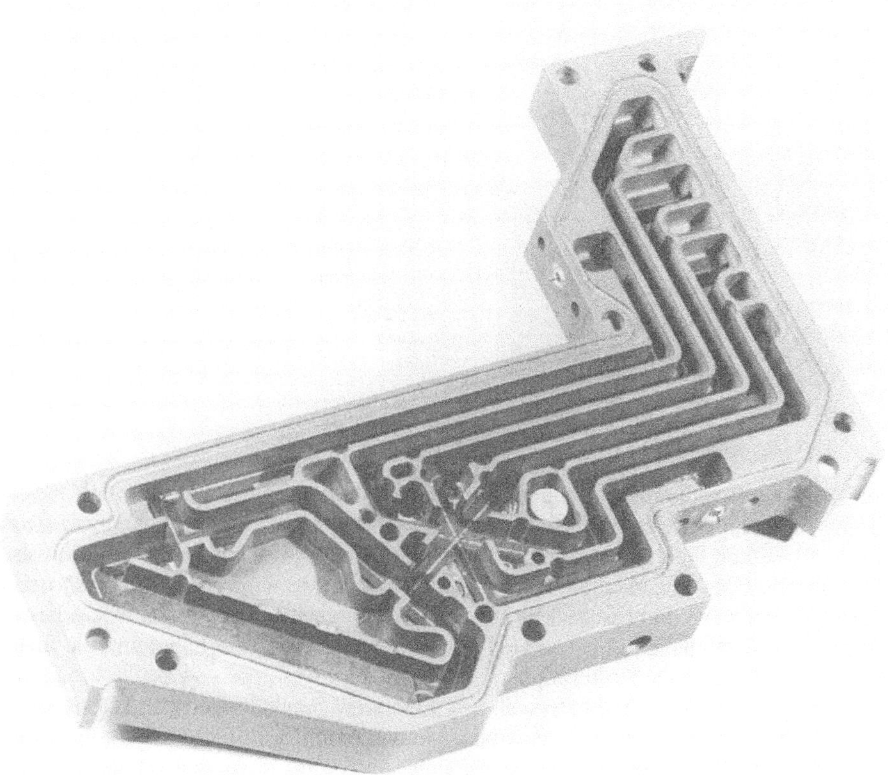

Figure 3. Plastic case of a millimeter wave module.

Figure 3 shows the main part of a plastic case of a millimeter wave module with several types of feedthroughs for high frequencies at the side wall as well as for DC signals at the base plate.
If all parts of a case are made from the same material global thermal mismatch cannot occur. If different materials are used local thermal mismatch may play an important role. E.g. feedthroughs that are optimized to be soldered into a metal case, will not automatically be best suited for plastic cases too. In general, the effects of thermal mismatch will be larger. Additional problems may occur due to the lower heat transfer coefficients of plastics compared to metals if modules with higher dissipation output are mounted.
Taking advance of the axial symmetry of the feedthrough, an axially symmetrical FE analysis of its vicinity should be performed to determine the strain and stress fields. The derived recommendations for material and geometric improvements should be confirmed by laser optic measurements.

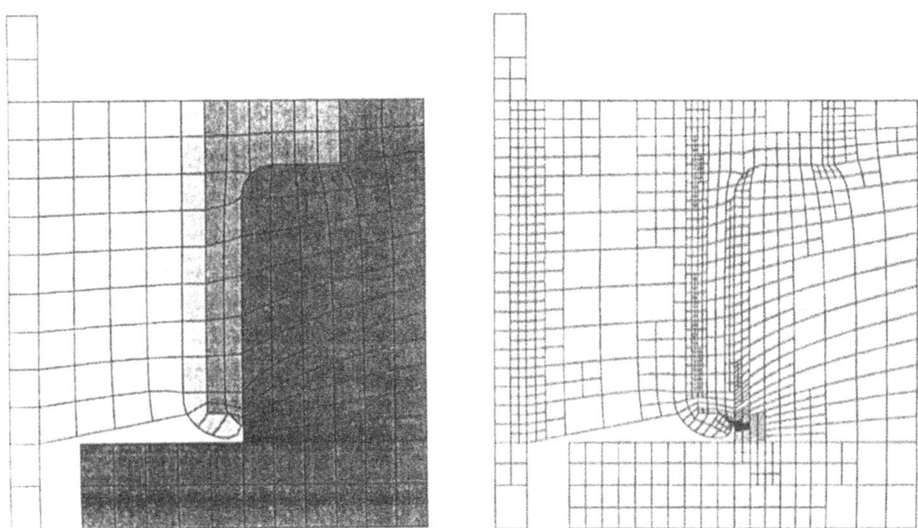

Figure 4. Mesh refinement technique.

A special meshing refinement technique was used [11]: Starting with a rough mesh containing the complete geometric information a simple linear elastic FE analysis was performed. Based on the deformation discontinuities (jumps) at the element edges it was decided wheather an element had to be subdivided or not. This procedure was performed in several steps and resulted in a FE mesh with very dense regions where large deformation gradients have to be expected. Figure 4 represents the original and the final meshes for the vicinity of a feedthrough.
In the second stage of FE analysis as well as temperature dependent elastic, plastic, and viscoplastic deformation behaviour was taken into account, especially for the tin lead solder alloy. A test cycle (see Figure 5) consisting of a cooling down phase, relaxation at the lower temperature level, heating up and a relaxation phase at the upper temperature level was investigated numerically, also for a structure with a circumferential interface crack.

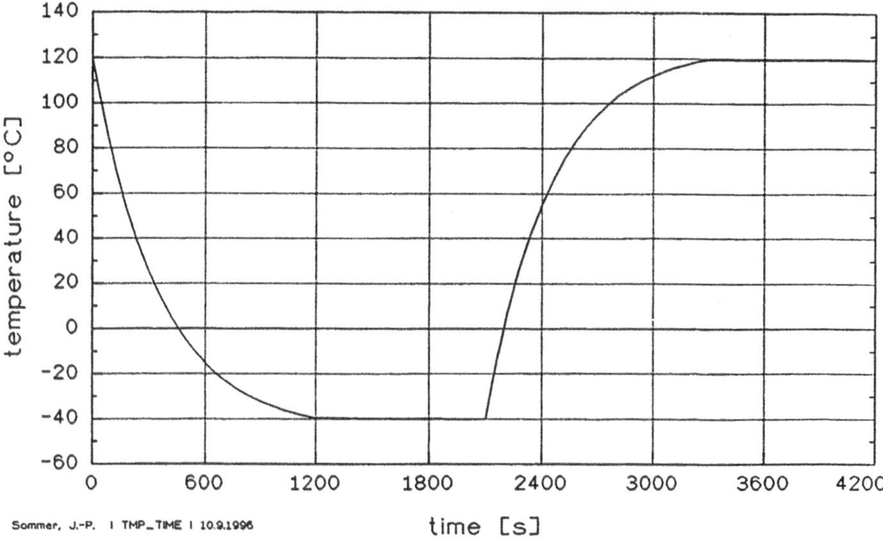

Figure 5. Test cycle for FE modelling.

As a result the Mises stress and the equivalent creep strain distributions at the end of each stage were obtained (see Figure 6). The effect of thermal mismatch could be studied explicitly. It was obvious that the parts of the cycle with a lower temperature level result in an essentially smaller relaxation than those with higher temperature. Therefore, most of the increase in creep strain takes place in the first cooling down step, and as a first rough approximation it should be sufficient to investigate only this phase of the complete cycle [12].
The Manson/Coffin-model

$$N_f = \frac{1}{2}\left[\frac{\Delta\gamma}{2\varepsilon_f}\right]^{\frac{1}{c}} \tag{1}$$

was developed for lifetime predictions in surface mount technology. The mean cycles to failure N_f can be derived, when the elastic part of the solder deformation is neglected. In (1) ε_f denotes the fatigue ductility coefficient, the exponent c is a constant, and $\Delta\gamma$ is the strain amplitude in a complete cycle. It was originally estimated for surface mount devices (SMD) by the equation

$$\Delta\gamma = \frac{L}{h} \cdot \Delta\alpha \cdot \Delta T \tag{2}$$

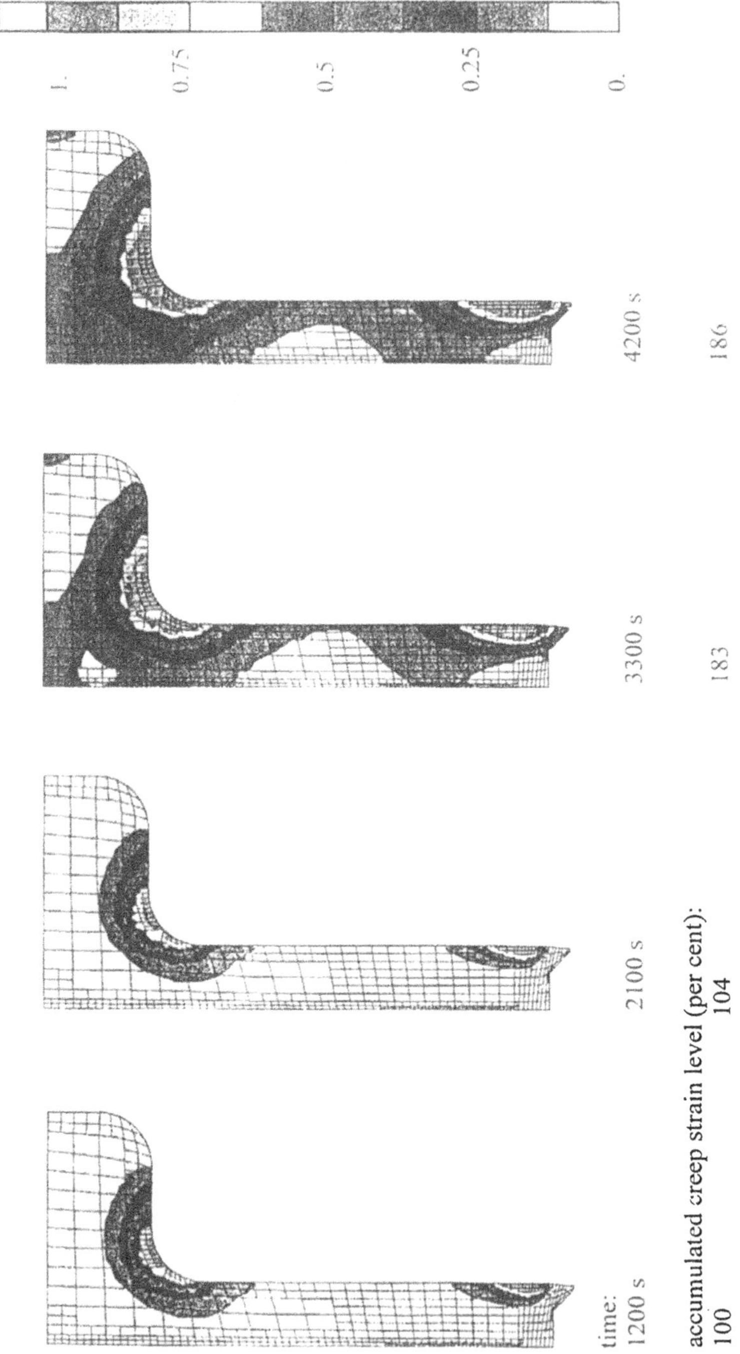

Figure 6. Creep strain distribution in the solder gap with an interface crack.

where 2L is the distance between the solder joints, h the solder joint hight, $\Delta\alpha$ the difference in the coefficients of thermal expansion of component and substrate and ΔT the temperature swing.

Engelmaier [13] has improved this model by modifying the exponent c and confirmed his model by various experimental studies. The authors propose to get delta gamma from the strain amplitude of a complete thermal cycle evaluated by a non-linear FE analysis [2]. This approach is valid not only for SMD [14].

5.2. CHOICE OF BEST SUITED CASE MATERIAL AND DESIGN FOR MILLIMETER WAVE DEVICES

The chosen case material must ensure that the heat energy can be dissipated and the thermal strength of the device will be limited. In general the heat dissipation has to be calculated already during the constructive phase of a package, e.g. on the basis of a thermal FE analysis.

The heat sources/sinks and the heat flux conditions at all surfaces have to be described as boundary conditions, computational series with variant parameters allow to detect sensitivities due to special inputs and to decide whether a modelled detail is essential or negligible.

If a plastic material does not ensure sufficiently large heat flux dissipation it is possible to build in a metallic insert and to attach the dies on it using a heat conductive adhesive. At the outside of the package the insert can be thermally coupled with a heat sink.

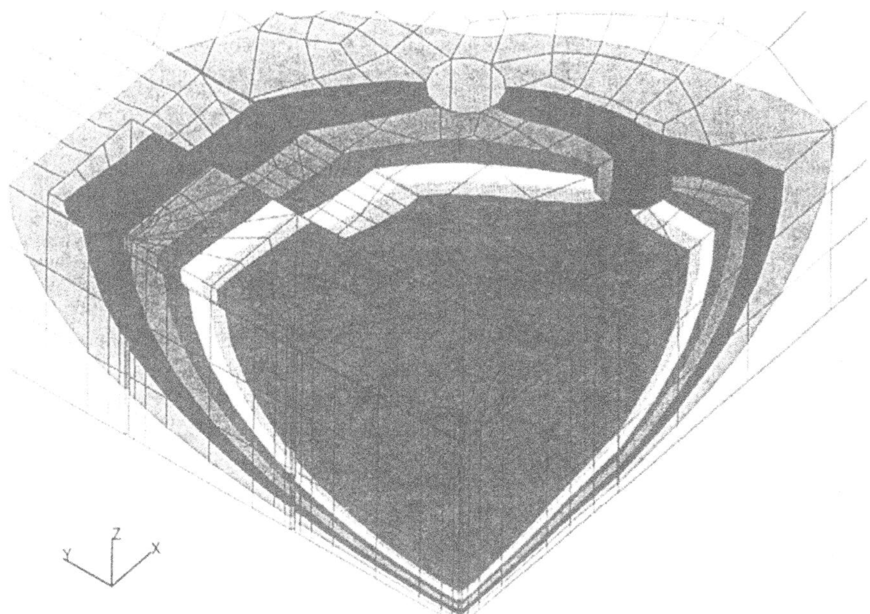

Figure 7. Temperature field near the GaAs transistor.

As an example, a MMMIC module was investigated. Normalizing the heat dissipation of each GaAs transistor by 1 Watt the heat conduction conditions of several package constructions and materials were compared: Plastic (polyether sulphone filled with ceramic powder) without and with an insert made of copper wolfram alloy, metal impact moulding material on iron base, several kinds of heat sinks at different locations at the outside of the package. Finally, the influence of convection was investigated by variation of the heat transition coefficients in the third kind boundary conditions.
A three dimensional FE model was created taking advantage of symmetry in two planes. Therefore, either the central transistor or a symmetric pair has been thermally investigated. Figure 7 represents a typical temperature field in the vicinity of a GaAs transistor. Using the superposition principle the effect of 5 active transistors could be derived too. Finally, a recommendation for the thermal layout could be formulated. It is planned to verify the numerical results by laser optic high precision measurements as soon as the first modules have been built.

6. Conclusion

Using the finite element method it is possible to investigate the thermal and/or mechanical behaviour of millimeter wave modules already in very early phases in layout and construction. By variation of selected parameters computational series can be carried out to determine sensitivities due to special inputs and to decide whether a modelled detail is essential or negligible. Wherever it can be performed computations should be combined with corresponding measurements. In addition, the more or less direct coupling of experimental techniques with numerical simulation tools (e.g. via image processing) decreases the expense in time and provides good means for advanced reliability assessment.

7. References

1. Sommer, J.-P., Dudek, R., and Michel, B.: Thermomechanische Simulation von mm-Wellen-Gehäusen. Forschungsbericht zum BMBF-Projekt "Systemtechnologien für Mikro- und Millimeterwellen-Schaltungen", Fraunhofer-IZM Berlin, 1995.
2. Michel, B., and Sommer, J.-P.: Application of Solid Mechanics in the Fields of Microelectronics and Micro System Technology, 2nd European Solid Mechanics Conference, Sept. 12-16, 1994, Proc. Euromech, Genoa, Italy, P. N9.
3. Michel, B., Sommer, J.-P., and Großer, V.: Application of Fracture Mechanics to Micromechanics and MicrosystemTechnology, Advances in Fracture Research and Structural Integrity, ed. V.V. Panasyuk et.al., Pergamon Press 1994, pp. 683-690.
4. Michel, B., Sommer, J.-P., Krause, F., Winkler, T., and Faust, W.: Bruchmechanische Untersuchungen zur Zuverlässigkeitsbewertung mikrotechnischer Aufbauten, Jahrestagung des AK "Bruchvorgänge" im Deutschen Verband für Materialforschung und -prüfung (DVM), Köln, 14.-16.2.1995.
5. Michel, B., Schubert, A., Dudek, R., and Großer, V.: Experimental and Numerical Investigations of Thermomechanically Stressed Micro-Components, Micro System Technologies 1 (1994) 1, Springer Verlag, pp. 14-22.
6. Michel, B.: Mechanische und thermische Probleme an Verbundwerkstoffen, Vortrag Arbeitsgemeinschaft "Verbundwerkstoffe" Berlin (BAV), Bundesanstalt für Materialforschung und -prüfung, Berlin, 25. 4. 1995.

7. Sommer, J.-P., Dudek, R., and Michel, B.: Numerische Beanspruchungsanalyse an mikrotechnischen Baugruppen mit Kunststoffgehäuse, Problemseminar "Deformation und Bruchverhalten von Kunststoffen", Merseburg, 28.-30.6.1995.
8. Großer, V., Sommer, J.-P., Faust, W., Bombach, C., and Michel, B.: Mechanisch-thermische Versagensdetektion an Leiterplatten mittels numerischer und laseroptischer Verfahren, SMT/ES&S/Hybrid '95, Nürnberg, 3.-5.5. 1995, Proc. S. 17-26.
9. Sommer, J.-P., Dudek, R., and Michel, B.: Deformationsanalyse mit Finite-Elemente- und laseroptischen Verfahren in der Mikrotechnik, Deutsche ISHM-Konferenz 1994, Int. Society for Hybrid Microelectronics, München, 24./25. 10. 1994
10. Vogel, D., Sommer, J.-P., and Michel, B.: Speckle Photographic Strain Field Measurement - A Helpful Aid in Fracture Modelling and Analysis of Microstructures, 10th European Conf. on Fracture (ECF 10), Berlin, Sept. 20-23, 1994, Proc. publ. in Structural Integrity: Experiments, Models and Applications, EMAS Ltd., Warley, 1994 (ed. C. Berger and K.-H. Schwalbe).
11. Michel, B., Kühnert, R., Auersperg, J., and Tränkner, K.: Experimentelle und numerische Deformationsanalyse an Komponenten der Mikrosystemtechnik, Deutscher Verband für Schweißtechnik, DVS-Bericht, Essen, Nr. 158, 122-125 (1994).
12. Dudek, R., Großer, V., and Michel, B.: Simulation und Bewertung der mechanisch-thermischen Zuverlässigkeit von Komponenten und Baugruppen der Mikrotechnik, SMT '94, Tutorial "Mechanisch-thermische Zuverlässigkeit von Komponenten der Mikrotechnik", Nürnberg, 17.-19.5.1994.
13. Engelmaier, W.: Solder Attachment Reliability, Accelerated Testing, and Result Evaluation, in: Lau, H. (ed.), Solder Joint Reliability, pp. 548-587, Van Nostrand Reinhold, New York, 1991.
14. Dudek, R., and Michel, B.: Thermomechanical Reliability Assessment in SM- and COB-Technology by Combined Experimental and Finite Element Method, Proc. Int. Reliability Physics Symp., San José, California, April 11-14, 1994, pp. 458-465.

AUTHOR INDEX

The manufacturer's authorised representative in the EU is Springer Nature Customer Service Centre GmbH, Europaplatz 3, 69115 Heidelberg, Germany. If you have any concerns regarding our products, please contact ProductSafety@springernature.com

Printed and bound by CPI Group (UK) Ltd, Croydon, CR0 4YY

24/04/2026

02096308-0009